Springer Theses

Recognizing Outstanding Ph.D. Research

AF212641

For further volumes:
http://www.springer.com/series/8790

Aims and Scope

The series "Springer Theses" brings together a selection of the very best Ph.D. theses from around the world and across the physical sciences. Nominated and endorsed by two recognized specialists, each published volume has been selected for its scientific excellence and the high impact of its contents for the pertinent field of research. For greater accessibility to non-specialists, the published versions include an extended introduction, as well as a foreword by the student's supervisor explaining the special relevance of the work for the field. As a whole, the series will provide a valuable resource both for newcomers to the research fields described, and for other scientists seeking detailed background information on special questions. Finally, it provides an accredited documentation of the valuable contributions made by today's younger generation of scientists.

Theses are accepted into the series by invited nomination only and must fulfill all of the following criteria

- They must be written in good English.
- The topic should fall within the confines of Chemistry, Physics, Earth Sciences, Engineering and related interdisciplinary fields such as Materials, Nanoscience, Chemical Engineering, Complex Systems and Biophysics.
- The work reported in the thesis must represent a significant scientific advance.
- If the thesis includes previously published material, permission to reproduce this must be gained from the respective copyright holder.
- They must have been examined and passed during the 12 months prior to nomination.
- Each thesis should include a foreword by the supervisor outlining the significance of its content.
- The theses should have a clearly defined structure including an introduction accessible to scientists not expert in that particular field.

Johannes Giesecke

Quantitative Recombination and Transport Properties in Silicon from Dynamic Luminescence

Doctoral Thesis accepted by
University of Konstanz, Germany

Springer

Author
Dr. Johannes Giesecke
Fraunhofer ISE
Freiburg
Germany

Supervisors
Prof. Giso Hahn
Universität Konstanz
Konstanz
Germany

Prof. Andrés Cuevas
Australian National University
Canberra, ACT
Australia

ISSN 2190-5053 ISSN 2190-5061 (electronic)
ISBN 978-3-319-36027-0 ISBN 978-3-319-06157-3 (eBook)
DOI 10.1007/978-3-319-06157-3
Springer Cham Heidelberg New York Dordrecht London

© Springer International Publishing Switzerland 2014
Softcover reprint of the hardcover 1st edition 2014
This work is subject to copyright. All rights are reserved by the Publisher, whether the whole or part of the material is concerned, specifically the rights of translation, reprinting, reuse of illustrations, recitation, broadcasting, reproduction on microfilms or in any other physical way, and transmission or information storage and retrieval, electronic adaptation, computer software, or by similar or dissimilar methodology now known or hereafter developed. Exempted from this legal reservation are brief excerpts in connection with reviews or scholarly analysis or material supplied specifically for the purpose of being entered and executed on a computer system, for exclusive use by the purchaser of the work. Duplication of this publication or parts thereof is permitted only under the provisions of the Copyright Law of the Publisher's location, in its current version, and permission for use must always be obtained from Springer. Permissions for use may be obtained through RightsLink at the Copyright Clearance Center. Violations are liable to prosecution under the respective Copyright Law.
The use of general descriptive names, registered names, trademarks, service marks, etc. in this publication does not imply, even in the absence of a specific statement, that such names are exempt from the relevant protective laws and regulations and therefore free for general use.
While the advice and information in this book are believed to be true and accurate at the date of publication, neither the authors nor the editors nor the publisher can accept any legal responsibility for any errors or omissions that may be made. The publisher makes no warranty, express or implied, with respect to the material contained herein.

Printed on acid-free paper

Springer is part of Springer Science+Business Media (www.springer.com)

To Maila—Joy, Wrath, and Sunshine

To Mollie—Joy, Wrath and Sunshine

Supervisor's Foreword

Semiconductor devices are present almost everywhere in our modern daily life. Examples are integrated circuits, light emitting diodes, laser diodes, or solar cells. The quality of the material used to fabricate those devices plays an important role in guaranteeing a reliable and an effective operation. The lifetime of excess charge carriers in semiconductors is one of the most important parameters for evaluating the quality of the material, as the carriers are needed for the functioning of the device. The generated charge carriers can recombine e.g., at crystal defects in the bulk of the crystal or at its surface, reducing the carrier density. Therefore, the determination of excess charge carrier lifetime in a semiconductor is of great interest for a broad variety of applications.

Recombination of excess charge carriers can take place via various mechanisms. Radiant recombination (or band-to-band recombination) of electrons and holes by emitting a photon is the inverse of carrier generation by absorption of photons. In indirect semiconductors like crystalline silicon, a phonon is involved to conserve momentum, making this process more unlikely than in direct semiconductors. Other recombination channels include Auger recombination and recombination via traps in the bulk or at the surface (referred to as Shockley-Read-Hall recombination).

For determination of excess charge carrier lifetime, many different techniques have been developed over the years. Many of them use direct or indirect measurement of photoconductivity over time after carrier generation via an excitation pulse. An alternative is the direct detection of photons emitted after generation via radiant recombination, called luminescence. In combination with cameras based on charge-coupled devices (CCD), spatially resolved images of luminescence intensity can be determined in relatively short time frames, which is of particular interest for materials exhibiting spatial inhomogeneities. For indirect semiconductors the luminescence signal is weak, and care has to be taken that it can be measured reliably.

In recent years, several measurement setups have been developed at various institutions, and all of them using luminescence in combination with various techniques to yield spatially resolved measurements of variables of interest in wafers or semiconductor devices, such as solar cells. Examples of relevant variables include excess carrier lifetime and series resistance. Although luminescence, in general, is a well understood phenomenon, up to now there has been no systematic compendium available, bringing together all information needed to gain a deep understanding of luminescence methods both from a physics and an application point of view.

Johannes Giesecke's thesis is based on a very thorough understanding of semiconductor physics and the underlying physical principles. Starting with the basic continuity equation he develops a very stringent formalism which gives new insights, for example into excess charge carrier lifetime and the different ways it can be derived. An important part is the description of dynamic luminescence techniques, partly coupled with additional bias light, resulting in a new way to determine the actual lifetime. In addition to the excellent, both physically and mathematically clear derivation of the underlying principles, it also presents a variety of new applications with high practical relevance. Examples are the determination of doping concentration independently of charge carrier mobility and the injection dependent measurement of excess charge carrier lifetimes down to carrier concentrations of 10^8 cm^{-3}. The latter in particular is of great interest for fundamental investigations of the underlying recombination mechanisms.

Another strength of the thesis is the analysis of experimental influences and errors on the obtained data in the measurement setup used for this study. In this way, reliable results can be presented.

In combination with an elaborate base of literature references, this thesis presents a very valuable and concentrated source of information on new measurement methods based on luminescence and their underlying principles. It is a valuable source for everybody interested in luminescence-based measurements of crystalline silicon wafers and devices. The exciting new field of camera-based luminescence measurement techniques has just recently opened up, and there are still plenty of things to be discovered. We believe that this thesis is an ideal way to be introduced to it. Hopefully you find it as instructive as we do.

Konstanz, February 2014 Prof. Giso Hahn

Abstract

This thesis elaborates techniques to accurately determine recombination and transport properties of silicon from harmonically time-modulated luminescence.

Harmonically modulated carrier lifetime techniques based on luminescence combine the general advantages of luminescence-based metrology with the benefits that the immediacy of a time-resolved lifetime approach brings about. While luminescence-based metrology offers a highly sensitive lifetime tool virtually unaffected by carrier trapping, the dynamic approach implies a minimum of otherwise required a priori assumptions about electronic material properties as well as vanishing artifacts imposed on measured amplitudes. The latter may be largely affected by optical properties, by interface morphology, or by self-absorption.

In the course of this work, several advances of luminescence-based harmonically modulated lifetime metrology were achieved. This involves theoretical, conceptual, and experimental advances:

A theoretical explanation of the mismatch between quasi-steady-state harmonically modulated lifetime and steady-state effective lifetime for high surface recombination velocities was found. Further, a theory of light-biased decay time reveals the differential nature of harmonically modulated lifetime techniques, which is crucial in the case of injection-dependent recombination.

Conceptual advances comprise a class of phase-sensitive carrier lifetime techniques complementing a self-consistent lifetime determination, an approach to accurately determine injection-dependent carrier lifetime over a very broad injection range, a specification of how to perform a spatially resolved carrier lifetime calibration of photoluminescence images, and an approach to access bulk lifetime and surface recombination velocity of silicon ingots. Further, a class of techniques to resolve net dopant concentration via modulated photoluminescence and a transit time approach to accurately determine minority carrier mobility were developed.

Experimental advances comprise harmonically modulated carrier lifetime measurements in the microsecond range, carrier lifetimes via dynamic electroluminescence, and a setup adaption enabling accurate carrier lifetime measurements in any photovoltaic processing stage—from an ingot to a finished solar cell.

Acknowledgments

Thanks to my internal supervisor Wilhelm Warta (the present work was elaborated at the Fraunhofer Institut für Solare Energiesysteme) for the cautious guidance, and above all for the remarkable amount of freedom I enjoyed—I greatly appreciate it.

Thanks to the supervising professors and examiners of my thesis, my first supervisor Giso Hahn, and my second supervisor Andrés Cuevas from the Australian National University, who joined us on the occasion of my doctoral examination on October 9, 2013 in Konstanz.

Thanks to Stefan Glunz, Martin Hermle, Elisabeth Schäffer, Jonas Schön, Martin Schubert, Wilhelm Warta, and numerous others for living and sustaining the spirit of Fraunhofer ISE, thereby making it the exceptional workplace that it is.

Thanks to Martin Schubert for the kind integration into the material characterization group at Fraunhofer ISE, and to all the very pleasant colleagues I have had the chance to work with.

Thanks to Ronald Sinton for an inspiring discussion on the limitations of dynamic lifetime measurement techniques. It eventually triggered the building of a substantial part of the theoretical foundation of this thesis.

Thanks to Markus Glatthaar and to David Kiliani for encouraging me to get involved with the Python programming language.

Thanks to an estimated two dozen of colleagues I shared my office with over the years. Thank you for the exceptionally quiet and concentrated working atmosphere.

Thanks to the numerous technical staff members at Fraunhofer ISE, who readily helped with the design and processing of samples, thanks to all colleagues who provided most appropriate samples for miscellaneous purposes, and thanks to the library staff for competent literature research.

Last, but never least, thanks to all involved in truly caring for my daughter Maila—to my wife Birte Gundelach, to many of our relatives, and to day care. Your time and effort eventually established the reconciliation of our work and family lives.

Contents

Abbreviations and Symbols

Abbreviations

The excessive use of abbreviations was avoided in this dissertation—aiming at a better legibility. Yet, some abbreviations (defined here) were deemed inevitable, either because they represent common terminology, or due to their use in figures.

ADC	Analog digital converter
ASTM	American Society for Testing and Materials
BSF	Back Surface Field
CCD	Charge-coupled device
CMOS	Complementary metal oxide semiconductor
CZ	Czochralski (crystal growth process)
d2a	Differential-to-actual (carrier lifetime methodology)
EL	Electroluminescence
FZ	Float zone (crystal growth process)
FWHM	Full width at half maximum
GUI	Graphical user interface
hli/lli	High-level injection/low-level injection
LED	Light emitting diode
mc	Multicrystalline (silicon material class)
PC	Photoconductance
PL	Photoluminescence
PLIR	Photoluminescence intensity ratio
PSG	Phosphosilicate glass
QSSL	Quasi-steady-state luminescence
QSSPC	Quasi-steady-state photoconductance
R&D	Research & Development
SRH	Shockley-Read-Hall (recombination statistics)
umg	Upgraded metallurgical grade (silicon material class)

Symbols

The following list of symbols provides a nomenclature of recurring symbols in this dissertation. Auxiliary variables of variable meaning as well as locally used symbols may exclusively be defined close to their occurrence in the text.

Fundamental Physical Constants

$c = 3 \cdot 10^{10}\,\text{cm s}^{-1}$	Vaccuum velocity of light
$h = 6.626 \cdot 10^{-34}\text{AVs}^2 = 4.136 \cdot 10^{-15}\,\text{eV s}$	Planck constant
$\hbar = h/2\pi$	.
$k_B = 1.381 \cdot 10^{-23}\text{AVs K}^{-1} = 8.62 \cdot 10^{-5}\,\text{eV K}^{-1}$	Boltzmann constant
$m_e = 9.109 \cdot 10^{-31}\text{kg}$	Free electron mass
$q = 1.602 \cdot 10^{-19}\text{A s}$	Elementary charge
$\epsilon_0 = 8.854 \cdot 10^{-14}\text{AVs}^{-1}\,\text{cm}^{-1}$	Vacuum permittivity

Roman Symbols

a_n, b_n	Solutions of transcendental equations for n-th excess carrier decay mode (transient Luke/Cheng approach) [cm^{-1}]
B	Coefficient of radiative recombination [$\text{cm}^3\ \text{s}^{-1}$]
$\mathcal{C}$	Modulated versus steady-state effective lifetime ratio
c_Φ	Scaling factor of luminescence intensity [cm^6]
c_φ	Scaling factor of local luminescence intensity [cm^6]
D_a	Ambipolar (minority) carrier diffusion coefficient [$\text{cm}^2\ \text{s}^{-1}$]
$D_{e/h}$	Minority carrier (electron/hole) diffusion coefficient [$\text{cm}^2\ \text{s}^{-1}$]
$D(\varepsilon)$	Density of electronic states (per energy and volume) [$\text{eV}^{-1}\ \text{cm}^{-3}$]
$D_\gamma(\varepsilon_\gamma)$	Photon density of states (per energy and volume) [$\text{eV}^{-1}\ \text{cm}^{-3}$]
d	Substrate thickness [cm]
$\mathbf{E}$	Electrical field [V cm^{-1}]
E	Internal energy [eV]
F	Free energy [eV]
$f(\varepsilon, \mu_c)$	Fermi distribution (occupation probability)
G	Excess carrier generation rate [$\text{cm}^{-3}\ \text{s}^{-1}$]
I	Detector photocurrent ($\propto$ detected light intensity) [A]
$\mathbf{j}_Q$	Charge current density [A cm^{-2}]
j_γ	Transmitted (and absorbed) photon current density [$\text{cm}^{-2}\ \text{s}^{-1}$]
$j_{\gamma,0}$	Maximal harmonically modulated photon current density [$\text{cm}^{-2}\ \text{s}^{-1}$]
$j_{\gamma,i}$	Impinging photon current density [$\text{cm}^{-2}\ \text{s}^{-1}$]
$\mathbf{K}$	Reciprocal Bravais lattice vector [cm^{-1}]
$\mathbf{k}$	Wave vector or reciprocal position vector [cm^{-1}]
L	Minority carrier (ambipolar) diffusion length [cm]
$m^*_{e/h}$	Electron/hole effective masses [kg]

N	Particle number
$N_{A/D}$	Acceptor/donor concentration [cm^{-3}]
$N_{C/V}$	Effective conduction/valence band densities of states [cm^{-3}]
N_t	Total trap (or trapping site) concentration [cm^{-3}]
n	Index of excess carrier decay mode order also locally used as a refractive index (real part)
n_0	Net dopant concentration [cm^{-3}]
Δn	Excess carrier concentration [cm^{-3}]
$\Delta n_{e/h}$	Excess electron/hole concentration [cm^{-3}]
$n_{e/h}$	Concentrations of free electrons/holes [cm^{-3}]
$n^0_{e/h}$	Dark concentrations of free electrons/holes [cm^{-3}]
n_i	Intrinsic free carrier concentration of silicon [cm^{-3}]
$n_{i,eff}$	Effective n_i (accounting for band gap narrowing) [cm^{-3}]
n_t	Occupied trap (or trapping site) concentration [cm^{-3}]
n_γ	Photon density [cm^{-3}]
$n_\#$	Number of harmonically modulated modulation cycles
$P(\vartheta)$	Light transmission probability in polar angle ϑ
p	Pressure [eV cm^{-3}]
$p(\vartheta)$	Probability density distribution of light transmission
$p_\Phi(z)$	Relative luminescence detection probability
Q	Electrical charge [As]
$Q_\Phi(\varepsilon_\gamma)$	Luminescence detector quantum efficiency
$\mathbf{R}$	Bravais lattice vector [cm]
$\mathcal{R}_\Phi$	Luminescence intensity peak curvature ratio
R	Excess carrier recombination rate [cm^{-3} s^{-1}]
R_B	Luminescence intensity ratio according to Baek
R_C	Compensation ratio (of dopant species)
R_γ	Interface reflectivity at photon energy ε_γ
$R_\Phi(\varepsilon_\gamma)$	luminescence self-absorption function
$\mathbf{r}$	Position vector [cm]
dr_{abs}	Photon net absorption rate in $[\varepsilon_\gamma, \varepsilon_\gamma + d\varepsilon_\gamma]$ [cm^{-3} s^{-1}]
$dr_\uparrow$	Photon absorption rate in $[\varepsilon_\gamma, \varepsilon_\gamma + d\varepsilon_\gamma]$ [cm^{-3} s^{-1}]
$dr_{\downarrow sp}$	Photon spontaneous emission rate in $[\varepsilon_\gamma, \varepsilon_\gamma + d\varepsilon_\gamma]$ [cm^{-3} s^{-1}]
$dr_{\downarrow st}$	Photon stimulated emission rate in $[\varepsilon_\gamma, \varepsilon_\gamma + d\varepsilon_\gamma$ [cm^{-3} s^{-1}]]
$\mathcal{S}_\Phi(r)$	Relative radial sensitivity of luminescence detection
$\mathcal{S}_R(r)$	Relative radial sensitivity of QSSPC sensor
S	Surface recombination velocity [cm s^{-1}]
S_0	Illuminated interface recombination velocity [cm s^{-1}]
S_d	Unilluminated interface recombination velocity [cm s^{-1}]
S^T	Entropy [eV K^{-1}]
$\mathcal{T}$	Modulation time [s]
T	Temperature [K]

$T_{\Phi}(\varepsilon_{\gamma})$ Luminescence optical transmission function
t time [s]
Δt Time shift between modulated $\Phi(t)$ and $G(t)$ [s]
$\Delta t'$ Detector syncronization time [s]
U Voltage [V]
V Volume [cm^3]
w Phong parameter (interface optical roughness)
x,y Variables indicating the lateral substrate position [cm]
z Substrate depth along the global surface normal [cm]

Greek Symbols

α_{γ} Intrinsic absorption coefficient at photon energy ε_{γ} [cm^{-1}]
β_n Weight of n-th decay mode of excess carrier density
ϵ Dielectric constant
ε Energy [eV]
$\varepsilon_{C/V}$ Energy of conduction/valence band edge [eV]
ε_F Fermi energy [eV]
$\varepsilon_{FC/FV}$ Quasi-Fermi energy of electrons/holes [eV]
ε_G Band gap energy [eV]
$\Delta\varepsilon_G$ Band gap narrowing [eV]
ε_i Intrinsic Fermi energy [eV]
$\varepsilon_{\mathbf{k}}$ Electron energy at wave vector $\mathbf{k}$[eV]
ε_t Energy level of defect state (trap) [eV]
ε_{Γ} Phonon energy [eV]
ε_{γ} Photon energy [eV]
η Energy conversion efficiency
$\eta_{e/h}$ Electrochemical potential of electrons/holes [eV]
$\eta_{\sum}$ Separation of quasi-Fermi energies [eV]
ϑ Polar angle of light transmission
κ_n Inverse decay time of n-th decay mode of excess carrier density [s^{-1}]
$\varkappa$ Factor relating average excess carrier generation rate to electronically active bulk thickness
λ Light wavelength [nm]
λ_n Solution of transcendental equation for n-th excess carrier decay mode (general Green's function approach) [cm^{-1}]
λ_{γ} Irradiation wavelength [nm]
μ_a Ambipolar mobility [$cm^2\,V^{-1}\,s^{-1}$]
μ_c Chemical potential [eV]
$\mu_{e/h}$ Electron/hole mobilities [$cm^2\,V^{-1}\,s^{-1}$]
μ_{Γ} Phonon chemical potential [eV]
μ_{γ} Photon chemical potential [eV]
$v_{e/h}$ Thermal electron/hole velocities [$cm\,s^{-1}$]

ξ	Bias light parameter
ρ	Electrical resistivity [VA^{-1} cm]
ρ_Q	Charge density [As cm^{-3}]
σ	Electrical conductivity [$AV^{-1}\,cm^{-1}$]
$\Delta\sigma$	Excess photoconductance [$AV^{-1}\,cm^{-1}$]
σ_{c_Φ}	Standard deviation of scaling factor c_Φ [cm^6]
$\sigma_{e/h}$	Electron/hole conductivity [$AV^{-1}\,cm^{-1}$]
$\sigma_{e/h}^{cap}$	Electron/hole capture cross sections [cm^2]
σ_I	Fluctuation of detector photocurrent [A]
τ	Excess carrier (bulk) recombination lifetime [s]
τ_{aug}	Auger recombination lifetime [s]
τ_{d2a}	Effective lifetime as of *d2a* analysis [s]
τ_{diel}	Dielectric relaxation time [s]
τ_{eff}	Effective lifetime [s]
$\tau_{eff,d}$	Dynamic effective lifetime ($\tau_{n=1}$) [s]
$\tau_{eff,h}$	Harmonically modulated effective lifetime [s]
$\tau_{eff,i}$	Local effective lifetime at lateral position (x_i, y_i) [s]
$\tau_{eff,s}$	Steady-state effective lifetime [s]
τ_g	Escape time of a trapping site [s]
τ_m	Differential (measurable) decay time [s]
τ_n	Relaxation time of *n*-th mode of excess carrier decay [s]
τ_{rad}	Radiative recombination lifetime [s]
τ_{srh}	Shockey-Read-Hall recombination lifetime [s]
τ_t	Capture time of a trapping site [s]
τ_{tr}	Transit time [s]
τ_σ	Self-consistent effective lifetime [s]
Φ	Luminescence intensity
ϕ	Electrical potential [V]
φ_i	Local luminescence intensity at lateral position $[x_i, y_i]$
Ω	Solid angle of luminescent emission/detection
ω	Angular modulation frequency $2\pi/T$ [s^{-1}]

Chapter 1
Preamble

1.1 The Imperative of Renewable Energy

Fossil fuels and nuclear energy sources are scarce, and their historic and present worldwide large scale utilization entails risks and poses miscellaneous severe environmental, social, and economic threats. The economist and politician Hermann Scheer compiled seven momentous energy-related world crises [1]: These are a world climate crisis, a crisis caused by the exhaustion of and the dependence on fossil and nuclear energy sources, a crisis of energy poverty in developing countries with dramatic environmental, social, and political consequences, a political crisis due to nuclear proliferation and nuclear armament, a water crisis brought about by the competition for water between citizens and the energy industry, an agricultural crisis caused by an increasing dependence of farmers on fossil fuels, and a health crisis related to energy emissions and exposure to radioactivity.

Renewable energy sources like wind power, hydrodynamic power, solar heat, photovoltaics, and (to some extent) biomass represent sustainable energy sources provided by the sun. As such, they are inexhaustible on any timescale relevant to mankind. Exceeding the world's current primary energy demand [2] by four orders of magnitude, they truly deserve to be referred to as abundant. The above addressed benefits together with a negligible cost of ownership, their peripheral installation capability, virtually negligible carbon and water footprints and the flexibility of sustainable agricultural and land integration render renewable energy sources imperative for the pursuit of energy autonomy and for the resolution of the above addressed world crises.

1.2 Photovoltaic Energy Conversion

Photovoltaic energy conversion represents the most immediate way to convert solar power into most flexible electrical power [3]. Exploiting the internal photoelectric effect (photovoltaic effect), the conversion process is distinguished by a comparatively low entropy gain.

J. Giesecke, *Quantitative Recombination and Transport Properties in Silicon from Dynamic Luminescence*, Springer Theses, DOI: 10.1007/978-3-319-06157-3_1,

© Springer International Publishing Switzerland 2014

A photovoltaic energy return on investment less than unity is a persistent popular belief which has been proven wrong since the 1970s. Life cycle assessments have robustly indicated energy returns on investment greater than 10 already a decade ago [4, 5]—maintaining a steady increase fostered by scaling effects and technological progress [6, 7]. Depending on the solar cell technology and the local solar irradiation, some energy payback times have substantially gone below a year, and energy return on investment has already reached a factor of 50 in some cases [8].

In terms of levelized cost of electricity, photovoltaic energy conversion has lacked competitiveness compared to alternative renewable energy conversion techniques for a long time. Therefore, the deployment of photovoltaics has largely been driven by subsidies and feed-in tariffs. However, at an impressive learning rate of 22 % [9], this is likely to change in the near future. With grid parity being within reach in many countries, photovoltaics is becoming increasingly competitive not only for isolated applications. Maintaining the current learning rate, photovoltaic energy conversion is likely to become the world's cheapest energy source in the long run.

1.3 Subject of This Work

The abundance of the chemical element silicon in the earth crust and the high level of knowledge about its advantageous electronic properties as a semiconductor may have caused its broad dissemination as a base material of solar cells. As solar cell technology advances to increasing energy conversion efficiencies, accurate and precise knowledge of recombination and transport properties is vital in order to further optimize solar cell concepts.

Luminescence-based characterization techniques of silicon recombination and transport properties have experienced a considerable boost in recent years. Much research has focused on camera-based steady-state imaging techniques aiming at the determination of spatially resolved recombination and transport properties, while exploiting the benefit of fast spatially resolved information [10, 11].

The present work focuses on other assets of luminescence—seizing a dynamic carrier lifetime approach distinguished by harmonic time modulation of excess carrier generation. Phase-sensitive harmonically modulated approaches aiming at the determination of recombination and transport properties of semiconductors trace back at least to the late 1940s, when Tolstoi and Feofilov determined phosphorescence lifetimes of photoexcited phosphors from phase information [12]. Their self-consistent approach of hysteresis-elimination spread among researchers in the 1950s, e.g. to adapt the time constant of an RC circuit in order to match the phase lag between a chopped [13] or harmonically modulated [14] optical irradiation and the *excess conductance* of a semiconductor substrate. In the 1970s, methods to investigate semiconductor recombination properties via harmonic time modulation of both light-induced [15, 16] and electron beam-induced [17] short circuit currents had been presented. In the 1980s, contactless lock-in based carrier lifetime approaches based on modulated photoluminescence (by Guidotti et al. [18, 19]) and on modulated free carrier absorption (by Sanii et al. [20, 21], later pursued by Glunz and Warta [22]) emerged. Harmonically modulated photoluminescence approaches have later been refined, extended, and adapted to photovoltaic applications by Trupke and Bardos [23–25], by Brüggemann and Reynolds [26], and by Mandelis and coworkers [27–29].[1]

The present work aims at novel techniques and applications to accurately and precisely access recombination and transport properties via harmonically modulated luminescence. The essential impetus of the present work was to exploit the very advantageous properties of dynamic luminescence in order to obtain a maximum of information about recombination and transport properties with a minimum of a priori assumptions having to rely upon.

1.4 Outline

Chapter 2 is devoted to the fundamental working principle of a solar cell. It briefly motivates the relevance of carrier recombination and carrier transport to solar cell operation.

Chapter 3 is concerned with the dynamics of charge carriers in silicon. Essential concepts like carrier recombination lifetime, surface recombination, and carrier diffusion are introduced. Subsequently, they are used to define and solve the relevant transport equations for the cases of interest—focusing on harmonic modulation of excess carrier generation rate.

The theory of spontaneous emission of radiation is outlined in Chap. 4 covering silicon luminescence. The link from luminescent emission to electronic properties is drawn, and it is briefly sketched and exemplified how this link has been exploited in several measurement techniques in the past. Here again, the main emphasis is on dynamic measurement techniques.

Chapter 5 is concerned with dynamic lifetime measurement techniques. Dynamic measurement techniques are classified in terms of their characteristic timescales, and different perceptions of a quasi-steady-state condition are discussed. Particular attention is given to a self-consistent dynamic approach to effective carrier lifetime.

Chapter 6 details key constraints of dynamic carrier lifetime. Considerations for dynamic lifetime measurements on substrates with an excess carrier generation much confined to highly recombinative interfaces are detailed [34–36], and a theory of light-biased decay time (differential lifetime) is derived [37].

Chapter 7 gives an account of the evolution of the harmonically modulated luminescence experimental setup in the course of this work. A detailed description of the state of the art experimental setup is followed by the description of three key innovations, involving considerations for carrier lifetime measurements in the microsecond

---

[1] Harmonic time modulation of optical irradiation has long been known and used to resolve relaxation times of optical transitions not only in semiconductor physics, but also in biology, biochemistry, and in the medical sciences [30–33].

range [38], carrier lifetime measurements on metalized substrates and solar cells [39], and dynamic electroluminescence [40].

Chapter 8 details conceptual advances regarding the determination of recombination properties. A class of three closely related phase-sensitive basic approaches to injection-dependent effective carrier lifetime via harmonically modulated luminescence is presented [37, 41–43]—complementing Trupke's self-consistent approach [25]. Furhter applications involve a technique to measure injection-dependent effective carrier lifetime down to a very low injection level [44], a spatially resolved carrier lifetime calibration of steady-state photoluminescence images via quasi-steady-state harmonically modulated photoluminescence measurements [45, 46], and a combined approach to access bulk lifetime and surface recombination velocity of silicon ingots [36].

Chapter 9 details conceptual advances regarding the determination of transport properties. These involve a class of photoluminescence techniques sensitive to net dopant concentration [47, 48], which can also be used for a determination of majority carrier mobility, and a transit time approach to minority carrier mobility [49].

Chapter 10 summarizes key findings and addresses possible further investigations as well as potential implementations of the gained knowledge in the future.

References

1. H. Scheer, *Energy Autonomy: The Ecomomic, Social and Technological Case for Renewable Energy* (Earthscan, London, 2007)
2. R. Dudley, *BP Statistical Review of World Energy, 2011*. Tech. rep, BP (2011)
3. D.M. Chapin, C.S. Fuller, G.L. Pearson, A new silicon p-n junction photocell for converting solar radiation into electrical power. J. Appl. Phys. **25**, 676–677 (1954)
4. E.A. Alsema, Energy pay-back time and CO_2 emissions of PV Systems. Prog. Photovoltaics: Res. Appl. **8**, 17–25 (2000)
5. V. Fthenakis, E. Alsema, Photovoltaics energy payback times, greenhouse gas emissions and external costs: 2004–early 2005 status. Prog. Photovoltaics: Res. Appl. **14**, 275–280 (2006)
6. W.C. Sinke, W. van Hooff, G. Coletti, B. Ehlen, G. Hahn, S. Reber, J. John, G. Beaucarne, E. van Kerschaver, M. de Wild-Scholten, A. Metz, Wafer-based crystalline silicon modules at 1 €/WP: final results from the crystal clear integrated project, in *Proceedings of the 24th EU PVSEC* (Hamburg, 2009), pp. 845–856
7. C. Olson, M. de Wild-Scholten, M. Scherff, P. J. Ribeyron, Life cycle assessment of heterojunction solar cells and modules, in *Proceedings of the 26th EU PVSEC* (Hamburg, 2011), pp. 3822–3825
8. V. Fthenakis, How long does it take for photovoltaics to produce the energy used? PE Mag. **1**, 16–17 (2012)
9. D. Gielen, *Renewable Energy Technologies: Cost Analysis Series—Solar Photovoltaics, 2012*, Tech. rep, IRENA (2012)
10. T. Fuyuki, H. Kondo, T. Yamazaki, Y. Takahashi, Y. Uraoka, Photographic surveying of minority carrier diffusion length in polycrystalline silicon solar cells by electroluminescence. Appl. Phys. Lett. **86**, 262108 (2005)
11. T. Trupke, R.A. Bardos, M.C. Schubert, W. Warta, Photoluminescence imaging of silicon wafers. Appl. Phys. Lett. **89**, 044107 (2006)
12. N.A. Tolstoi, P.P. Feofilov, Novel method to investigate the processes of rise and decay of luminescence (translation). Bull. Russ. Acad. Sci. Phys. **13**, 211–217 (1949)

13. A.R. Engler, C.J. Kevane, Direct reading minority carrier lifetime measuring apparatus. Rev. Sci. Instr. **28**, 548–551 (1957)
14. E. Harnik, A. Many, N.B. Grover, Phase shift method of carrier lifetime measurements in semiconductors. Rev. Sci. Instr. **29**, 889–891 (1958)
15. H. Reichl, H. Bernt, Lifetime measurements in silicon epitaxial materials. Solid-State Electron. **18**, 453–458 (1975)
16. G. Schwab, H. Bernt, H. Reichl, Measurement of minority carrier lifetime profiles in silicon. Solid-State Electron. **20**, 91–94 (1977)
17. O. von Roos, Recombination lifetimes and surface recombination velocities of minority carriers in N-P junctions. A new method for their determination by means of a stationary amplitude-modulated electron beam. J. Appl. Phys. **50**, 3738–3742 (1979)
18. D. Guidotti, J.S. Batchelder, J.A. Van Vechten, A. Finkel, Nondestructive depth profiling of carrier lifetimes in full silicon wafers. Appl. Phys. Lett. **48**, 68–70 (1986)
19. D. Guidotti, J.S. Batchelder, A. Finkel, P.D. Gerber, J.A. Van Vechten, Modeling of luminescence phase delay for nondestructive characterization of Si wafers. J. Appl. Phys. **66**, 2542–2553 (1989)
20. F. Sanii, R.J. Schwartz, R.F. Pierret, W.M. Au, The measurement of bulk and surface recombination by means of modulated free carrier absorption, in *Proceedings of the 20th IEEE PVSC* (Las Vegas, 1988), pp. 575–580
21. F. Sanii, F.P. Giles, R.J. Schwartz, J.L. Gray, Contactless nondestrcutive measurement of bulk and surface recombination using frequency-modulated free carrier absorption. Solid-State Electron. **35**, 311–317 (1992)
22. S.W. Glunz, W. Warta, High-resolution lifetime mapping using modulated free-carrier absorption. J. Appl. Phys. **77**, 3243–3247 (1994)
23. T. Trupke, R.A. Bardos, F. Hudert, P. Würfel, J. Zhao, A. Wang, M.A. Green, Effective excess carrier lifetimes exceeding 100 milliseconds in float zone silicon determined from photoluminescence, in *Proceedings of the 19th EU PVSEC* (Paris, 2004), pp. 758–761
24. T. Trupke, R.A. Bardos, Photoluminescence: a surprisingly sensitive lifetime technique, in *Proceedings of the 31st IEEE PVSC* (Orlando, 2005), pp. 903–906
25. T. Trupke, R.A. Bardos, M.D. Abbott, Self-consistent calibration of photoluminescence and photoconductance lifetime measurements. Appl. Phys. Lett. **87**, 184102 (2005)
26. R. Brüggemann, S. Reynolds, Modulated photoluminescence studies for lifetime determination in amorphous-silicon passivated crystalline-silicon wafers. J. Non-Cryst. Solids **352**, 1888–1891 (2006)
27. A. Mandelis, J. Batista, D. Shaughnessy, Infrared photocarrier radiometry of semiconductors: physical principles, quantitative depth profilometry, and scanning imaging of deep subsurface electronic defects. Phys. Rev. B **67**, 205208 (2003)
28. A. Melnikov, A. Mandelis, J. Tolev, P. Chen, S. Huq, Infrared lock-in carrierography (photocarrier radiometric imaging) of Si solar cells. J. Appl. Phys. **107**, 114513 (2010)
29. Q. Sun, A. Melnikov, A. Mandelis, Quantitative self-calibrating lock-in carrierographic lifetime imaging of silicon wafers. Appl. Phys. Lett. **101**, 242107 (2012)
30. R.B. Cundall, R.B. Dale, *Time-Resolved Fluorescence Spectroscopy in Biochemistry and Biology* (Plenum Press, New York, 1983)
31. P.C. Schneider, R.M. Clegg, Rapid acquisition, analysis, and display of fluorescence lifetime-resolved images for real-time applications. Rev. Sci. Instr. **68**, 4107–4119 (1997)
32. A. Squire, P.J. Verveer, P.I.H. Bastiaens, Multiple frequency fluorescence lifetime imaging microscopy. J. Microsc. **197**, 136–149 (2000)
33. E.B. van Munster, T.W.J. Gadella, Fluorescence lifetime imaging microscopy (FLIM), in *Microscopy Techniques* (Springer, Berlin, 2005), pp. 143–175
34. R.A. Sinton, T. Trupke, Limitations on dynamic excess carrier lifetime calibration methods. Prog. Photovoltaics: Res. Appl. **20**, 246–249 (2012)
35. J.A. Giesecke, W. Warta, Understanding carrier lifetime measurements at nonuniform recombination. Appl. Phys. Lett. **104**, 082103 (2014)

36. J.A. Giesecke, R.A. Sinton, M.C. Schubert, S. Riepe, W. Warta, Determination of bulk lifetime and surface recombination velocity of silicon ingots from dynamic photoluminescence. IEEE J. Photovoltaics **3**, 1311–1318 (2013)
37. J.A. Giesecke, S.W. Glunz, W. Warta, Understanding and resolving the discrepancy between differential and actual minority carrier lifetime. J. Appl. Phys. **113**, 073706 (2013)
38. J.A. Giesecke, W. Warta, Microsecond carrier lifetime measurements in silicon via quasi-steady-state photoluminescence. Prog. Photovoltaics: Res. Appl. **20**, 238–245 (2012)
39. J.A. Giesecke, B. Michl, F. Schindler, M.C. Schubert, W. Warta, Minority carrier lifetime of silicon solar cells from quasi-steady-state photoluminescence. Sol. Energy Mater. Sol. Cells **95**, 1979–1982 (2011)
40. J.A. Giesecke, M.C. Schubert, W. Warta, Carrier lifetime from dynamic electroluminescence. IEEE J. Photovoltaics **3**, 1012–1015 (2013)
41. J.A. Giesecke, M.C. Schubert, D. Walter, W. Warta, Minority carrier lifetime in silicon wafers from quasi-steady-state photoluminescence. Appl. Phys. Lett. **97**, 092109 (2010)
42. J.A. Giesecke, M.C. Schubert, W. Warta, Self-sufficient minority carrier lifetime in silicon from quasi-steady-state photoluminescence. Phys. Stat. Sol. A **209**, 2286–2290 (2012)
43. J.A. Giesecke, S.W. Glunz, W. Warta, Determination of actual carrier lifetime from differential measurements. Energy Procedia **38**, 59–65 (2013)
44. J.A. Giesecke, T. Niewelt, M. Rüdiger, M. Rauer, M.C. Schubert, W. Warta, Broad range injection-dependent minority carrier lifetime from photoluminescence. Sol. Energy Mater. Sol. Cells **102**, 220–224 (2012)
45. J.A. Giesecke, M.C. Schubert, B. Michl, F. Schindler, W. Warta, Minority carrier lifetime imaging of silicon wafers calibrated by quasi-steady-state photoluminescence. Sol. Energy Mater. Sol. Cells **95**, 1011–1018 (2011)
46. J.A. Giesecke, B. Michl, F. Schindler, M.C. Schubert, W. Warta, Spatially resolved carrier lifetime calibrated via quasi-steady-state photoluminescence. Energy Procedia **8**, 64–70 (2011)
47. J.A. Giesecke, M.C. Schubert, W. Warta, Measurement of net dopant concentration via dynamic photoluminescence. J. Appl. Phys. **112**, 063704 (2012)
48. J.A. Giesecke, D. Walter, F. Kopp, P. Rosenits, M.C. Schubert, W. Warta, Simultaneous determination of carrier lifetime and net dopant concentration of silicon wafers from photoluminescence, in *Proceedings of the 35th IEEE PVSC* (Honolulu, 2010), pp. 847–851
49. J.A. Giesecke, F. Schindler, M. Bühler, M.C. Schubert, W. Warta, Accurate determination of minority carrier mobility in silicon from quasi-steady-state photoluminescence. J. Appl. Phys. **113**, 213705 (2013)

Chapter 2
Introduction to Solar Cell Operation

This chapter discusses the very fundamental working principle of a solar cell. It is intended to briefly motivate the relevance of carrier recombination and carrier transport to solar cell operation without going into technological detail.

2.1 Doped Semiconductors

2.1.1 The Band Gap

An adequate description of electrons in a crystal (represented by a periodic assembly of atoms) would require the solution of a many-electron problem with a Hamiltonian involving both the interaction between electrons and atomic nuclei as well as the interaction among electrons. However, in order to qualitatively motivate some fundamental properties of metals and semiconductors, it may suffice to focus on weakly bound electrons (Bloch electrons) in an effective one-electron potential representing the interaction with a matrix of atomic nuclei.[1] For all Bravais lattice vectors $\mathbf{R}$ of the crystal lattice, the potential $\mathcal{U}(\mathbf{r})$ must obey

$$\mathcal{U}(\mathbf{r}+\mathbf{R}) = \mathcal{U}(\mathbf{r}). \tag{2.1}$$

Bloch's theorem states that for such a periodic effective one-electron potential $\mathcal{U}(\mathbf{r})$, the eigenstates $\psi_{\mathbf{k}}$, which are solutions of the Schrödinger equation

$$\left(-\frac{\hbar^2}{2m}\nabla^2 + \mathcal{U}(\mathbf{r})\right)\psi_{\mathbf{k}}(\mathbf{r}) = \varepsilon_{\mathbf{k}}\psi_{\mathbf{k}}(\mathbf{r}) \tag{2.2}$$

[1] The main lines of reasoning in this section draw on full derivations given in [1].

J. Giesecke, *Quantitative Recombination and Transport Properties in Silicon from Dynamic Luminescence*, Springer Theses, DOI: 10.1007/978-3-319-06157-3_2,
© Springer International Publishing Switzerland 2014

to eigenvalues $\varepsilon_{\mathbf{k}}$, can be chosen as the product of a plane wave and a function $u_{\mathbf{k}}(\mathbf{r})$ with the periodicity of the Bravais lattice

$$\psi_{\mathbf{k}}(\mathbf{r}) = e^{i\mathbf{k}\cdot\mathbf{r}} u_{\mathbf{k}}(\mathbf{r}). \tag{2.3}$$

With reciprocal lattice vectors $\mathbf{K}$, the periodicity of the Bravais lattice allows expression of the eigenstates $\psi_{\mathbf{k}}$ and the potential $\mathcal{U}(\mathbf{r})$ in terms of Fourier components $c_{\mathbf{k}-\mathbf{K}}$ and $\mathcal{U}_{\mathbf{K}}$ as

$$\psi_{\mathbf{k}}(\mathbf{r}) = \sum_{\mathbf{K}} c_{\mathbf{k}-\mathbf{K}} e^{i(\mathbf{k}-\mathbf{K})\cdot\mathbf{r}} \tag{2.4}$$

$$\mathcal{U}(\mathbf{r}) = \sum_{\mathbf{K}} \mathcal{U}_{\mathbf{K}} e^{i\mathbf{K}\cdot\mathbf{r}}. \tag{2.5}$$

In the limit of vanishing $\mathcal{U}_{\mathbf{K}}$, insertion of this expression for $\psi_{\mathbf{k}}$ into Eq. 2.2 reveals a parabolic electron band structure with degenerate eigenvalues at Bragg planes (Brillouin zone boundaries). For a weak but finite $\mathcal{U}_{\mathbf{K}}$, perturbation theory yields a splitting of otherwise degenerate energy levels at a magnitude of $2 \cdot |\mathcal{U}_{\mathbf{K}}|$ [1]. This finding implies a vanishing density of electronic states at certain energy levels.

According to the Pauli exclusion principle, the ground state of a gas of weakly bound Bloch electrons (valence electrons) in a crystalline solid implies occupation of all one-electron levels below a so-called Fermi energy ε_F, which is distinguished by requiring the total number of one-electron levels with energies less than ε_F to be equal to the number of Bloch electrons in a crystalline solid [1]. The Pauli exclusion principle is also reflected in the Fermi-Dirac distribution representing the occupation probability

$$f(\varepsilon_{\mathbf{k}}) = \frac{1}{e^{\frac{\varepsilon_{\mathbf{k}}-\varepsilon_F}{k_B T}} + 1} \tag{2.6}$$

of free (or quasi-free) one-electron levels. If—for reasons outlined above—there is a considerable difference between the highest occupied level and the lowest unoccupied level in the ground state of the electron gas (at temperature $T = 0$), this energy difference is referred to as a *band gap*. Adjacent bands are denoted as valence band (below the band gap) and conduction band (above the band gap).

2.1.2 Classification of Semiconductors

Both electronic and optical properties of crystalline solids are essentially affected by the existence and width of a band gap and by the characteristics of the band structure near the band gap. Insulating solids are distinguished by a band gap much greater than $k_B T$. Intrinsic semiconductors feature a band gap in the range of $k_B T$, whereas metals feature a vanishing band gap. The above perception of the term semiconductor [1] is somewhat inadequate with regard to most semiconductors of technological relevance (e.g. for silicon, the band gap is more than 40 times greater than room temperature $k_B T$). Therefore, here the term semiconductor shall denote crystalline solids with a band gap in the energy range of visible light.

2.1.3 Interaction with Light—Indirect Semiconductors

Electrons in solids may interact with photons. Such optical transitions enable excitation of electrons from their ground state to a previously unoccupied state. The interaction between light and matter must obey conservation of energy, momentum, and angular momentum. For indirect semiconductors such as silicon and germanium, the locations of conduction band minimum and valence band maximum in **k**-space differ such that the interaction between electrons and photons must involve another excitation (e.g. a phonon) in order to satisfy conservation of momentum. The characteristic of an indirect semiconductor largely affects the optical properties of silicon, as the prerequisite of excitation or anihilation of phonons renders both absorption of light as well as radiative recombination very unlikely compared to a direct semiconductor of comparable band gap. This fact is essential when relating recombination lifetime as inferred from measurements of radiative recombination (luminescence) to other recombination causes such as scattering of electrons at impurities and crystal defects.

2.1.4 Doping

In a semiconductor with n_v valence electrons per atom, adding atoms with $n_v \pm 1$ valence electrons induces shallow electronic states in the band gap. Such states are much confined ($\sim k_B T$) to an energy range close to the conduction band edge for $n_v + 1$ (donors, n-type doping, e.g. phosphorous doping of silicon) whereas they are much confined to an energy range close to the valence band edge for $n_v - 1$ (acceptors, p-type doping, e.g. boron doping of silicon).

Adding of donors (free electrons) or acceptors (free vacancies or *holes*) to a semiconductor brings about a significant change of electrical conductivity, which can be explained by a high degree of room temperature ionization of the electronic states induced via doping—or in other words by a high free carrier concentration. Acceptor levels allow thermal excitation of electrons from the adjacent valence band, whereas donor electrons are easily thermally excited into the adjacent conduction band.

With a density of electronic states per energy and volume in a solid $D(\varepsilon_{\mathbf{k}})$, the dark densities of electrons and holes are

$$n_e^0 = \int_{\varepsilon_C}^{\infty} d\varepsilon_{\mathbf{k}}\, f(\varepsilon_{\mathbf{k}}) D(\varepsilon_{\mathbf{k}}) \quad (2.7)$$

$$n_h^0 = \int_{-\infty}^{\varepsilon_V} d\varepsilon_{\mathbf{k}} \left[1 - f(\varepsilon_{\mathbf{k}})\right] D(\varepsilon_{\mathbf{k}}). \quad (2.8)$$

Here, ε_C and ε_V denote edges of the conduction and valence band, respectively. For a symmetric shape of $D(\varepsilon_{\mathbf{k}})$ with respect to the band gap (intrinsic semiconductor), this implies a Fermi energy close to the center of the band gap, leading to an intrinsic free carrier concentration in silicon of $n_i \approx 10^{10}\,\mathrm{cm}^{-3}$ at room temperature [2–5]. However, with a substantial amount of electronic states near either conduction band or valence band, the symmetry of $D(\varepsilon_{\mathbf{k}})$ is destroyed, leading to a shift of Fermi energy toward the respective band edge in order to satisfy Eqs. 2.7 and 2.8. For $T = 0$, the Fermi energy must lie between the respective band edge and the adjacent energy level of the donors or acceptors—leading to a vanishing concentration of free carriers. Yet, at an increasing temperature the Fermi level advances toward the center of the bandgap, such that at room temperature free carrier concentration (ionized dopant concentration) nearly corresponds to dopant concentration. At typical dopant concentrations of 10^{15}–$10^{16}\,\mathrm{cm}^{-3}$ this brings about an increase of free carrier density and hence of electrical conductivity by 5–6 orders of magnitude.

The density of free carriers due to ionized dopant atoms is denoted as majority carrier density, whereas the density of the complementary charge carrier species is denoted as minority carrier density. The shift of Fermi energy in a doped semiconductor away from the minority carrier band leads to an inverse proportionality between dark minority carrier density and dopant concentration.

2.2 Quasi-Fermi Energy—Electrochemical Potential

While the band population via free electrons (and—equivalently—holes) may be described via Fermi-Dirac statistics according to Eq. 2.7, this description is complicated in the presence of additional generation of excess carrier density Δn beyond dark equilibrium. The concept of so-called Quasi Fermi levels aims at an adequate description of such a state of electrochemical non-equilibrium.

2.2.1 Quasi-Fermi Levels

Due to a square root shape of the density of states $D(\varepsilon_{\mathbf{k}}) \propto \sqrt{\pm\left(\varepsilon_{\mathbf{k}} - \varepsilon_{C/V}\right)}$, free carrier densities in the dark may be accurately expressed as[2]

$$n_e^0 = N_C e^{\frac{\varepsilon_F - \varepsilon_C}{k_B T}}, \tag{2.9}$$

$$n_h^0 = N_V e^{\frac{\varepsilon_V - \varepsilon_F}{k_B T}}, \tag{2.10}$$

with N_C and N_V denoting effective densities of states of the conduction and valence band

$$N_C = 2\left(\frac{2\pi m_e^* k_B T}{h^2}\right)^{\frac{3}{2}}, \tag{2.11}$$

$$N_V = 2\left(\frac{2\pi m_h^* k_B T}{h^2}\right)^{\frac{3}{2}}, \tag{2.12}$$

and with effective masses of electrons and holes [1] $m^*_{e/h}$ and the Planck constant h.

In the case of excess carrier generation beyond thermal equilibrium, the population of valence and conduction band through excess carriers is no longer adequately described via Fermi-Dirac statistics. Whereas in the dark state, the product of minority and majority carrier densities must correspond to intrinsic carrier density $n_e^0 n_h^0 = n_i^2$, now both carrier species feature enhanced densities, requiring a shift of Fermi energy toward both band edges. This dilemma is resolved by the concept of quasi-Fermi energies, which allow the description of a state of electrochemical non-equilibrium of a semiconductor under illumination via Fermi-Dirac statistics. Quasi-Fermi energies ε_{FC} and ε_{FV} of electrons and holes replace Fermi energy ε_F in the Fermi-Dirac distribution (cf. Eq. 2.6), respectively. Electron and hole densities can thus be expressed as[3]

$$n_e = n_e^0 + \Delta n = N_C e^{\frac{\varepsilon_{FC} - \varepsilon_C}{k_B T}}, \tag{2.13}$$

$$n_h = n_h^0 + \Delta n = N_V e^{\frac{\varepsilon_V - \varepsilon_{FV}}{k_B T}}. \tag{2.14}$$

The separation of quasi-Fermi energies of electrons and holes relates to the implied voltage U of a semiconductor under illumination $\varepsilon_{FC} - \varepsilon_{FV} = qU$. The following derivation reveals the relation between quasi-Fermi levels and the so-called electrochemical potential of carriers.

[2] This approximation is deemed accurate for $|\varepsilon_F - \varepsilon_{C/V}| \gtrsim 3k_B T$.

[3] Here, $\Delta n_e = \Delta n_h = \Delta n$ is assumed.

2.2.2 The Electrochemical Potential

The change of internal energy dE of a statistical ensemble such as a gas of electrons or holes corresponds to the balance of changes of all relevant energy forms[4]

$$dE = TdS^T - pdV + \mu_c dN + \phi dQ. \tag{2.15}$$

Here, changes of thermal energy, compressional energy, chemical energy, and electrical energy are considered. The system is distinguished by the intensive variables temperature T, pressure p, chemical potential μ_c, and electrical potential ϕ. Further, the extensive variables giving rise to a change of internal energy are the entropy S^T, the volume V, the particle number N, and the charge Q. The energy which a solar cell delivers to a load is given by the sum of the changes of free energy $dF_{e/h}$ of electrons and holes, respectively.

$$dF = dE - d\left(TS^T\right) = S^T dT - pdV + \mu_c dN + \phi dQ \tag{2.16}$$

With constant semiconductor volume and temperature during solar cell operation, this expression simplifies accordingly such that

$$dF = (\eta_e + \eta_h)\, dN, \tag{2.17}$$

with the electrochemical potential

$$\eta_{e/h} = \mu_{c,e/h} \mp q\phi, \tag{2.18}$$

and with the positive elementary charge q. With a free energy increment according to Eq. 2.17, the electrochemical potential represents the essential potential of excess carriers in semiconductors. Its gradient gives rise to net currents of excess carriers.

Consider the mean energy per particle of a gas of $N_{e/h}$ free electrons/holes at temperature T

$$\langle \varepsilon_{e/h} \rangle = \pm\varepsilon_{C/V} + \frac{3}{2} k_B T = \frac{TS^T_{e/h} + p_{e/h}V + \eta_{e/h}}{N_{e/h}}. \tag{2.19}$$

Making use of the modified Sackur-Tetrode equation expressing the entropy of an ideal gas [6]

$$S^T_{e/h} = k_B N_{e/h} \left(\frac{5}{2} + \ln \frac{N_{C/V}}{n_{e/h}}\right) \tag{2.20}$$

and exploiting the equation of state of an ideal gas

[4] This derivation proceeds along the lines of a derivation given in [6].

$$p_{e/h}V = N_{e/h}k_BT, \tag{2.21}$$

one obtains

$$n_e = N_C e^{\frac{\eta_e - \varepsilon_C}{k_BT}}, \tag{2.22}$$

$$n_h = N_V e^{\frac{\eta_h + \varepsilon_V}{k_BT}}. \tag{2.23}$$

Comparison to Eqs. 2.13 and 2.14 reveals

$$\eta_e = \varepsilon_{FC} \tag{2.24}$$

$$\eta_h = -\varepsilon_{FV}. \tag{2.25}$$

Therefore, the implied voltage U of a semiconductor under illumination can be expressed via the sum of electrochemical potentials

$$qU = \varepsilon_{FC} - \varepsilon_{FV} = \eta_e + \eta_h = \eta_\Sigma. \tag{2.26}$$

With a chemical potential of the form

$$\mu_{c,e/h} = k_BT \ln \frac{n_{e/h}}{a_{e/h}} \tag{2.27}$$

with constant densities $a_{e/h}$,[5] according to Eq. 2.18 the electrochemical equilibrium condition (vanishing separation of quasi-Fermi energies $\eta_\Sigma = 0$) yields

$$a_e a_h = n_e^0 n_h^0 = n_i^2. \tag{2.28}$$

Consequently, the separation of quasi-Fermi energies η_Σ can be expressed in terms of the product of minority and majority carrier density as

$$\eta_\Sigma = k_BT \ln \frac{n_e n_h}{n_i^2}. \tag{2.29}$$

2.3 The pn-Junction

The essential task of a photovoltaic device is to separate electrical charge released upon absorption of light—preferrably at the highest possible conversion efficiency from photon energy to electrical energy (as the product of voltage and electrical charge). This essential task requires efficient mechanisms of both light absorption

[5] $a_{e/h}$ enable a dimensionless logarithmic argument and incorporate a constant offset of $\mu_{c,e/h}$.

and charge separation. In semiconductor photovoltaic devices, this task is commonly accomplished by a pn-junction within an absorber. This is the interface region between n- and p-doped semiconductor volumes. The light absorbing semiconductor bulk supplies electrical charge, while the pn-junction provides a means to separate it.

The charge current released by an illuminated pn-junction device with a separation of quasi-Fermi energies $qU = \eta_{\sum}$ gives rise to the delivery of work to an external load. In the following, this charge current is motivated—substantiating the relevance of transport properties such as carrier mobilities, diffusion coefficients, and dopant concentrations.

2.3.1 Dark pn-Junction

Electron and hole charge current density in semiconductors can be expressed in terms of the gradient of the electrochemical potential. With the electrical conductivity

$$\sigma_{e/h} = q n_{e/h} \mu_{e/h} \tag{2.30}$$

and with $\mu_{e/h}$ denoting electron and hole carrier mobilities, electron or hole charge current density is

$$\mathbf{j}_{Q,e/h} = \pm \frac{\sigma_{e/h}}{q} \nabla \eta_{e/h}. \tag{2.31}$$

Insertion of the electrochemical potential of Eq. 2.18 with a chemical potential given by Eq. 2.27 yields charge currents

$$\mathbf{j}_{Q,e/h} = \pm \frac{\sigma_{e/h}}{q} \left(\frac{k_B T}{n_{e/h}} \nabla n_{e/h} \pm q\mathbf{E} \right), \tag{2.32}$$

with a local electrical field $\mathbf{E} = -\nabla\phi$. In the dark, i.e. in electrochemical equilibrium, net current must vanish everywhere in the semiconductor, which is equivalent to the condition

$$\frac{k_B T}{n_{e/h}} \nabla n_{e/h} = \pm q\mathbf{E}. \tag{2.33}$$

Thus, in dark equilibrium, the gradient of electron and hole densities must be balanced by a local electrical field. Integrating Eq. 2.32 across the pn-junction from the n-contact to the p-contact (indicated via superscript here) yields

$$\phi^n_{e/h} - \phi^p_{e/h} = \pm \frac{k_B T}{q} \ln \frac{n^n_{e/h}}{n^p_{e/h}}. \tag{2.34}$$

The quantity $\phi^n_{e/h} - \phi^p_{e/h}$—which is also referred to as a dark diffusion voltage [6]—accounts for the compensation of electrical and chemical potential differences across the pn-junction such that in electrochemical equilibrium the voltage between the terminals of a (dark) pn-junction device is zero.

The large difference between free electron and hole densities across the pn-junction leads to diffusion of carriers from the region where they represent majority carriers to the region where they represent minority carriers. This diffusion process brings about a buildup of local charge within a so-called space charge region.

2.3.2 Illuminated pn-Junction

Illumination of a pn-junction with photon energies above the band gap leads to band-to-band photon absorption and generation of excess charge carrier density Δn. This process increases the number of free electrons and holes by the same amount. Dark majority and minority carrier concentrations commonly differ by a factor of

$$\frac{n^0_{maj}}{n^0_{min}} = \left(\frac{n^0_{maj}}{n_i}\right)^2 \ggg 1. \tag{2.35}$$

Therefore, the relative increase of minority carrier concentration upon illumination is much greater than the relative increase of majority carrier concentration, and the space charge and the diffusion voltage of the pn-junction are substantially reduced as a consequence of excess carrier generation—as can already be concluded from Eq. 2.34.

$$\left|\phi^n_{e/h} - \phi^p_{e/h}\right|_{\Delta n=0} > \left|\phi^n_{e/h} - \phi^p_{e/h}\right|_{\Delta n>0} \tag{2.36}$$

In the metal contacts of a pn-junction, there is no band gap and therefore excess carrier density must smoothly approach zero toward the terminals at the side faces of a pn-junction. This again implies that the difference of chemical potentials between the terminals of a pn-junction must not change upon excess carrier generation

$$\left|\ln n^n_{e/h} - \ln n^p_{e/h}\right|_{\Delta n=0} = \left|\ln n^n_{e/h} - \ln n^p_{e/h}\right|_{\Delta n>0}. \tag{2.37}$$

Thus, the voltage delivered at the metal electrodes of an illuminated pn-junction corresponds to the amount by which the diffusion voltage of the pn-junction (cf. Eq. 2.34) is reduced through a finite excess carrier density. This state of electrochemical non-equilibrium is equivalent to a finite separation of quasi-Fermi energies $0 \neq \eta_{\sum} = qU$. Figure 2.1 depicts the space charge, the band edges, and the quasi-Fermi levels of electrons and holes in a pn-junction device both, in (dark) electrochemical equilibrium, and under illumination or carrier injection.

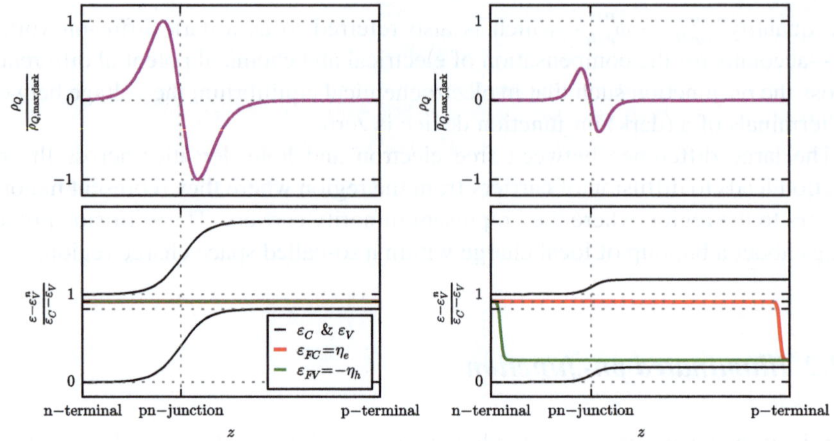

Fig. 2.1 Depiction of local charge density ρ_Q, valence and conduction band edges, and the quasi-Fermi energies (± electrochemical potentials) of electrons and holes across a pn-junction. The state of electrochemical equilibrium without illumination (*left*) is compared to the non-equilibrium state under illumination—featuring a finite separation of quasi-Fermi energies $\eta_\Sigma = qU$ corresponding to the implied voltage U. ε_V^n denotes the energy of the valence band edge at the n-terminal

2.3.3 Charge Current of a pn-Junction

As yet, the implied voltage of a pn-junction device under illumination has been motivated. However, what has not been motivated so far is the nature of the charge current out of the pn-junction device, which gives rise to the delivery of work in an external load. With electrical conductivity as specified in Eq. 2.30 and along the lines of Eq. 2.31, the total charge current density in a pn-junction device corresponds to the sum of charge current densities of electrons and holes

$$\mathbf{j}_Q = \frac{1}{q}\sum_{e/h} \pm\sigma_{e/h}\nabla\eta_{e/h}. \tag{2.38}$$

The charge current released at the terminals of a pn-junction device is a majority carrier drift current. As such, the solar cell current is induced by an electrical field—albeit not explicitly by the field of the space charge region, which does not deliver work on carriers [6]. In order to motivate the driftlike characteristic of a solar cell current, the nature of drift and dielectric relaxation has to be reflected upon. As one may conclude from Eqs. 2.17 and 2.18, minimization of free energy in a solar cell is equivalent to minimization of the sum of chemical potentials (via diffusion/random motion) and of electrical potentials (via field-driven drift/dielectric relaxation) of charge carriers, respectively.

2.3.3.1 Dielectric Relaxation

Considering the dissipation of a local charge density ρ_Q, the combination of the conservation of charge[6]

$$\frac{\partial \rho_Q}{\partial t} + \nabla \mathbf{j}_Q = 0 \tag{2.39}$$

with the charge current density

$$\mathbf{j}_Q = \sigma \mathbf{E} \tag{2.40}$$

and with the differential representation of Gauss's law

$$\epsilon\epsilon_0 \nabla \mathbf{E} = \rho_Q \tag{2.41}$$

yields the ordinary equation

$$\left(\frac{d}{dt} + \frac{\sigma}{\epsilon\epsilon_0}\right) \rho_Q = 0. \tag{2.42}$$

Its solution for an initial value $\rho_Q(t = 0) = \rho_{Q,0}$ and with a dielectric relaxation time $\tau_{diel} = \frac{\epsilon\epsilon_0}{\sigma}$ is

$$\rho_Q(t) = \rho_{Q,0} e^{-\frac{t}{\tau_{diel}}}. \tag{2.43}$$

With electrical conductivity as of Eq. 2.30, τ_{diel} must be much shorter for majority carriers than for minority carriers. Therefore, local charge imbalances in doped (and nondegenerate) semiconductors dissipate via dielectric relaxation of majority carriers. With the dielectric constant of silicon $\epsilon \approx 12$, the vacuum permittivity [4] $\epsilon_0 = 8.854 \times 10^{-14}\,\mathrm{A\,s\,V^{-1}\,cm^{-1}}$, a carrier mobility of roughly $\mu \approx 10^3\,\mathrm{cm^2\,V^{-1}\,s^{-1}}$, and the elementary charge $q = 1.602 \times 10^{-19}\,\mathrm{A\,s}$, τ_{diel} may be roughly approximated as

$$\tau_{diel} \approx \frac{7 \times 10^3}{n_{maj}}\,\mathrm{cm^{-3}s}, \tag{2.44}$$

which is below the range of picoseconds for typical dopant concentrations (e.g. $\tau_{diel} \approx 5 \times 10^{-13}\,\mathrm{s}$ for $1\,\Omega\,\mathrm{cm}$ p-type material). Within this timescale, the characteristic travelling distance of drift-induced transport can be conservatively estimated by solar cell thickness ($d \approx 2 \times 10^{-2}\,\mathrm{cm}$). Comparison with carrier diffusion distances within this timescale of typically $\sqrt{D\tau_{diel}} \lesssim 5 \times 10^{-6}\,\mathrm{cm} \ll d$ (diffusion coefficient D) allow the conclusion that drift and dielectric relaxation of local excess charge

[6] Derivation proceeds along the lines of [6].

proceeds instantaneously on relevant timescales of carrier diffusion through the solar cell. This justifies the assumption of local quasi charge neutrality $\Delta n = \Delta n_e \approx \Delta n_h$ outside the space charge region.

2.3.3.2 Diffusion Currents

Let us assume constant dopant concentrations $n^0_{e/h}$ on both sides of a pn-junction such that $\nabla n_{e/h} = \nabla \Delta n$ holds locally, and let the chemical potential drop due to changes of electron and hole concentration across the pn-junction be entirely compensated by an inverse electrical potential drop in the space charge region. Let us further assume a vanishing electrical potential gradient beyond the space charge region—i.e. vanishing majority carrier drift. In this hypothetical scenario, any gradient of the electrochemical potential would have to be due to a gradient of excess carrier density, and any electrical current would have to be caused by diffusion. With the assumptions stated above, and accounting for electrical conductivity according to Eq. 2.30, Eq. 2.31 would then become

$$\mathbf{j}_{Q,e/h} = \pm k_B T \mu_{e/h} \nabla \Delta n. \tag{2.45}$$

Consequently, (with $\nabla \Delta n_e \approx \nabla \Delta n_h$) the sum of charge current densities of electrons and holes would have to locally compensate (except for the minor difference between their mobilities). Thus, diffusion currents from the pn-junction to the device terminals do not give rise to a charge release at the terminals of a pn-junction device. Diffusion currents can be detrimental as recombination currents associated with recombination in the bulk and interfaces of an absorber, and at the same time they are vital for solar cell operation as they ensure minority carrier transport to the pn-junction. However, diffusion currents alone cannot explain solar cell operation. Rather, it is a majority carrier drift induced by an electrical potential gradient, which gives rise to charge release at the terminals of a pn-junction device.

2.3.3.3 Selectivity of the pn-Junction

The capability of the pn-junction to actually separate the charge of excess carriers is denoted here as *selectivity*. Two essential features account for the selectivity of a pn-junction: the electrical field of the space charge region, and the large difference between majority and minority carrier conductivities.

The electrical field of the space charge region enables substantial carrier transport against a very pronounced chemical potential gradient. Such net carrier transport is essentially rendered energetically favorable due to elastic backscattering of majority carriers in the space charge region's field. Optical generation of excess carriers then causes a considerable pressure of highly conductive excess majority carriers (and of their corresponding charge) at the device electrodes, respectively. In spite of the relatively high chemical potential of excess majority carriers, this pressure cannot relax via diffusion across the pn-junction, as this is prevented by the electrical field. Conversely, when connecting the electrodes of an illuminated solar cell via a load, the excess majority carrier charge pressure may relax through this external circuit. Thus, the resulting majority carrier *drift* current is eventually driven by a faint electrical potential gradient due to the connection of the device electrodes via a load. It appears noteworthy that, although the electrical field of the space charge region does not deliver work on charge carriers, it is nonetheless an electrical field which drives—or rather pulls—majority carrier currents out of the electrodes of a solar cell: this is the electrical field established between the electrodes or terminals of a solar cell by connecting them through a load—sustained and fed by the steady-state generation of excess carriers.

2.3.3.4 Drift or Diffusion

As outlined above, solar cell operation is enabled by the interplay of drift and diffusion currents. While the kinetics of transport of minority carriers to the pn-junction is governed by their random motion with a diffusion coefficient D, the kinetics of the transport of majority carriers to the terminals of a photovoltaic device is governed by drift or dielectric relaxation. Both carrier mobilities $\mu_{e/h}$ and concentrations $n_{e/h}$ are crucial determinants of this transport process.

2.4 Solar Cell Efficiency—Loss Mechanisms

The efficiency η of photovoltaic energy conversion is defined as the ratio

$$\eta = \frac{P_{el}}{P_{ph}} \tag{2.46}$$

between the electrical power $P_{el} = U_{mpp} I_{mpp}$ delivered to a load by a photovoltaic device and the light power P_{ph} to which the latter is exposed. Here, U_{mpp} and I_{mpp} denote the steady-state (maximum-power-point) operation voltage and current delivered by a photovoltaic device.

2.4.1 Idealized Reversible Conversion Process

Photovoltaic energy conversion efficiency is inherently costrained to $\eta < 1$, as follows from thermodynamic considerations. An idealized treatment by Landsberg [7, 8] considers the internal energy, the entropy, and the radiation pressure of isotropic blackbody radiation of temperature T_p (pump) surrounding a photovoltaic converter at a temperature T_c in thermal equilibrium with a thermal sink of temperature $T_s = T_c$.

The resulting Landsberg efficiency is

$$\eta_L = 1 - \frac{4}{3}\frac{T_s}{T_p} + \frac{1}{3}\left(\frac{T_s}{T_p}\right)^4. \tag{2.47}$$

It represents a reversible process of energy conversion (i.e. no generation of entropy), and therefore provides an upper limit of photovoltaic energy conversion efficiency. Landsberg efficiency η_L differs from the efficiency η_C of an idealized Carnot engine (acting between the gas of blackbody radiation at temperature T_p and a converter at temperature $T_c = T_s$)

$$\eta_C = 1 - \frac{T_s}{T_p} \tag{2.48}$$

in that it not only accounts for the entropy of radiation, but also for radiation of the converter. Minor inaccuracies of the Landsberg efficiency limit stem from the fact that it assumes perfect blackbody radiation of both the sun (pump) and the converter. Assuming a converter Temperature of $T_c = T_s = 300$ K and an effective temperature of solar black body radiation of $T_p \approx 5{,}800$ K [9], Carnot efficiency is $\eta_C = 0.948$, whereas Landsberg efficiency is $\eta_L = 0.931$.

Real solar energy conversion features efficiencies below the Landsberg limit. Landsberg efficiency represents the upper efficiency limit of an energy conversion process with respect to the second law of thermodynamics, as any energy conversion process of practical relevance is irreversible (i.e. involves generation of entropy) and accompanied by power losses.

2.4.2 *Single Junction-Detailed Balance Limit*

In 1961, Shockley and Queisser [10] identified a limit of the solar energy conversion efficiency via semiconductors, which essentially accounts for irreversibility, and which incorporates all loss mechanisms to be outlined in the following. The principle of detailed balance, which requires radiative recombination of a light absorbing semiconductor, sets an upper limit to the maximally achievable electrical power $U_{mpp}I_{mpp}$ of a photovoltaic device exposed to sunlight.

Shockley and Queisser first proposed an idealized calculus, assuming the solar (6,000 K [10]) black body spectrum to be absorbed and to entirely contribute to solar cell current for photon energies above the semiconductor bandgap ε_G, whereas photon energies below ε_G would not be absorbed at all. This calculus yielded an ultimate efficiency limit of $\eta = 0.44$ peaking at a band gap of $\varepsilon_G = 1.1$ eV, and suggesting silicon (with a band gap slightly above the latter value) as an optimal single junction photovoltaic device. Accounting for radiative recombination of absorbed light without any concentration of sunlight, the resulting detailed balance limit encountered at $\varepsilon_G = 1.1$ eV is $\eta = 0.30$.[7]

2.4.3 Classification of Power Losses

One way to classify the power loss mechanisms of a solar cell is to distinguish between thermal, optical, and electrical power losses. The categories overlap to some extent. Yet, this classification proves to be convenient for the specification of the subject of this work.

1. **Thermal** losses denote dissipation of photon energy in excess of a solar cell band gap upon photon absorption. Once excited, electrons and holes quickly relax to the lowest unoccupied states in their respective bands via generation of phonons. In an indirect semiconductor, this relaxation takes place on a timescale much less than excess carrier recombination. Thermal losses imply a major loss mechanism in single-junction devices such as silicon solar cells. They can be reduced via band gap engineering, involving an appropriate design and sequence of materials with different band gaps (e.g. compound semiconductors or quantum wells) tuned to the solar spectrum.
2. **Optical** losses denote deviations from total light absorption by a photovoltaic absorber. They incorporate losses due to reflection, transmission, and parasitic (i.e. non-band-to-band) absorption. High-efficiency solar cells provide miscellaneous concepts to minimize optical power losses, ranging from antireflective coating and light trapping techniques to approaches of up- and/or downconversion of remanent solar irradiation.
3. **Electrical** losses denote losses due to recombination or the imperfection of transport of excess carriers. Recombinative losses involve radiative recombination, recombination due to the interaction among excess carriers (Auger recombination), and recombination at crystal defects and impurities causing electronic states in the band gap (Shockley-Read-Hall (SRH) recombination [11, 12]). In indirect semiconductors, recombinative losses are mostly associated with phonon generation. Thus, they imply thermal losses as well. Transport related loss mechanisms involve resistive losses (finite series resistance of pn-junction device) as well as shunt losses (finite parallel resistance of pn-junction devices accompanied by an ohmic reflow of excess charge).

There is a substantial gap between the above stated detailed balance limit of photovoltaic energy conversion efficiency via single junction devices of $\eta = 0.30$ and the highest efficiencies reached on laboratory type silicon solar cells of $\eta = 0.250 \pm 0.005$ [13–15]. This gap can be attributed to optical and electrical causes. As for

[7] Note that the maximum of the detailed balance limit (without concentration) as a function of band gap energy ε_G is shifted to $\sim$1.3 eV as opposed to the maximum at $\sim$1.1 eV encountered for the ultimate efficiency limit.

optical causes, Shockley and Queisser assumed a Heaviside-like quantum efficiency. In reality, capture of light near the band gap energy ε_G has proven to be among the most demanding challenges on the path toward the Shockley-Queisser-Limit [13]. As for electrical causes, recombination at metal contacts probably represents the most dominant loss mechanism [16, 17].

2.5 Carrier Recombination and Transport

The above-mentioned category of electrical power losses is closely related to both recombination and transport properties of a solar cell. While concepts to adequately and quantitatively describe recombination rates and essential carrier transport properties are detailed in the following chapter, this section is intended to qualitatively sketch the relevance of carrier recombination and transport to solar cell operation.

Upon optical generation of excess electrons and holes in a semiconductor solar cell, excess carriers pile up in the absorber—bringing about a separation of quasi-Fermi levels $\eta_{\sum}$ of electrons and holes equivalent to an implied device voltage $U \approx \frac{\eta_{\sum}}{q}$, which enables the delivery of work $dW = \eta_{\sum} dN$ to an external load by releasing dN carriers of charge q. The steady-state excess carrier density of a solar cell in operation results from a balance between excess carrier generation on the one hand and both excess carrier recombination and charge extraction on the other hand.

2.5.1 Recombination

To some extent, excess charge carrier recombination is inherent to the operation of a semiconductor solar cell. This particularly applies to radiative recombination and Auger recombination. However, SRH recombination through electronic states in the band gap is by no means inherent to solar cell operation. Rather, it largely depends on material quality and on the formation or passivation of electronic states in the band gap during the solar cell production process. Therefore, the prevention and the efficient passivation of such electronic states is the essential task of all solar cell technologists concerned about excess charge carrier recombination.

SRH recombination occurs both at interfaces as well as in the semiconductor bulk. Whereas bulk recombination may be limiting for multicrystalline and possibly upgraded metallurgical grade silicon wafers [18] as well as for p-type materials featuring substantial recombination related to impurity-acceptor complexes [19, 20], the dominant recombination rate of most high efficiency solar cell concepts represents emitter or interface recombination, and particularly recombination at contacts between a semiconductor and metal electrodes [16, 17].

The optimization of the solar cell production process with regard to excess charge carrier recombination calls for accurate and precise measurement techniques of recombination properties, and for an accurate localization of recombination rates in bulk or interfaces.

2.5.2 Transport

Equation 2.38 indicates how electrical charge is delivered to the metal contacts of a solar cell. As can be seen already from this very simple and schematic representation, transport properties such as carrier mobilities, dopant concentrations, and the minority carrier diffusion coefficient largely affect solar cell operation. Therefore, accurate knowledge of these properties is vital for the further improvement of solar cell technologies.

Particularly carrier mobilities in crystalline silicon are often assumed to be accurately known, and as such they are occasionally used in measurements and device simulations as if they featured no uncertainty. Yet, considerable deviations of mobility data and models have been reported in literature (even for monocrystalline silicon) [21–23], and in spite of speculations about potential measurement artifacts, the cause to these deviations has not yet been resolved. With novel techniques to measure both majority and minority carrier mobility, the present work provides new experimental evidence in an unresolved controversy on the one hand. On the other hand, the techniques to be presented open up new pathways to access carrier mobilities in multicrystalline and particularly in upgraded metallurgical grade silicon.

References

1. N.W. Ashcroft, D.N. Mermin, *Solid State Physics* (Saunders College, Philadelphia, 1976)
2. M.A. Green, Intrinsic concentration, effective densities of states, and effective mass in Silicon. J. Appl. Phys. **67**, 2944–2954 (1990)
3. A.B. Sproul, M.A. Green, Improved Value for the Silicon intrinsic carrier concentration from 275 to 375 K. J. Appl. Phys. **70**, 846–854 (1991)
4. A.B. Sproul, M.A. Green, Intrinsic carrier concentration and minority-carrier mobility of Silicon from 77 to 300 K. J. Appl. Phys. **73**, 1214–1225 (1993)
5. P.P. Altermatt, A. Schenk, F. Geelhaar, G. Heiser, Reassessment of the intrinsic carrier density in Crystalline Silicon in view of band-gap narrowing. J. Appl. Phys. **93**, 1598–1604 (2003)
6. P. Würfel, *Physics of Solar Cells: From Basic Principles to Advanced Concepts* (Wiley-VCH, Weinheim, 2009)
7. P.T. Landsberg, G. Tonge, Thermodynamic energy conversion efficiencies. J. Appl. Phys. **51**, R1–R20 (1980)
8. P. Baruch, A. De Vos, P.T. Landsberg, J.E. Parrott, On some thermodynamic aspects of photovoltaic Solar Energy conversion. Solar Energy Mater. Solar Cells **36**, 201–222 (1995)
9. D.R. Williams, Sun Fact Sheet, NASA (2012). http://nssdc.gsfc.nasa.gov/planetary/factsheet/sunfact.html

10. W. Shockley, H.J. Queisser, Detailed balance limit of efficiency of p-n junction solar cells. J. Appl. Phys. **32**, 510–519 (1961)
11. R.N. Hall, Germanium rectifier characteristics. Phys. Rev. **83**, 228 (1951)
12. W. Shockley, W.T. Read, Statistics of the recombinations of holes and electrons. Phys. Rev. **87**, 835–842 (1952)
13. J. Zhao, A. Wang, M.A. Green, F. Ferrazza, 19.8 % efficient "Honeycomb" textured multicrystalline and 24.4 % monocrystalline silicon solar cells. Appl. Phys. Lett. **73**, 1991–1993 (1998)
14. M.A. Green, The path to 25 % silicon solar sell efficiency: history of silicon cell evolution. Prog. Photovoltaics: Res. Appl. **17**, 183–189 (2009)
15. M.A. Green, K. Emery, Y. Hishikawa, W. Warta, E.D. Dunlop, Solar cell efficiency tables (Version 41). Prog. Photovoltaics: Res. Appl. **21**, 1–11 (2012)
16. J. Benick, B. Hoex, M.C.M. Van de Sanden, W.M.M. Kessels, O. Schultz, S.W. Glunz, High efficiency n-type Si solar cells on AlO-passivated Boron Emitters. Appl. Phys. Lett. **92**, 253504 (2008)
17. P.J. Cousins, D.D. Smith, H.C. Luan, J. Manning, T.D. Dennis, A. Waldhauer, K.E. Wilson, G. Harley, W.P. Mulligan, Generation 3: improved performance at lower cost, in *Proceedings of the 35th IEEE PVSC* (Honolulu, 2010), pp. 275–278
18. B. Michl, M. Rüdiger, J.A. Giesecke, M. Hermle, W. Warta, M.C. Schubert, Efficiency limiting bulk recombination in multicrystalline silicon solar cells. Solar Energy Mater. Solar Cells **98**, 441–447 (2012)
19. D. Macdonald, T. Roth, P.N.K. Deenapanray, K. Bothe, P. Pohl, J. Schmidt, Formation rates of Iron-Acceptor pairs in crystalline Silicon. J. Appl. Phys. **98**, 083509 (2005)
20. K. Bothe, J. Schmidt, Electronically activated Boron-Oxygen-related recombination centers in crystalline Silicon. J. Appl. Phys. **99**, 013701 (2006)
21. D.B.M. Klaassen, A unified mobility model for device simulation—I. Model equations and concentration dependence. Solid State Electron. **35**, 953–959 (1992)
22. D.B.M. Klaassen, A unified mobility model for device simulation—II. Temperature dependence of carrier mobility and lifetime. Solid State Electron. **35**, 961–967 (1992)
23. A.B. Sproul, M.A. Green, A.W. Stephens, Accurate determination of minority carrier- and lattice scattering-mobility in Silicon from photoconductance decay. J. Appl. Phys. **72**, 4161–4171 (1992)

Chapter 3
Dynamics of Charge Carriers

This chapter introduces the essential concepts of carrier recombination and transport in silicon. The differential equations of interest are derived and solutions are given—particularly focusing on harmonically modulated excess carrier generation.

3.1 The Continuity Equation

The dynamics of an ensemble of excess charge carriers in a semiconductor may be affected by their local generation, recombination, by diffusion, and by local electrical and magnetic fields. With indistinguishable charge carriers, their dynamics is adequately described by a time- and space-dependent excess carrier density $\Delta n(t, \mathbf{r})$. This section motivates the conditions accounting for the dynamics of excess charge carriers, leading to the so-called *continuity equation*—which is sometimes referred to as a *diffusion equation*.[1]

3.1.1 Recombination Lifetime

Excess charge carrier recombination and generation can be most adequately described via local recombination and generation rates $R(\mathbf{r})$ and $G(\mathbf{r})$, denoting the number of events of electron-hole pair recombination and generation per volume and time, respectively.

[1] In spite of their frequent occurence in literature, both terms are ambiguous. While the term *continuity equation* is likewise used to express both local and temporal conservation of particles, probability, or electrical charge (cf. Eq. 3.13), the term *diffusion equation* is also used as a synonym for the heat equation which is equivalent to Fick's second law of diffusion. It will be outlined in this chapter that the *continuity equation* describing the dynamics of excess carriers in a semiconductor is a special case of the general inhomogeneous diffusion equation.

J. Giesecke, *Quantitative Recombination and Transport Properties in Silicon from Dynamic Luminescence*, Springer Theses, DOI: 10.1007/978-3-319-06157-3_3,
© Springer International Publishing Switzerland 2014

3.1.1.1 Definition for Exponential Excess Carrier Decay

Let us first assume an idealized homogeneous system of spatially uniform (constant) excess carrier density $\Delta n(\mathbf{r})$, generation rate $G(\mathbf{r})$, and recombination rate $R(\mathbf{r})$. Further, let there be an abrupt switch-off of a steady-state excess carrier generation rate such that $G(t < t_0) = G_0$ and $G(t \geq t_0) = 0$, as depicted in Fig. 3.1.

According to the above definition of a recombination rate, the derivative of excess carrier density with respect to time must read

$$\frac{d\Delta n}{dt} = -R. \tag{3.1}$$

As for many decay processes in nature, it may appear manifest to assume an exponential characteristic of the decay $\Delta n(t)$. According to Eq. 3.1, this would require a recombination rate proportional to excess carrier density $R \propto \Delta n$. This assumption proves to be highly inaccurate for excess charge carrier recombination of semiconductors, and particularly of an indirect semiconductor like silicon. In this context, it is common to speak of an injection dependence of recombination properties, which typiclally occurs for SRH recombination at crystal defects (to name but one affected class of recombination rates). The injection dependence of recombination properties brings about numerous implications that are essential to some of the developments emerging from this work. At first, injection-dependence shall be neglected in a simple, straightforward, and very intuitive lifetime definition. Subsequently, a lifetime definition which accounts for injection dependence of recombination properties shall be given.

With a recombination rate $R = \frac{\Delta n}{\tau}$ with the constant of proportionality of $R \propto \Delta n$ corresponding to the inverse of a so-called excess charge carrier recombination lifetime τ, the differential Eq. 3.1 together with the initial value $\Delta n(t_0) = \Delta n_0$ represents a linear first order initial value problem of unique solution

$$\Delta n(t) = \Delta n_0 e^{-\frac{t}{\tau}}. \tag{3.2}$$

Therefore, in this special case, excess carrier recombination lifetime τ corresponds to the time it takes for excess carrier density to decay from an initial value Δn_0 to $\Delta n(t = \tau) = e^{-1}\Delta n_0$, as also depicted in Fig. 3.1.

3.1.1.2 General Definition

For the general case of an injection-dependent recombination rate, manifesting itself in a deviation from an exponential decay, the above definition of lifetime τ is not appropriate. For a definition which accounts for injection dependence of lifetime, we shall again consider the case of an idealized homogeneous system upon an abrupt switch-off of steady-state excess carrier generation rate G. In Eq. 3.1, any deviation from the proportionality $R \propto \Delta n$ may be lumped into an injection-dependent excess

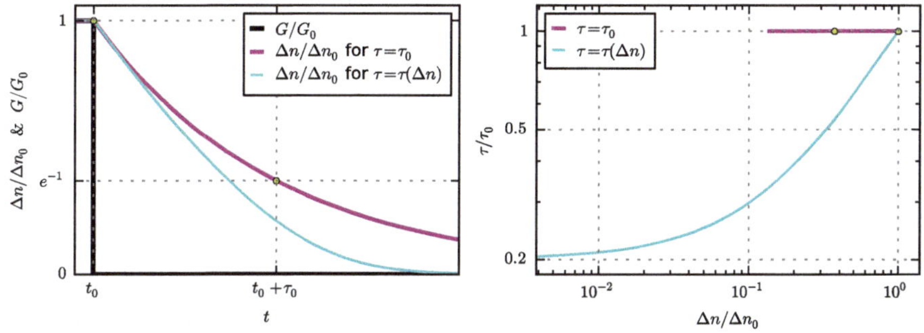

Fig. 3.1 On the definition of excess carrier recombination lifetime. The *left plot* shows two scenarios of excess carrier decay in a spatially uniform semiconductor after an abrupt switch-off of steady-state excess carrier generation rate G: a constant excess carrier recombination lifetime $\tau = \tau_0$, and an injection-dependent lifetime $\tau = \tau(\Delta n)$. The *right plot* depicts a lifetime interpretation of the excess carrier decay shown on the left according to Eq. 3.4

charge carrier recombination lifetime $\tau(\Delta n)$, featuring an arbitrary dependence on Δn. Therefore, Eq. 3.1 can be written as

$$\frac{d\Delta n}{dt} = -\frac{\Delta n}{\tau(\Delta n)}. \tag{3.3}$$

According to Eq. 3.3, for times $t \geq t_0$ injection-dependent lifetime $\tau(\Delta n)$ is defined as

$$\tau\left(\Delta n\left(t\right)\right) = -\frac{\Delta n\left(t\right)}{\left.\frac{d\Delta n}{dt}\right|_t}. \tag{3.4}$$

Thus, for a semiconductor featuring a spatially uniform excess carrier density Δn, the corresponding excess charge carrier recombination lifetime $\tau(\Delta n)$ corresponds to the negative ratio between excess carrier density and its derivative with respect to time upon abrupt switch-off of a steady-state generation rate.

3.1.1.3 Nonuniform Excess Carrier Density

From a rigorous point of view, Eq. 3.4 does not represent a generally valid definition of excess charge carrier recombination lifetime because of the constraint of spatial uniformity of recombination rate, generation rate, and of excess carrier density. However, it is virtually impossible to uphold a rigorous lifetime definition which appropriately accounts for nonuniform spatial distributions of these three quantities in a semiconductor. The cause for this complication is diffusion of excess carriers. If diffusion (spatial redistribution of excess carriers) did not occur, every location of a semiconductor would have a unique lifetime according to Eq. 3.4. Yet, diffusion precludes this in cases of recombination at interfaces, spatial inhomogeneity of recombination rate (e.g. locally confined recombination centers), or spatial inhomogeneity of generation rate.

This uncertainty of the definition of lifetime is attempted to be resolved by the concept of a so-called **effective lifetime** (cf. Sect. 3.1.4). The aim of this concept is to find a most appropriate lifetime value for an entity of vicinal locations, distinguished by the condition that excess carriers could reach any location within this entity during their lifetime.

Even though the concept of an effective lifetime may be very useful, it must be noted that there is no unique and exact effective lifetime for any situation. The recursive nature of the above perception of effective lifetime adds uncertainty for one thing. Moreover, particularly concerning dynamic scenarios of excess carrier generation, different definitions of effective lifetime may yield values deviating by several orders of magnitude—to be detailed later.

3.1.1.4 Classification of Bulk Recombination Mechanisms

The local excess carrier recombination rate $R(\mathbf{r})$ always represents a superposition of recombination rates associated with different causes

$$R = \sum_i R_i = \Delta n \sum_i \tau_i^{-1} \tag{3.5}$$

Such causes have been mentioned in Sect. 2.4. On the one hand, they involve the intrinsic bulk recombination mechanisms of radiative (or direct [1]) recombination and Auger recombination. On the other hand, they also involve the extrinsic SRH recombination mechanism due to electronic states in the band gap, which manifests itself both as bulk and interface recombination.

The distinction between different recombination mechanisms is crucial in order to interpret lifetime measurements in terms of isolated recombination mechanisms. The analysis of emitter saturation current density inferred from injection-dependent lifetime measurements [2, 3] or defect level lifetime spectroscopy [4] are only two classes of applications among a multitude.

Here, recombination mechanisms shall only be briefly outlined.[2] Elaborate treatments are to be found elsewhere [4–6]. For finite temperatures $T \neq 0$, any of the recombination mechanisms discussed below also occur at $\eta_\Sigma = 0$, i.e. in electrochemical equilibrium. Involving both free electrons and holes, recombination rates are in any case proportional to the product of their densities[3] $n_e = n_e^0 + \Delta n_e$ and

[2] Note that the focus of the present work lies in the measurement and interpretation of total recombination rate rather than in an interpretation in terms of an isolated recombination mechanism.

[3] Note that the notation for carrier densities according to Würfel [5] is adhered throughout this work, as it facilitates indexing.

$n_h = n_h^0 + \Delta n_h$ (assuming $\Delta n_e = \Delta n_h$, cf. Sect. 3.1.6).[4] However, the **excess** carrier recombination rates described in the following are all proportional to $n_e n_h - n_i^2$.

1. **Radiative** recombination denotes the inverse process of the optical generation of free electrons and holes. Involving both electrons and holes, the radiative recombination process is proportional to free carrier densities n_e and n_h. Therefore, radiative recombination of excess carriers is

$$R_{rad} = B \cdot \left(n_e n_h - n_i^2\right), \tag{3.6}$$

with the coefficient of radiative recombination B. As radiative recombination in an indirect semiconductor like silicon must involve phonon generation or annihilation, B also accounts for phonon density, which brings about a pronounced dependence on temperature T. Further, B depends on free carrier density due to screening of Coulomb attraction between electrons and holes toward relatively high free carrier densities [7, 8]. Radiative recombination and luminescence denote the same physical process, which is essential for the measurement techniques developed in the course of this work. Therefore, the laws of radiative recombination in a semiconductor are detailed in Sect. 4.1.

2. **Auger** recombination denotes direct recombination of an electron and a hole via energy transfer to another free carrier. This three-particle interaction is referred to as an *eeh* or *ehh* process, depending on whether the excitation energy is transferred to an electron or to a hole. Auger recombination rates are therefore proportional to the involved carrier densities, respectively. With Auger coefficients $C_{e/h}$ [9, 10] for the *eeh* and *ehh* processes, recombination rate reads

$$R_{aug} = \left(n_e n_h - n_i^2\right)\left(C_e n_e + C_h n_h\right), \tag{3.7}$$

leading to a corresponding Auger lifetime

$$\tau_{aug} = \frac{1}{\left(\Delta n + n_e^0 + n_h^0\right)\left(C_e\left(n_e^0 + \Delta n\right) + C_h\left(n_h^0 + \Delta n\right)\right)}. \tag{3.8}$$

While this traditional theory of Auger recombination proved to be consistent for relatively high free carrier densities, it was found to fail toward relatively low free carrier densities. Hangleiter and Häcker [11] attributed this to a Coulomb enhancement of Auger recombination toward low free carrier densities due to exciton formation, to be resolved by multiplication of Auger coefficients with so-called Coulomb enhancement factors g_{eeh} and g_{ehh}. A new parametrization of these Coulomb enhancement factors was recently found by Richter et al. [6].

[4] If not otherwise specified, the denotation Δn implies equality of excess electron and hole densities.

3. **Shockley-Read-Hall** (SRH) recombination denotes phonon-mediated recombination via electronic states within the band gap. It is founded on statistical considerations about free carrier capture of a defect level ε_t of density N_t in the band gap of a nondegenerate semiconductor by Hall [12] and by Shockley and Read [13]. Thermal carrier velocities $\nu_{e/h}$ and carrier capture cross sections of traps $\sigma_{e/h}^{\text{cap}}$ define a capture time constant of electrons $\tau_{e0} = \left(N_t \nu_e \sigma_e^{\text{cap}}\right)^{-1}$ and of holes $\tau_{h0} = \left(N_t \nu_h \sigma_h^{\text{cap}}\right)^{-1}$, respectively. With so-called SRH densities

$$n_{e1} = n_i e^{\frac{\varepsilon_t - \varepsilon_i}{k_B T}} \tag{3.9}$$

$$n_{h1} = n_i e^{\frac{\varepsilon_i - \varepsilon_t}{k_B T}} \tag{3.10}$$

and with ε_i denoting the intrinsic fermi energy, the recombination rate associated with a trap of energy ε_t and density N_t was found to be

$$R_{srh} = \frac{n_e n_h - n_i^2}{\tau_{e0}\left(n_h + n_{h1}\right) + \tau_{h0}\left(n_e + n_{e1}\right)} \tag{3.11}$$

With $n_e n_h - n_i^2 = \Delta n \left(\Delta n + n_e^0 + n_h^0\right)$, the corresponding SRH recombination lifetime $\tau_{srh} = \frac{\Delta n}{R_{srh}}$ must read

$$\tau_{srh} = \frac{\tau_{e0}\left(n_h^0 + n_{h1} + \Delta n\right) + \tau_{h0}\left(n_e^0 + n_{e1} + \Delta n\right)}{n_e^0 + n_h^0 + \Delta n}. \tag{3.12}$$

3.1.1.5 Surface Recombination

At interfaces of a semiconductor, surface recombination may occur. Any interface of a semiconductor—regardless whether external or internal (e.g. a grain boundary in multicrystalline silicon)—represents a disruption of crystal symmetry, inducing a band of electronic states in the band gap [5, 14, 15]. Further, crystal interfaces are typically exposed to an enhanced mechanical stress, which renders them a likely source of crystal defects, and by nature they agglomerate impurities. The recombination activity of interfaces may also be affected by band bending.

Surface recombination is recombination at defect states in the band gap of a semiconductor, and as such it can be referred to as SRH recombination. Yet, as SRH theory is subject to the condition of a single trap level, interface recombination cannot be adequately described by Eq. 3.11. An extended SRH formalism attempts to integrate the contributions of a continuum of surface states across the band gap to form a surface recombination rate R_s [4].

3.1.2 Motivation of the Continuity Equation

The conservation of particles, probability, or electrical charge in space and time can be expressed by a relation that is sometimes referred to as the continuity equation. It states that the temporal change of a density $n_x(\mathbf{r}, t)$ of an extensive physical quantity x like particle number, probability, or electrical charge contained within a volume increment $[\mathbf{r} + d\mathbf{r}]$ must be balanced by a flow of that same extensive physical quantity into or out of the volume increment of interest:

$$\frac{\partial n_x(\mathbf{r}, t)}{\partial t} + \nabla \mathbf{j}_x(\mathbf{r}, t) = 0 \tag{3.13}$$

The concept of conservation of particles can be applied to excess electrons and holes of density Δn in a semiconductor, except for a possible imbalance between excess carrier generation and recombination rate. Therefore, the continuity equation of an ensemble of excess electrons and holes in a semiconductor reads

$$\frac{\partial \Delta n(\mathbf{r}, t)}{\partial t} + \nabla \mathbf{j}_{\Delta n}(\mathbf{r}, t) = G(\mathbf{r}, t) - R(\mathbf{r}, t) \tag{3.14}$$

As detailed in Sect. 3.1.1, the recombination rate can be expressed in terms of an excess carrier recombination lifetime $R = \frac{\Delta n}{\tau(\Delta n)}$. In Eqs. 2.31 and 2.32, charge current densities of electrons and holes were introduced. These expressions can be transformed to particle current densities through division of charge current density by negative elementary charge $-q$ for electrons and by positive elementary charge q for holes. Further accounting for both, the conductivity $\sigma_{e/h}$ as of Eq. 2.30 as well as the Einstein relation between minority carrier mobility $\mu_{e/h}$ and the diffusion coefficient $D_{e/h}$

$$\mu_{e/h} = \frac{q D_{e/h}}{k_B T}, \tag{3.15}$$

one obtains the particle current

$$\mathbf{j}_{e/h}(\mathbf{r}, t) = -D_{e/h} \nabla n_{e/h}(\mathbf{r}, t) \mp \frac{\sigma_{e/h}}{q} \mathbf{E}(\mathbf{r}, t). \tag{3.16}$$

In Sect. 2.3, majority carrier drift currents induced by dielectric relaxation of a local excess charge were used to motivate the release of charge from an illuminated pn-junction device. Even though this field-induced transport process is vital for solar cell operation, the solution of the continuity equation $\Delta n(\mathbf{r}, t)$ is only weakly affected by drift currents. The reason for this is once again dielectric relaxation of the field caused by excess charge (cf. Sect. 2.3), which takes place on a timescale much shorter than the timescale of interest for investigations of recombination properties.

Yet, there are configurations in which local fields substantially affect the dynamics of excess carriers. For instance, in the case of large steady-state gradients of excess carrier density, differing electron and hole mobilities may cause a buildup of local charge. This effect can be taken into account by an effective or *ambipolar* diffusion coefficient D_a (cf. Sect. 3.1.5), which eventually incorporates the effect of local electrical fields $\mathbf{E}$ on the dynamics of excess carriers such that electron and hole current densities can be expressed in a form much resembling Fick's first law of diffusion

$$\mathbf{j}_{\Delta n}(\mathbf{r}, t) = -D_a \nabla \Delta n(\mathbf{r}, t), \tag{3.17}$$

assuming spatially uniform dark free carrier density such that the gradient of total free carrier density coincides with the gradient of excess carrier density $\nabla n_e = \nabla n_h = \nabla \Delta n$. Insertion of Eq. 3.17 into Eq. 3.14 yields

$$\frac{\partial \Delta n(\mathbf{r}, t)}{\partial t} - D_a \nabla^2 \Delta n(\mathbf{r}, t) + \frac{\Delta n(\mathbf{r}, t)}{\tau} = G(\mathbf{r}, t). \tag{3.18}$$

The explicit indication of the injection dependence of carrier lifetime τ has been omitted here for reasons of simplicity. However, such injection dependence is crucial with regard to the solution of the continuity equation for dynamic generation rate in general and for harmonically modulated generation rate in particular, and it will therefore play a crucial role throughout this entire thesis (cf. Sect. 6.2).

Equation 3.18 is the inhomogeneous continuity equation describing the dynamics of excess carriers in silicon. Mathematically, this second-order parabolic partial differential equation is equivalent to a special case of the heat equation describing the temperature distribution $T(\mathbf{r}, t)$ in a solid with both, a built-in heat source (analogous to excess carrier generation rate) and dissipation of heat to an infinitely large external sink at $T = 0$ (analogous to excess carrier recombination, e.g. via thermal radiation).

3.1.2.1 One-Dimensional Treatment

Aiming at straightforward and mostly analytical solutions, the treatment of the continuity equation shall be constrained to a one-dimensional treatment here. Involving only one space variable z which denotes substrate depth as the perpendicular distance from an irradiated interface, the continuity equation becomes:

$$\frac{\partial \Delta n(z, t)}{\partial t} - D_a \frac{\partial^2 \Delta n(z, t)}{\partial z^2} + \frac{\Delta n(z, t)}{\tau} = G(z, t). \tag{3.19}$$

With this constraint, the rigorous validity of the solutions provided in Sect. 3.2 is restricted to laterally uniform material, i.e. to material featuring invariance of electronic properties in the lateral substrate dimensions x and y. Note that, particularly—but not only—in multicrystalline silicon, such conditions are seldom encountered. However, there are numerous applications of harmonically modulated lifetime metrology that are virtually unaffected by a local failure of the above-mentioned assumption. Moreover, the combination of dynamic lifetime measurement results with steady-state photoluminescence may (under preconditions to be addressed e.g. in Sect. 8.3) yield useful results even in locations of substantial lateral excess carrier diffusion (e.g. grain boundaries in multicrystalline material).

3.1.3 Boundary Conditions: Interface Recombination

In order to explicitly solve the continuity equation (Eq. 3.19), initial values must be given and/or boundary conditions must be imposed on the continuity equation. As the continuity equation describes the excess carrier density $\Delta n\,(z,t)$ in a semiconductor, boundary conditions define carrier currents into and through the interfaces confining the semiconductor bulk of interest. Such currents could either be related to electrical injection or extraction of carriers at interfaces, or they could be caused by excess carrier recombination at (or near) interfaces.[5] Interface recombination is most appropriately represented by a surface recombination velocity S, which is defined through the boundary condition

$$D_a \left.\frac{\partial \Delta n}{\partial z}\right|_{z=z_{\mp}} = \pm S \Delta n\,(z = z_{\mp})\,. \tag{3.20}$$

Here, the subscript $\mp$ denotes whether an interface at $z_{\mp}$ confines the bulk toward lower (z_-) or higher (z_+) values of the variable z, which indicates the position along an axis perpendicular to the interface.

Being of second order in z and of first order in t the unique solution of the continuity equation requires two boundary conditions and an initial value. For a silicon wafer of thickness d, the boundary conditions are thus

$$D_a \left.\frac{\partial \Delta n}{\partial z}\right|_{z=0} = S_0 \Delta n\,(z = 0) \tag{3.21}$$

$$D_a \left.\frac{\partial \Delta n}{\partial z}\right|_{z=d} = -S_d \Delta n\,(z = d)\,. \tag{3.22}$$

In the case of a substrate thickness d much greater than the minority carrier diffusion length (e.g. in a silicon ingot) and an excess carrier generation much confined to the interface $z = 0$, the substrate can be considered as semi-infinite, thereby reducing the second boundary condition (Eq. 3.22) to $\Delta n\,(z \to \infty) = 0$.

The concept of surface recombination is often expanded to excess carrier recombination in close proximity of interfaces (e.g. solar cell emitters or space-charge-regions

[5] Throughout this work, boundary conditions are solely represented by interface recombination. In the case of electrical current injection to be discussed in Sect. 7.4, carrier recombination and injection through the interface can be accurately separated in most cases. Thus, the problem can be treated with the same boundary conditions as introduced here.

at the pn-junction), yielding a so-called effective surface recombination velocity. Throughout this work, there shall be no distinction between rigorous and effective surface recombination velocities.

3.1.4 Effective Carrier Lifetime

In Sect. 3.1.1, the futility of upholding a rigorous and generally valid definition of excess carrier recombination lifetime in the presence of locally varying bulk recombination, significant surface recombination, and possibly nonuniform excess carrier generation has been underlined. Excess carrier diffusion—the very process that enables solar cell operation—prohibits a rigorous definition. In spite of such complications, a concept of an effective lifetime is necessary—allowing the interpretation of a specific type of lifetime measurements in terms of quantities of interest such as bulk lifetime or surface recombination velocity.

3.1.4.1 Aim of the Concept of Effective Lifetime

It must be noted that there is no unique definition of the concept of effective lifetime. Rather, it can be distinguished by the aim of obtaining a most appropriate and analytically traceable aggregated lifetime value for an entity of vicinal locations, allowing excess carriers to reach any location within this entity during their *lifetime* via diffusion.

3.1.4.2 Different Perceptions

The concept of effective lifetime is often perceived as a set of instructions on how to obtain recombination properties of interest (e.g. bulk lifetime or surface recombination velocities) from lifetime measurements. Yet, such an analytical framework largely depends on the precise perception of effective lifetime. There are cases of large discrepancies between effective lifetimes according to different perceptions of the term [16]. Therefore, it is imperative to be aware of the underlying perception when interpreting effective lifetimes in terms of recombination properties of interest. The two most prevalent perceptions shall be outlined in the following—with a focus on their differences. Here, they are referred to as a **steady-state perception** and a **dynamic perception**.

3.1.4.3 Steady-State Perception of Effective Lifetime

The most straightforward perception of effective lifetime is the arithmetic average of the duration of excitation of an ensemble of excess carriers in a specified semiconductor volume V. In the case of a steady-state excess carrier generation rate, this is the ratio between cumulative excess carrier density and generation rate

$$\tau_{eff,s} = \frac{\int_V d\mathbf{r}\, \Delta n(\mathbf{r})}{\int_V d\mathbf{r}\, G(\mathbf{r})}. \tag{3.23}$$

According to this perception, steady-state effective lifetime of a laterally homogeneous semiconductor substrate of thickness d is

$$\tau_{eff,s} = \frac{\int_0^d dz\, \Delta n(z)}{\int_0^d dz\, G(z)}. \tag{3.24}$$

Insertion of the steady-state solution $\Delta n(z)$ of the continuity equation (cf. Sect. 3.2.2) yields analytical expressions, which are compared to the ones derived from a dynamic perception for special cases in the following.

3.1.4.4 Dynamic Perception of Effective Lifetime

Another often encountered and occasionally confused perception of effective lifetime relates to **decay modes** of excess carrier density after a very short pulse of monochromatic optical excess carrier generation

$$G(z,t) = \alpha_\gamma j_\gamma(t)\, e^{-\alpha_\gamma z}. \tag{3.25}$$

Here, α_γ denotes the band-to-band absorption coefficient, which corresponds to an inverse light penetration depth if competing absorption processes such as free carrier absorption are negligible.[6] With a light pulse of a transmitted areal density of N_γ photons at time $t_0 = 0$, the transmitted photon current density can be expressed in terms of a delta function as

$$j_\gamma(t) = N_\gamma \delta(t). \tag{3.26}$$

The resulting initial value of excess carrier density is

$$\Delta n(z,0) = \int dt\, G(z,t) = \alpha_\gamma N_\gamma e^{-\alpha_\gamma z}. \tag{3.27}$$

[6] Inverse light penetration depth and the band-to-band absorption coefficient do not coincide if significant competing light absorption processes exist. Note that—although in literature the notation of the band-to-band absorption coefficient may differ from the notation of the total absorption coefficient [17, 18]—α_γ denotes the band-to-band absorption coefficient throughout this work.

With such an initial value $\Delta n\,(z,0)$, the homogeneous continuity equation (cf. Eq. 3.19 with $G\,(z,t) = 0$) can be solved via separation of variables [19–21], leading to a solution consisting of a superposition of odd and even harmonic functions. With a transformation of the point of origin to the wafer center $z' = z - \frac{d}{2}$, the solution of the continuity equation is

$$\Delta n\left(z', t\right) = e^{-\frac{t}{\tau}} \sum_{n=1}^{\infty} \left(A_n e^{-a_n^2 D_a t} \cos a_n z' + B_n e^{-b_n^2 D_a t} \sin b_n z' \right). \qquad (3.28)$$

The coefficients A_n and B_n in this series of modes are determined by the initial value $\Delta n\,(z,0)$, whereas the coefficients a_n and b_n are determined by the boundary conditions given in Eqs. 3.21 and 3.22. If a_n and b_n are chosen such that any mode in the above series satisfies the boundary conditions, the series itself must satisfy the boundary conditions as well. Assuming identical surface recombination velocities on both wafer sides $S_0 = S_d = S$, insertion of the modes of the solution given in Eq. 3.28 into the boundary conditions leads to the following equations determining the coefficients a_n and b_n:

$$\cot \frac{a_n d}{2} = \frac{a_n D_a}{S} \qquad (3.29)$$

$$\tan \frac{b_n d}{2} = -\frac{b_n D_a}{S} \qquad (3.30)$$

Carrier decay time measurements are commonly conducted in a depth-integrated manner. As the depth integral of the odd mode vanishes, we shall focus on the even mode in the following—as previously argued by Luke and Cheng [20].[7] According to Eq. 3.28 every mode of the solution of the continuity equation features a characteristic decay time.

$$\frac{1}{\tau_n} = \frac{1}{\tau} + a_n^2 D_a \qquad (3.31)$$

The decay time τ_n can be perceived as an effective lifetime of the n-th mode of the solution of the continuity equation [22]. Commonly, the decay time of the most slowly decaying fundamental mode $n = 1$ is denoted as *effective lifetime*. Here, it shall be denoted as **dynamic** effective lifetime

$$\frac{1}{\tau_{eff,d}} = \frac{1}{\tau} + a_1^2 D_a. \qquad (3.32)$$

---

[7] Note that for decay time measurements featuring a weighted depth-dependent sensitivity with respect to excess carrier density (e.g. due to luminescence reabsorption or high-level injection measurement conditions) or depth-resolved decay time measurement techniques, the odd mode may not generally be negligible.

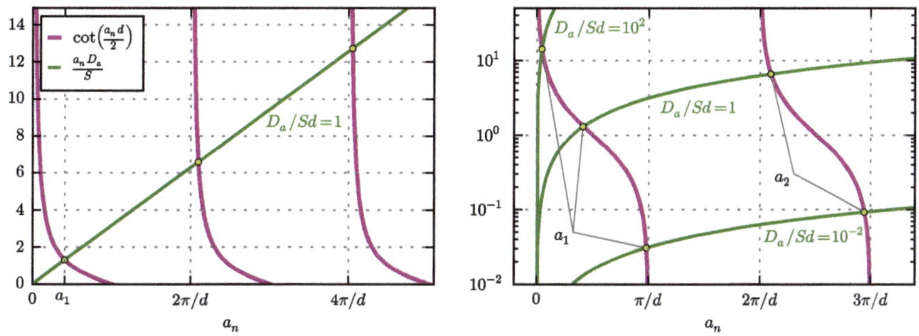

Fig. 3.2 Graphical illustration of the solutions a_n of Eq. 3.29 for three different surface recombination velocity scenarios, distinguished by a ratio $\frac{D_a}{Sd}$ of 10^2, 1, and 10^{-2}, respectively. *On the left* $\cot\frac{a_n d}{2}$ and $\frac{a_n D_a}{S}$ are plotted as functions of a_n for a ratio $\frac{D_a}{Sd}=1$. *On the right* these functions are plotted on a semi-logarithmic scale in order to properly depict intercepts a_n for high and low values of $\frac{D_a}{Sd}$. The coefficient a_1, which is crucial for the effective lifetime of the fundamental decay mode (cf. Eq. 3.32), approaches 0 in the limit $\frac{D_a}{Sd}\to\infty$, and $\frac{\pi}{d}$ in the limit $\frac{D_a}{Sd}\to 0$

Equation 3.29 is crucial for this dynamic perception of effective lifetime. Solutions of Eq. 3.29 are graphically illustrated in Fig. 3.2.

Limiting special cases of dynamic effective lifetime are distinguished by extreme values of the ratio $\frac{D_a}{Sd}$. For large surface recombination velocities, linear expansion of the function $\cot\frac{a_n d}{2}$ around its roots at $a_n = \frac{2n-1}{d}\pi$ allows an analytic expression of the intercept

$$a_n = \frac{(2n-1)\,\pi}{d\left(\frac{2D_a}{Sd}+1\right)}. \tag{3.33}$$

The accuracy of this expression declines with increasing order n, as can be understood from the illustration of intercepts given in Fig. 3.2. In the limit $\frac{D_a}{Sd}\ll 1$, the intercept simplifies such that $a_1=\frac{\pi}{d}$, leading to a dynamic effective lifetime of

$$\frac{1}{\tau_{eff,d}} = \frac{1}{\tau} + \frac{\pi^2}{d^2}D_a. \tag{3.34}$$

For very low surface recombination velocities corresponding to the case $\frac{D_a}{Sd}\gg 1$, the solution a_1 of Eq. 3.29 takes very small values. Thus, the function $\cot\frac{a_1 d}{2}$ is accurately approximated by $\frac{2}{a_1 d}$, yielding $a_1^2=\frac{2S}{D_a d}$ and consequently

$$\frac{1}{\tau_{eff,d}} = \frac{1}{\tau} + \frac{2S}{d}. \tag{3.35}$$

3.1.4.5 Discussion of Different Perceptions

For a discussion of the differences between the steady-state and the dynamic perception of effective lifetime, let us consider a wafer featuring a uniform bulk lifetime τ and identical surface recombination velocities on both interfaces $S_0 = S_d = S$.

In the case of negligible bulk recombination and a small but dominant surface recombination $S \ll \alpha_\gamma D_a$, steady-state and dynamic expressions for effective lifetime coincide, as can be retraced by comparing the dynamic expression given in Eq. 3.35 and the derivation of a corresponding expression from the steady-state solution of the continuity equation provided in Appendix A.1.1. Notably, in this case effective lifetime expressions coincide even in the presence of heavily nonuniform excess carrier generation.

With increasing surface recombination velocity, deviations between steady-state and dynamic effective lifetime occur. In the limit $S \to \infty$, the discrepancy between steady-state and dynamic effective lifetime amounts to a minimum of a factor of

$$\frac{\tau_{eff,d}}{\tau_{eff,s}} \geq \frac{12}{\pi^2}. \tag{3.36}$$

A corresponding expression for steady-state effecive lifetime is stated in [16, 23] and derived in Appendix A.1.1. However, this minimal mismatch only holds for a uniform excess carrier generation rate $G(z)$. In practice, the mismatch is typically greater, because optical excess carrier generation is by nature nonuniform. Rather, in many optical lifetime techniques, excess carrier generation is required to be much confined to a region close to a substrate interface ($\alpha_\gamma d \gg 1$), bringing about a high degree of nonuniformity.[8] Figure 3.3 depicts the mismatch between dynamic effective lifetime $\tau_{eff,d}$ and steady-state effective lifetime $\tau_{eff,s}$ as a function of both thickness d and the band-to-band absorption coefficient α_γ in the limit $S \to \infty$.

3.1.4.6 Origin of Mismatch Between Dynamic and Steady-State Effective Lifetime

As can be seen in Fig. 3.3, the mismatch between dynamic and steady-state effective lifetime (in the limit $S \to \infty$) is a monotonously increasing function of both substrate thickness d and inverse light penetration depth α_γ. This mismatch is due to the fact that dynamic and steady-state effective lifetime expressions weight the decay modes of excess carrier density in a different manner. While dynamic effective lifetime by definition corresponds to the fundamental decay mode, steady-state effective lifetime accounts for higher (and faster) decay modes as well [16]. The impact of higher order decay modes on steady-state effective lifetime increases with increasing surface recombination velocity and with a decreasing depth of excess carrier generation. When maintaining a shallow steady-state excess carrier generation at a high surface

[8] The condition $\alpha_\gamma d \gg 1$ is also denoted as *shallow* excess carrier generation in the following.

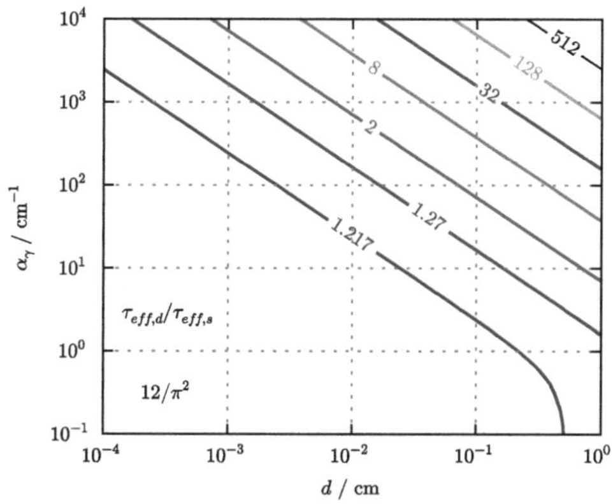

Fig. 3.3 Contour plot of the ratio between dynamic effective lifetime $\tau_{eff,d} = \frac{d^2}{\pi^2 D_a}$ and steady-state effective lifetime $\tau_{eff,s}$ according to Eq. 3.24 as a function of substrate thickness d and of the band-to-band absorption coefficient α_γ. Here, infinite surface recombination velocity and bulk lifetime, and an ambipolar minority carrier diffusion coefficient $D_a = 27\,\text{cm}^2\text{s}^{-1}$ were assumed. One may conclude that, if substrate thickness d substantially exceeds excess carrier generation depth α_γ^{-1}, the ratio $\frac{\tau_{eff,d}}{\tau_{eff,s}}$ is only a function of $\alpha_\gamma d$

recombination velocity, the fraction of excess carriers recombining at the interface is much greater than after an abrupt switch-off of excess carrier generation rate. In the latter case, recombination of carriers at the interface requires their diffusion to the interface—thereby limiting total excess carrier recombination rate.

3.1.4.7 Note on Lateral Carrier Diffusion

The dimension of an entity of vicinal locations to be assigned one effective lifetime value may depend on effective lifetime itself. Assuming e.g. an idealized highly recombinative point defect in an otherwise perfect crystal featuring only radiative and Auger recombination, the size of such an entity (and therefore the effective lifetime to be assigned to it) would critically depend on the distance between the point defect and a point of interest.

Although much of this work is restricted to a one-dimensional treatment of the continuity equation, lateral diffusion of the above kind may be incorporated in a concept of local effective lifetime. For instance, the local effective lifetime in one pixel of a lifetime map of a silicon wafer may be lowered due to recombination occurring in a location corresponding to another (vicinal) pixel. A concept of effective lifetime which incorporates such lateral diffusion effects is deemed superior in terms of the accuracy of predictions of solar cell performance. However, the impact of lateral carrier diffusion on effective lifetime largely depends on the perception of carrier lifetime. Generally, the impact of adjacent recombination centers is less pronounced according to a dynamic perception of effective lifetime [16] (cf. Sects. 6.1 and 8.4).

3.1.5 Ambipolar Diffusion

There are relevant configurations in which an electrical field affects the dynamics of excess charge carriers—even on the timescale of interest of relevant carrier lifetimes. This applies if excess carrier densities at intermediate to high-level injection conditions (comparable to net dopant concentration) coincide with a substantial excess carrier density gradient. In this configuration, the differing diffusion coefficients of electrons and holes lead to a local electrical field, accelerating the relatively inert holes while decelerating the relatively mobile electrons. It is possible to adequately describe the dynamics of such a system via decoupled equations of motion for both electrons and holes via an effective or *ambipolar* diffusion coefficient. Starting with the transport equations for electrons and holes (Eq. 3.16),[9] a vanishing local charge current $j_h - j_e = 0$ may be assumed in the absence of external electrical fields:

$$0 = q\left(D_e \nabla n_e(\mathbf{r}) - D_h \nabla n_h(\mathbf{r})\right) + (\sigma_e + \sigma_h)\,\mathbf{E}(\mathbf{r}) \tag{3.37}$$

Local charge and local electrical fields may result from a local difference between excess electron and hole densities $\Delta n_e \neq \Delta n_h$. However, such difference may be assumed to be small enough in order to assume nearly equal gradients of excess carrier densities $\nabla n_e \approx \nabla n_h$, which are (as in Eq. 3.17) denoted as $\nabla \Delta n$ in the following. Therefore, the electrical field from Eq. 3.37 is

$$\mathbf{E}(\mathbf{r}) = q\frac{D_h - D_e}{\sigma_e + \sigma_h}\nabla \Delta n(\mathbf{r}). \tag{3.38}$$

Insertion of this expression into Eq. 3.16 for either electrons or holes yields

$$\mathbf{j}_{e/h}(\mathbf{r}) = -\frac{\sigma_e D_h + \sigma_h D_e}{\sigma_e + \sigma_h}\nabla \Delta n(\mathbf{r}), \tag{3.39}$$

which has the form of Fick's first law of diffusion with a diffusion coefficient

$$D_a = \frac{\sigma_e D_h + \sigma_h D_e}{\sigma_e + \sigma_h}. \tag{3.40}$$

Expressing electrical conductivities as functions of free electron and hole densities $n_{e/h}$ according to Eq. 2.30, and exploiting the Einstein relation (cf. Eq. 3.15), the ambipolar diffusion coefficient reads

$$D_a = \frac{D_e D_h (n_e + n_h)}{D_e n_e + D_h n_h}. \tag{3.41}$$

Equivalently, the Einstein relation allows an ambipolar carrier mobility μ_a to be expressed in terms of electron and hole mobilities rather than diffusion coefficients.

---

[9] Treatment analogous to [5].

At low-level injection ($\Delta n \approx n_h \ll n_e$ for n-type doping, or $\Delta n \approx n_e \ll n_h$ for p-type doping), the ambipolar diffusion coefficient corresponds to the minority carrier diffusion coefficient, approaching an intermediate value between minority and majority carrier diffusion coefficients toward intermediate- and high-level injection.

Notably, significant deviations between the minority carrier diffusion coefficient and the ambipolar diffusion coefficient may already set in at excess carrier densities well below an order of magnitude less than net dopant concentration. Especially when conducting lifetime measurements under open circuit conditions—i.e. without electrical carrier extraction—ambipolar effects should be taken into account. In the present work, the consideration of ambipolar diffusion proved to be vital for the determination of minority carrier mobility from lifetime measurements in a highly transport limited measurement design (cf. Sect. 9.2).

3.1.6 Trapping

Apart from the case of persistent local electrical fields (e.g. related to ambipolar diffusion) the densities of excess electrons Δn_e and holes Δn_h were assumed to be equal throughout the above considerations. This was attributed to dielectric relaxation of majority carriers such that local charge neutrality was ensured on timescales above the picosecond range—as detailed in Sect. 2.3. Yet, substantial deviations $\Delta n_e \neq \Delta n_h$ may occur on any timescale and even in the absence of local charge.

The cause for an imbalance of excess electrons and holes lies in a temporary capture of excess minority carriers in (mostly shallow, i.e. close to the minority carrier band) electronic states in the band gap, followed by reemission into the minority carrier band instead of recombination. This phenomenon, which is also referred to as *trapping*, was originally observed by Haynes and Westphal as a measurement artifact in drift experiments aiming at the determination of minority carrier mobility [24, 25]. Haynes and Hornbeck provided experimental evidence of trapping in both p- and n-type silicon [26, 27], and they adapted a model originally proposed by Fan to explain trapping kinetics based on the continuity equation [28], to be sketched in the following.[10]

3.1.6.1 Trapping According to Fan, Hornbeck, and Haynes

The model of Fan, Hornbeck, and Haynes assumes a total density of traps N_t, which may become occupied via capture of excess minority carriers at a density $n_t \leq N_t$. Further, the rate at which excess minority carriers are captured is distinguished by the capture time τ_t, while the rate at which captured carriers escape back to the minority carrier band is distinguished by the escape time τ_g. These quantities enable a description of the population kinetics of both the traps as well as the minority

[10] The chosen nomenclature is adapted from [29] for the most part.

carrier band via the ordinary time-dependent continuity equation.

$$\frac{d\Delta n_e}{dt} = G - \frac{\Delta n_e}{\tau} - \frac{dn_t}{dt} \tag{3.42}$$

$$\frac{dn_t}{dt} = -\frac{n_t}{\tau_g} + \frac{\Delta n_e \left(1 - \frac{n_t}{N_t}\right)}{\tau_t} \tag{3.43}$$

Here, the density of excess electrons in p-type silicon Δn_e may likewise be replaced by the density of excess holes in n-type silicon Δn_h. The population kinetics of both minority and majority carrier bands may be substantially affected by trapping.

3.1.6.2 Implications for Lifetime Measurements

Macdonald and Cuevas found experimental evidence for trapping to be caused by both dislocations in multicrystalline silicon and boron-impurity pairs or complexes in boron doped (p-type) silicon [29, 30], substantiating the relevance of trapping for photovoltaic characterization at $\Delta n \lesssim N_t$. Steady-state excess electron and hole densities differ in the presence of trapping. This fact, which was originally stressed by Fan [28], is crucial and fundamentally limiting for some steady-state and quasi-steady-state[11] lifetime measurement techniques. Beyond, dynamic lifetime measurement techniques may be affected by trapping likewise. This also applies to a harmonically modulated phase-sensitive lifetime approach, which represents a special case of a decay time measurement (cf. Sect. 6.2).

3.1.6.3 Steady-State Impact of Trapping

For steady-state excess carrier generation, the model by Hornbeck and Haynes yields

$$\Delta n_e = G\tau \tag{3.44}$$

$$n_t = \frac{N_t \Delta n_e}{\Delta n_e + \frac{N_t \tau_t}{\tau_g}}. \tag{3.45}$$

The invariance of steady-state excess minority carrier density Δn_e with regard to trapping brings about a remarkable implication: excess majority carrier concentration has to balance the sum of excess minority carriers and trapped carriers via dielectric relaxation. Therefore, excess majority carrier density must satisfy $\Delta n_h = \Delta n_e + n_t$ in the case of p-type doping.

Dielectric relaxation of majority carrier density due to trapping imposes severe limitations upon any steady-state lifetime technique sensitive to the total sum of free

[11] There are diverging perceptions of a quasi-steady-state condition, as will be discussed in Sect. 5.1. Here, quasi-steady-state denotes negligible derivatives with respect to time.

carrier densities.[12] Such techniques infer excess carrier density from the difference between the total sum of free carrier densities at a finite excess carrier generation rate $G \neq 0$ versus in the dark ($G = 0$). Most prominently, this involves quasi-steady-state inductive eddy current techniques [31, 32], but amplitude-sensitive[13] techniques involving intraband free carrier absorption or emission of infrared light [33–35] or microwave reflection are affected in the very same manner.

Let the measured amplitude of either of the above-mentioned affected steady-state techniques be

$$\mathcal{A}_+ = a_e n_e + a_h n_h, \tag{3.46}$$

with total free electron and hole densities n_e and n_h and positive real prefactors a_e and a_h. Let us restrict ourselves to p-type material with electron trapping (without loss of generality). The resulting excess amplitude (difference between illuminated and dark amplitude) amounts to

$$\Delta\mathcal{A}_+ = a_e \Delta n_e + a_h \left(\Delta n_e + n_t\right) \tag{3.47}$$

With Eqs. 3.44 and 3.45, the mismatch between an apparent lifetime τ_{app} to be obtained upon neglect of trapping and true lifetime τ can be expressed as a function of the ratio between true excess carrier density and dark trap density $x = \frac{\Delta n_e}{N_t}$. It is

$$\frac{\tau_{app}}{\tau}(x) = \frac{a_e + a_h\left(1 + \frac{1}{x + \frac{\tau_t}{\tau_g}}\right)}{a_e + a_h}. \tag{3.48}$$

For excess carrier densities much greater than dark trap density, i.e. in the limit $\frac{\Delta n_e}{N_t} \to \infty$, Eq. 3.48 implies coinciding $\tau_{app} = \tau$. Yet, in the low-level injection limit $\frac{\Delta n_e}{N_t} \to 0$, $\frac{\tau_{app}}{\tau}(x)$ may substantially deviate from unity

$$\frac{\tau_{app}}{\tau}(x \to 0) = \frac{a_e + a_h\left(1 + \frac{\tau_g}{\tau_t}\right)}{a_e + a_h}. \tag{3.49}$$

Equation 3.49 indicates the relevance of the ratio $\frac{\tau_g}{\tau_t}$ for the low-level injection mismatch betwen apparent lifetime and true lifetime. For equal coefficients $a_e = a_h$ and $\tau_g \gg \tau_t$ it would amount to

$$\frac{\tau_{app}}{\tau}(x \to 0) \approx \frac{1}{2}\frac{\tau_g}{\tau_t}. \tag{3.50}$$

[12] Such techniques are denoted as *additive* here, with an amplitude subscript +.

[13] *Amplitude-sensitive* shall denote the interpretation of a signal amplitude in terms of carrier lifetime via models, assumptions, and calibrations, rather than to exploit the time dependence of a measured amplitude.

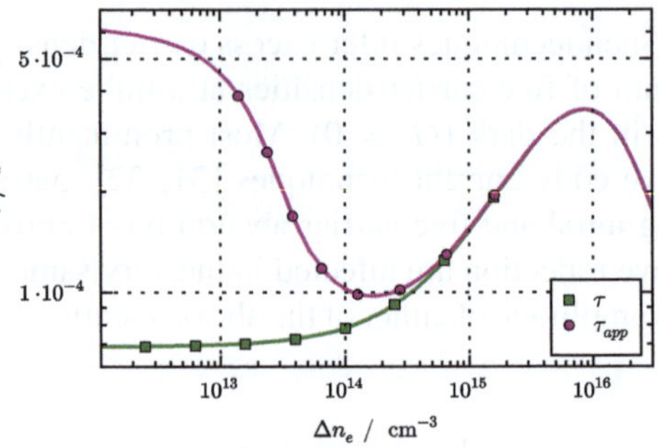

Fig. 3.4 Modeling of true lifetime τ versus apparent lifetime τ_{app} according to Eq. 3.48 for an injection-dependent QSSPC measurement with $a_e = q\mu_e$ and $a_h = q\mu_h$ [31, 32] (cf. Appendix B.1) The model parameters are $N_t = 10^{14}\,\mathrm{cm}^{-3}$, $\frac{\tau_g}{\tau_t} = 30$, and $\frac{\mu_e}{\mu_h} = 2.7$ (approximately corresponding to an acceptor concentration of $n_h^0 = 10^{16}\,\mathrm{cm}^{-3}$)

Macdonald and Cuevas encountered ratios of $10 \lesssim \frac{\tau_g}{\tau_t} \lesssim 100$ in multicrystalline p-type silicon [29]. Figure 3.4 illustrates the modeled mismatch between apparent lifetime and true lifetime as it would occur in a quasi-steady-state photoconductance measurement [31, 32].

Radiative recombination is sensitive to the product of majority carrier densities and minority carrier densities rather than to the difference between illuminated and dark total free carrier densities. This characteristic of steady-state luminescence-based lifetime techniques[14] is advantageous with regard to trapping, provided that net dopant concentration substantially exceeds the density of trapped carriers n_t. This was previously pointed out by Bardos et al. [36], and it can be explained by regarding Eq. 3.6. Again assuming p-type material with trapped electrons and neglecting dark electron density $n_e^0 \ll \Delta n_e$, one obtains the measured photoluminescence amplitude

$$\mathcal{A}_\times \propto \Delta n_e \left(\Delta n_e + \left(n_h^0 + n_t\right)\right). \tag{3.51}$$

With $\Delta n_e = \tau G$ (cf. Eq. 3.44), and with $n_t \leq N_t$ the ratio between apparent lifetime τ_{app} and true lifetime τ can be confined to

$$1 \leq \frac{\tau_{app}}{\tau} \leq \frac{N_t + n_h^0}{n_h^0} \tag{3.52}$$

Therefore, if $n_h^0 \gg N_t$ is satisfied, steady-state luminescence measurements are unaffected by trapping. Although a trapping artifact in steady-state luminescence measurements is most pronounced at low-level inection $\Delta n \ll n_h^0$, the injection dependence of the trapping artifact is much weaker than in the above discussed additive class of lifetime measurement techniques. This can be exploited for injection-dependent luminescence lifetime measurements at extremely low excess

[14] multiplicative techniques, with an amplitude subscript $\times$.

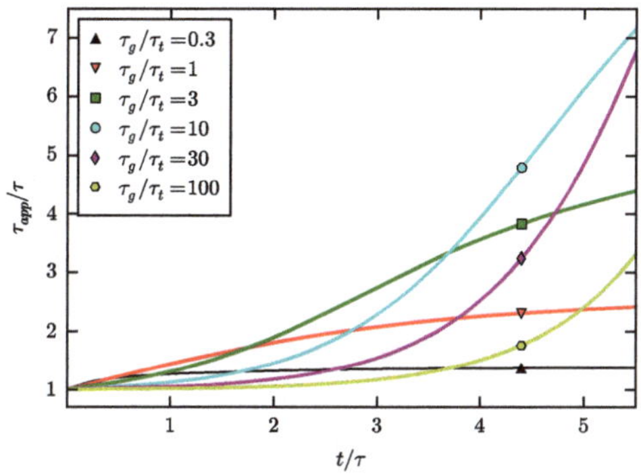

Fig. 3.5 Depiction of the mismatch between apparent lifetime τ_{app}—calculated from a numerical solution $\Delta n\,(t)$ according to Eq. 3.4—and true lifetime τ after an abrupt switch-off of excess carrier generation rate. $\Delta n\,(t)$ was modeled according to the trapping model by Fan, Hornbeck, and Haynes (cf. Eqs. 3.42 and 3.43) with an initial excess carrier density $\Delta n\,(t = 0) = N_t$, a lifetime $\tau = 10^{-5}$ s, and a capture time $\tau_t = \tau$

carrier densities, featuring a state-of-the-art sensitivity limit of $\Delta n_{min} \approx 10^8\,\text{cm}^{-3}$ at a dopant concentration of $n_h^0 \approx 1.5 \times 10^{16}\,\text{cm}^{-3}$ [37–39] (cf. Sect. 8.2).

3.1.6.4 Dynamic Impact of Trapping

For excess minority carrier densities $\Delta n \lesssim N_t$,[15] the population kinetics of the trap level may crucially interact with the population kinetics of the minority carrier band. The ensemble of captured carriers of density n_t may act as a significant source or generation rate of excess minority carriers upon decrease of external excess carrier generation. Likewise, the ensemble of unoccupied trap states of density $N_t - n_t$ may act as a significant sink for excess minority carriers upon onset of external excess carrier generation. Such processes complicate the determination of excess carrier recombination lifetime τ from decay time measurements at excess carrier densities $\Delta n \lesssim N_t$. Harmonically modulated lifetime measurements—making use of a superposition of infinitesimal carrier relaxation processes—could also be affected by such competing recombination and generation channels for sufficiently small injection $\Delta n \lesssim N_t$. In the following, the dynamic impact of trapping will be discussed and illustrated for the two cases of an abruptly vanishing steady-state excess carrier generation rate and a harmonically modulated excess carrier generation rate.

To illustrate the dynamic impact of trapping, Eqs. 3.42 and 3.43 were numerically solved via a finite element approach, assuming an injection-independent true excess carrier recombination lifetime $\tau = 10^{-5}$ s and an initial/maximal excess carrier density $\Delta n_{max} = N_t$.

Transient Techniques: For the case of an abruptly vanishing excess carrier generation rate, apparent lifetime τ_{app} was calculated from the numerical solution $\Delta n\,(t)$

[15] Dynamic measurement artifacts caused by trapping are not caused by the mismatch between Δn_e and Δn_h. Therefore, excess carrier densities are again denoted as Δn in the following.

according to Eq. 3.4 for different values of the ratio $\frac{\tau_g}{\tau_t}$. Figure 3.5 depicts the resulting mismatch $\frac{\tau_{app}}{\tau}$ as a function of the normalized time $\frac{t}{\tau}$ after an abrupt switch-off of generation rate, and for a capture time $\tau_t = \tau$. With regard to the resulting apparent lifetime curves, it can be stated that trapping may cause significant artifacts. The characteristics of the mismatch between apparent lifetime and true lifetime largely depend on the ratio $\frac{\tau_g}{\tau_t}$, but also on the ratio between true lifetime τ and both capture time τ_t and escape time τ_g.

Generally, the mismatch $\frac{\tau_{app}}{\tau}$ is a monotonously increasing function of decay time t. With an increasing ratio $\frac{\tau_g}{\tau_t}$, the significant onset of the mismatch is shifted to greater decay times t. Yet, the saturated magnitude of $\frac{\tau_{app}}{\tau}$ $(t \to \infty)$ is also a monotonously increasing function of $\frac{\tau_g}{\tau_t}$. Also, with an increasing/decreasing $\frac{\tau_t}{\tau}$, the onset of the mismatch is shifted to greater/shorter times t, respectively. In conclusion, the trapping error in transient lifetime analyses largely depends on their time frame. Given the ratios $10 \lesssim \frac{\tau_g}{\tau_t} \lesssim 100$ found by Macdonald and Cuevas for multicrystalline silicon wafers [29, 30], it appears that appropriate measurement conditions could be encountered at small decay times $t \ll \tau$ if $\tau_t \not\ll \tau$. However, the latter condition may be violated, and the restriction $t \ll \tau$ usually conflicts with both finite switch-off times of generation rate and appropriate measurement conditions for the fundamental decay mode of effective carrier lifetime (cf. Sect. 3.1.4).

Harmonically Modulated Techniques: As for lifetime measurements featuring a harmonically modulated generation rate, it is demonstrated in 3.2.3 that the time shift between maxima of excess carrier density and generation rate corresponds to excess carrier recombination lifetime,[16] provided that the timescale of a significant variation of generation rate (periodic time $\mathcal{T}$) is much greater than carrier lifetime.[17] In order to test the impact of trapping on such harmonically modulated lifetime measurements, the time shift resulting from a numerical solution $\Delta n\,(t)$ of Eqs. 3.42 and 3.43 for harmonically modulated excess carrier generation was systematically investigated.

Figures 3.6 and 3.7 depict the ratio between this timeshift τ_{app} and true lifetime τ (as assumed for the numerical simulation) as a function of both $\frac{\tau_t}{\tau}$ and $\frac{\tau_g}{\tau}$. Model assumptions were an injection-independent lifetime $\tau = 10^{-5}$ s, a sinusoidally modulated excess carrier generation rate peaking at $G_{max} = 10^{19}\,\mathrm{cm}^{-3}\mathrm{s}^{-1}$ and featuring periodic times $\mathcal{T}$ varying in the range between 10^{-3} and 10^{-1} s, and a dark trap density N_t varying in the range between $0.2\Delta n_{max}$ and $5\Delta n_{max}$. While Fig. 3.6 shows a variation of dark trap density N_t at a periodic time of $\mathcal{T} = 10^2$ s, Fig. 3.7 shows a variation of periodic time $\mathcal{T}$ at a dark trap density of $N_t = 10^{14}\,\mathrm{cm}^{-3}$.

The general conclusion of this model investigation is that there may be a pronounced and non-negligible impact of trapping in modulated lifetime measure-

[16] The equality of lifetime and harmonically modulated time shift only holds in case of injection-independent carrier lifetime. This is fully derived in Sect. 6.2.

[17] If such a quasi-steady-state condition is not fulfilled, time shifts become smaller than carrier lifetime. This is actually visible on the left of Fig. 3.7: at a ratio $\mathcal{T}/\tau = 100$ the time shift (without trapping artifact) is already measurably lower than true lifetime (cf. Fig. 3.8). Note that a self-consistent lifetime analysis would yield exact lifetime, even if a quasi-steady-state condition was violated (cf. Sect. 5.1).

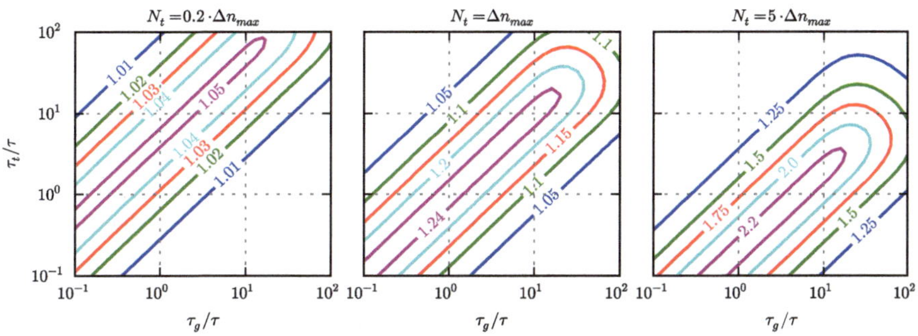

Fig. 3.6 Depiction of the ratio between apparent lifetime τ_{app} (as obtained from the time shift between modeled excess carrier density $\Delta n(t)$ and generation rate $G(t)$) and true lifetime τ. $\Delta n\,(t)$ was modeled according to the trapping model by Fan, Hornbeck, and Haynes (cf. Eqs. 3.42 and 3.43). Specific model parameters are mentioned in the text. Here, the periodic time of a harmonic time modulation is held constant at $\mathcal{T} = 10^{-2}$ s $= 10^3\,\tau$ and the dark trap density is varied as indicated

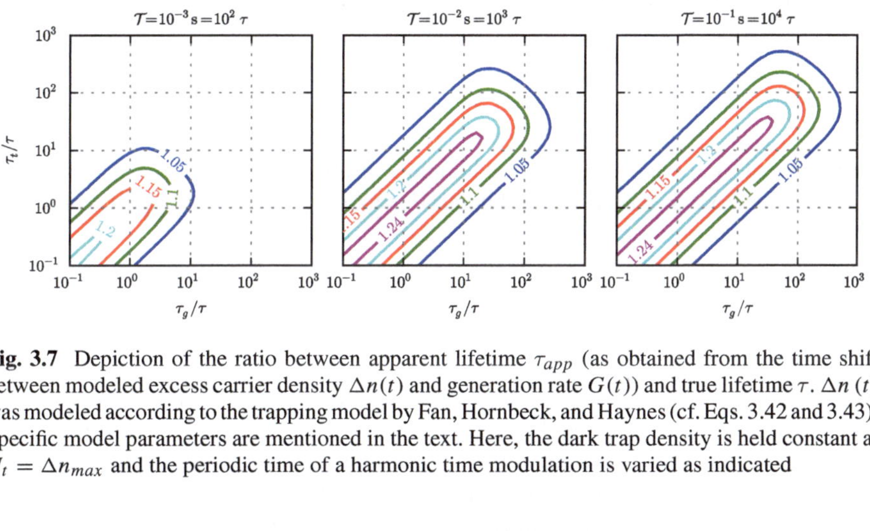

Fig. 3.7 Depiction of the ratio between apparent lifetime τ_{app} (as obtained from the time shift between modeled excess carrier density $\Delta n(t)$ and generation rate $G(t)$) and true lifetime τ. $\Delta n\,(t)$ was modeled according to the trapping model by Fan, Hornbeck, and Haynes (cf. Eqs. 3.42 and 3.43). Specific model parameters are mentioned in the text. Here, the dark trap density is held constant at $N_t = \Delta n_{max}$ and the periodic time of a harmonic time modulation is varied as indicated

ments, if excess carrier density Δn does not substantially exceed dark trap density N_t. The lower the ratio $\frac{\Delta n}{N_t}$, the more severe the artifact. Further, the lifetime mismatch induced by trapping appears to be equal for a fixed ratio $\frac{\tau_g}{\tau_t}$ if both τ_g and τ_t are sufficiently small ($\tau_g, \tau_t \not\gg \tau$). Figure 3.6 particularly reveals that the mismatch is most pronouced if the condition

$$\frac{N_t}{\Delta n}\frac{\tau_t}{\tau_g} = 1 \tag{3.53}$$

is satisfied. Figure 3.7 reveals another noteworthy detail: Toward the intermediate regime between quasi-steady-state[18] and transient measurement conditions, it appears that the mismatch between apparent and true lifetime becomes dependent on the periodic time $\mathcal{T}$ of the harmonic time modulation. Therefore, a trapping artifact in a modulated lifetime measurement could be reduced by augmenting the modulation frequency up to an intermediate regime between quasi-steady-state and transient time modulation. There is a direct analogy to the lifetime determination after an abrupt switch-off of excess carrier generation as discussed above. In that case (cf. Fig. 3.5) a trapping artifact was least pronounced for shortest measurement times, whereas here it is least pronounced for highest modulation frequencies. This phenomenon could also be one possible explanation for an (occasionally observed) dependence of carrier lifetime as obtained from a modulated technique on $\mathcal{T}$, which would not be straightforwardly explicable from the continuity equation.

3.2 Solutions of the Continuity Equation

This section elaborates solutions of the inhomogeneous continuity equation (Eq. 3.19) under conditions to be specified in Sect. 3.2.1, focusing on harmonically modulated excess carrier generation. Starting with ordinary solutions under steady-state conditions in Sect. 3.2.2 and ordinary modulated solutions neglecting any depth dependence in Sect. 3.2.3, both time- and depth-dependent solutions are outlined for both wafers and semi-infinite ingots in Sect. 3.2.4.

3.2.1 Excess Carrier Generation Rate

The present treatment of the continuity equation is not comprehensive in the sense that any solution would be covered. Besides the depth-dependent treatment resulting from the (occasionally violated) assumption of lateral invariance of electronic properties, the solutions provided here are based on some geometrical constraints to be addressed below. Also, these solutions mainly focus on a certain time dependence of excess carrier generation rate—distinguished by harmonic time modulation.

3.2.1.1 Depth Profile

As for the above mentioned geometrical constraints, the present treatment focuses on perpendicular irradiation of a planar semiconductor interface with monochromatic light of a room temperature penetration depth much less than the thickness of an investigated substrate (wafer, solar cell, ingot) $\alpha_\gamma^{-1} \ll d$. According to the Lambert-Beer law, collinear incidence of monochromatic light on a planar interface

[18] Here, the quasi-steady-state condition denotes a significant variation of excess carrier generation rate only on a timescale much greater than excess carrier recombination lifetime (cf. Sect. 5.1).

justifies the assumption of a mono-exponential decay of light intensity as a function of perpendicular substrate depth z. A nonplanar interface would scatter transmitted light, leading to deivations from a mono-exponential depth decay—particularly in substrates featuring an interface texture aiming at an enhanced absorption of light wavelengths close to the silicon band gap. For nonplanar interface morphologies featuring no intentional roughening (e.g. chemically polished multicrystalline wafers), light scattering was found to negligibly affect the depth profile of excess carrier generation rate (cf. Appendix E).

Further, it is assumed that any light transmitted through the interface of a semiconductor substrate is absorbed via excess carrier generation only. At monochromatic irradiation wavelengths of 791 and 880 nm, this is justified, as the probability of intraband absorption by free carriers is many orders of magnitude less than the probability of band-to-band absorption at these wavelengths [40–42] (cf. Appendix D).

With the assumptions stated above, according to the Lambert-Beer law excess carrier generation rate can be expressed as

$$G(z,t) = j_\gamma(t)\,\alpha_\gamma e^{-\alpha_\gamma z}, \tag{3.54}$$

with the possibly time-dependent transmitted (and absorbed) photon current density[19] $j_\gamma(t)$ and the band-to-band absorption coefficient α_γ at the irradiation wavelength.

For arbitrary non-monochromatic (continuous) spectra of irradiation light (of photon energy ε_γ), the depth profile of excess carrier generation rate is

$$G(z,t) = \int d\varepsilon_\gamma \, \frac{dj_\gamma\left(t,\varepsilon_\gamma\right)}{d\varepsilon_\gamma} \alpha_\gamma\left(\varepsilon_\gamma\right) e^{-\alpha_\gamma(\varepsilon_\gamma) z}. \tag{3.55}$$

Accordingly, solutions of the continuity equation may be determined via superposition of monochromatic solutions weighted by $\frac{dj_\gamma}{d\varepsilon_\gamma}$.

3.2.1.2 Time Profile

In the case of a steady-state optical irradiation (to be dealt with initially in Sect. 3.2.2) the impinging photon current density (cf. Eq. 3.54) is invariant in time. However, the present work mainly concerns harmonically modulated luminescence lifetime metrology. The most straightforward expression of a harmonic time modulation in this case is

$$j_\gamma(t) = j_{\gamma,0} e^{i\omega(t-t_0)} \tag{3.56}$$

[19] The transmitted photon current density corresponds to the impinging photon current density $j_{\gamma,i}$ multiplied by the substrate interface transmission $1 - R_\gamma$.

with the angular frequency ω linked to the periodic time T via $\omega T = 2\pi$. This expression will be used for a simplified treatment of the solution of the modulated continuity equation. To a large extent, this approach proves to yield correct conclusions, albeit being counterintuitive and to some extent inappropriate because of a possibly negative real part, which conflicts the physical conception of a photon current density, generation rate, excess carrier density, or of any intensity. An expression accounting for the latter conflict is

$$j_\gamma(t) = j_{\gamma,0} \sin^2 \frac{\pi}{T} (t - t_0). \tag{3.57}$$

Under conditions to be addressed in the following chapters, it may be justified to use a depth-averaged value of excess carrier generation rate. Integraton of Eq. 3.54 yields

$$G(t) = \frac{j_\gamma(t)}{d}, \tag{3.58}$$

for a substrate thickness $d \gg \alpha_\gamma^{-1}$.

3.2.1.3 Note on Electrical Injection of Excess Carriers

The depth-dependent solutions provided in the following are based on optical excess carrier generation upon illumination at a distinct photon energy ε_γ corresponding to a finite light penetration depth α_γ. In a depth-averaged treatment of the continuity equation, electrical injection of excess carriers into a pn-junction device via application of a forward bias (as applied in Sect. 7.4) is under some preconditions equivalent to optical excess carrier generation.

Preconditions are a sufficiently low surface recombination velocity of the substrate interface close to the pn-junction, a minority carrier diffusion length much greater than the light penetration depth, a sufficiently high shunt resistance of the pn-junction device, and to some extent a sufficient lateral homogeneity of electronic quality.

Electrical injection implies that minority carriers are driven into the pn-junction device. Thus, the effective penetration depth of electrical injection is virtually negligible, and injection is confined to the pn-junction. If the above preconditions are satisfied, the said confinement of carrier injection to the pn-junction has no impact on a depth-averaged solution of the continuity equation.

3.2.2 *Ordinary Steady-State Solution*

The steady-state solution of the continuity equation as given in Eq. 3.19 is required for numerous applications throughout this work. Steady-state excess carrier generation brings about a vanishing derivative with respect to time $\frac{\partial \Delta n}{\partial t} = 0$. Therefore, the resulting ordinary differential equation is

$$-D_a \frac{d^2 \Delta n(z)}{dz^2} + \frac{\Delta n(z)}{\tau} = G(z). \tag{3.59}$$

With an exponential ansatz, the general solution of the homogeneous differential Eq. 3.59 ($G = 0$) with a characteristic polynomial

$$\chi(\lambda) = -D_a \lambda^2 + \frac{1}{\tau} \tag{3.60}$$

takes the form

$$\Delta n_h(z) = A e^{\frac{z}{L}} + B e^{-\frac{z}{L}}, \tag{3.61}$$

with the inverse $L = \pm\sqrt{D_a \tau}$ of the root of the characteristic polynomial of the differential operator $\mathcal{D} = -D_a d_z^2 + \tau^{-1}$. This quantity can be physically interpreted as the ambipolar diffusion length of minority carriers.

A particular solution of the inhomogeneous differential Eq. 3.59 can again be derived from the characteristic polynomial Eq. 3.60. With an exponentially decaying excess carrier generation rate $G \propto e^{\upsilon z}$, a particular solution of the inhomogeneous Eq. 3.59 is given by

$$\Delta n_i \propto \frac{1}{\chi(\upsilon)} e^{\upsilon z}, \tag{3.62}$$

if $\chi(\upsilon) \neq 0$. Therefore, with an excess carrier generation rate

$$G(z) = j_\gamma \alpha_\gamma e^{-\alpha_\gamma z}, \tag{3.63}$$

the corresponding particular solution is

$$\Delta n_i = -\frac{j_\gamma \alpha_\gamma}{D_a \left(\alpha_\gamma^2 - L^{-2}\right)} e^{-\alpha_\gamma z}, \tag{3.64}$$

if $L \neq \alpha_\gamma^{-1}$.[20] Having now found a general solution $\Delta n_h(z)$ of the homogeneous steady-state continuity equation and one particular solution $\Delta n_i(z)$, which solves the steady-state continuity equation for an inhomogeneity as of Eq. 3.63, any particular solution

[20] For $L = \alpha_\gamma^{-1}$, the solution given in Eq. 3.64 is not defined. However, this appears to be an irrelevant restriction for virtually any practical application, as this particular solution is valid for an arbitrarily small $|L - \alpha_\gamma^{-1}|$.

$$\Delta n(z) = \Delta n_h(z) + \Delta n_i(z) \tag{3.65}$$

must satisfy the boundary conditions of a boundary value problem, respectively. In this manner, particular solutions of the steady-state continuity equation for wafers and ingots will be provided in the following.

3.2.2.1 Wafer

Inserting Eq. 3.65 into the boundary conditions for a silicon wafer of thickness d as given by Eqs. 3.21 and 3.22, one obtains the following system of equations:

$$\begin{pmatrix} 1 - \frac{D_a}{S_0 L} & 1 + \frac{D_a}{S_0 L} \\ \left(1 + \frac{D_a}{S_d L}\right) e^{\frac{d}{L}} & \left(1 - \frac{D_a}{S_d L}\right) e^{-\frac{d}{L}} \end{pmatrix} \begin{pmatrix} A \\ B \end{pmatrix} = \frac{\alpha_\gamma j_\gamma}{D\left(\alpha_\gamma^2 - L^{-2}\right)} \begin{pmatrix} 1 + \frac{\alpha_\gamma D_a}{S_0} \\ \left(1 - \frac{\alpha_\gamma D_a}{S_d}\right) e^{-\alpha_\gamma d} \end{pmatrix}. \tag{3.66}$$

Expression in terms of matrix elements

$$\begin{pmatrix} a_{11} & a_{12} \\ a_{21} & a_{22} \end{pmatrix} \begin{pmatrix} A \\ B \end{pmatrix} = \begin{pmatrix} b_1 \\ b_2 \end{pmatrix}. \tag{3.67}$$

yields the coefficients

$$A = \frac{b_1 a_{22} - b_2 a_{12}}{a_{11} a_{22} - a_{21} a_{12}} \tag{3.68}$$

$$B = \frac{b_1 a_{21} - b_2 a_{11}}{a_{21} a_{12} - a_{11} a_{22}}. \tag{3.69}$$

3.2.2.2 Ingot

In the case of a silicon ingot (denoted with an index I in the following), the electronically active area distinguished by the minority carrier diffusion length L is confined to depths z much less than the substrate thickness d. Therefore, the matrix elements a_{22} and b_2 of Eq. 3.67 vanish, while the matrix element a_{21} becomes infinitely large, leading to a simplification of coefficients

$$A_I = 0 \tag{3.70}$$

$$B_I = \frac{b_1}{a_{12}}. \tag{3.71}$$

Correspondingly, the steady-state solution for ingots with a surface recombination velocity $S_0 = S$ reads

$$\Delta n_I(z) = \frac{\alpha_\gamma j_\gamma}{D\left(\alpha_\gamma^2 - L^{-2}\right)} \left(\frac{\alpha_\gamma D + S}{D/L + S} e^{-\frac{z}{L}} - e^{-\alpha_\gamma z}\right) \tag{3.72}$$

3.2.3 Ordinary Harmonically Modulated Solution

The most essential analytical relations that this work draws on are the conclusions of an ordinary treatment of the continuity equation upon a harmonic time modulation. In this treatment, finite surface recombination velocities, nonuniform excess carrier generation, and minority carrier diffusion are considered via an effective lifetime.

Except for some special cases, it would be highly negligent not to take excess carrier diffusion and surface recombination into account for an ordinary time-modulated treatment of the continuity equation. In fact—in combination with the boundary conditions Eqs. 3.21 and 3.22—the diffusion term in Eq. 3.19 can be perceived as a surface recombination rate. Finite surface recombination velocities $S > 0$ cause a negative second derivative of excess carrier density with respect to substrate depth z. Thus, for $S > 0$, the depth integral satisfies

$$\frac{1}{d}\int_0^d dz \left(-D_a \frac{\partial^2 \Delta n(z,t)}{\partial z^2}\right) > 0, \tag{3.73}$$

which justifies the perception of the diffusion term in the continuity equation as an effective surface recombination rate. This allows a depth-independent expression of total recombination rate

$$\frac{\Delta n(t)}{\tau_{eff}} = \frac{1}{d}\int_0^d dz \left(\frac{\Delta n(z,t)}{\tau} - D_a \frac{\partial^2 \Delta n(z,t)}{\partial z^2}\right). \tag{3.74}$$

It is yet to be discussed how the effective lifetime τ_{eff} in the latter expression relates to dynamic or steady-state effective lifetime.[21] For harmonically modulated measurements, it will prove to be neither exactly identical to the one nor to the other. In Sect. 6.1, an exact expression of harmonically modulated effective lifetime in the case of quasi-steady-state measurement conditions (cf. Sect. 5.1) will be derived. The distinction between transient, harmonically modulated, and steady-state effective lifetime expressions is most crucial in the case of thick substrates with high surface recombination velocities and shallow excess carrier generation. For now we shall consider the case of coinciding steady-state and dynamic effective lifetime expressions. The ordinary differential equation of interest is therefore

[21] The depth profile of excess carrier density in Eq. 3.74 may be additionally weighted due to the sensitivity of the applied measurement technique.

$$\frac{d\Delta n\left(t\right)}{dt}+\frac{\Delta n\left(t\right)}{\tau_{\mathit{eff}}}=G\left(t\right), \tag{3.75}$$

with a harmonically modulated inhomogeneity $G(t)$.

3.2.3.1 Complex Ansatz

A mathematically straightforward treatment of an oscillatory physical system would consider the real part of the solution of Eq. 3.75 for an inhomogeneity of the form of a complex oscillation

$$G\left(t\right)=G_{0}e^{i\omega\left(t-t_{0}\right)}, \tag{3.76}$$

with an angular frequency $\omega=\frac{2\pi}{T}$. In spite of the fact that this system is unphysical in the sense that it features negative real parts of excess carrier generation rate, this treatment yet yields the same essential conclusions as a slightly more complicated treatment with a real positive inhomogeneity $G\left(t\right)$. In order to find a solution of the initial value problem represented by Eq. 3.75 together with an initial value $\Delta n\left(t_{0}\right)$, one may proceed analogously to the steady-state treatment sketched in Sect. 3.2.2. The characteristic polynomial of the homogeneous ordinary time-dependent continuity equation is

$$\chi\left(\lambda\right)=\lambda+\frac{1}{\tau_{\mathit{eff}}}. \tag{3.77}$$

A particular solution of the initial value problem is

$$\Delta n_{i}\left(t\right)=\frac{G_{0}}{i\omega+\tau_{\mathit{eff}}^{-1}}e^{i\omega\left(t-t_{0}\right)}. \tag{3.78}$$

The general solution of the homogeneous equation takes the form

$$\Delta n_{h}\left(t\right)=Ae^{-\frac{t-t_{0}}{\tau_{\mathit{eff}}}}, \tag{3.79}$$

with a coefficient A to be determined from an initial value $\Delta n\left(t_{0}\right)$. With this, the solution of Eq. 3.75 for any initial value is again given by the sum of Δn_{i} and Δn_{h}. As we are interested in $\Delta n\left(t-t_{0}\gg\tau_{\mathit{eff}}\right)$, the homogeneous part of the solution turns out to be irrelevant here, yielding

$$\Delta n\left(t-t_{0}\gg\tau_{\mathit{eff}}\right)=\Delta n_{i}\left(t\right)=\frac{G_{0}}{i\omega+\tau_{\mathit{eff}}^{-1}}e^{i\omega\left(t-t_{0}\right)}. \tag{3.80}$$

Let us now focus on very slow variations of excess carrier generation rate compared to effetive lifetime $\omega^{-1} \gg \tau_{eff}$.[22] In this case, the following linear Taylor series expansions are accurate:

$$\frac{1}{i\omega + \tau_{eff}^{-1}} \approx \tau_{eff}\left(1 - i\omega\tau_{eff}\right) \approx \tau_{eff} e^{-i\omega\tau_{eff}} \tag{3.81}$$

This enables quasi-steady-state excess carrier density to be expressed as

$$\Delta n\left(t\right) = \tau_{eff} G_0 e^{-i\omega\left(t - t_0 - \tau_{eff}\right)}, \tag{3.82}$$

allowing the preliminary conclusion that quasi-steady-state effective lifetime simply corresponds to the time shift between excess carrier density and excess carrier generation rate.[23]

3.2.3.2 Real Ansatz

In order to account for the fact that excess carrier density and generation rate are by definition positive, the solution of Eq. 3.75 for a real inhomogeneity $G\left(t\right)$ shall be given and discussed here. With a periodic time $\mathcal{T} = \frac{2\pi}{\omega}$, a real positive harmonically modulated excess carrier generation rate may take the form

$$\begin{aligned} G\left(t\right) &= G_0 \sin^2\left(\frac{\omega}{2}\left(t - t_0\right)\right) \\ &= \frac{G_0}{4}\left(2 - e^{i\omega(t - t_0)} - e^{-i\omega(t - t_0)}\right). \end{aligned} \tag{3.83}$$

For times $t - t_0 \gg \tau_{eff}$, the solution of Eq. 3.75 is then

$$\begin{aligned} \Delta n\left(t\right) &= \frac{G_0}{4}\left(2\tau_{eff} - \frac{1}{\tau_{eff}^{-1} + i\omega} e^{i\omega(t - t_0)} - \frac{1}{\tau_{eff}^{-1} - i\omega} e^{-i\omega(t - t_0)}\right) \\ &= \frac{G_0}{2}\left(\tau_{eff} - \frac{1}{\tau_{eff}^{-2} + \omega^2}\left(\frac{1}{\tau_{eff}}\cos\left(\omega\left(t - t_0\right)\right) + \omega\sin\left(\omega\left(t - t_0\right)\right)\right)\right). \end{aligned} \tag{3.84}$$

---

[22] This form of a quasi-steady-state condition will recur throughout this work.

[23] This relation will prove inaccurate (even if the above stated quasi-steady-state condition is satisfied) for injection-dependent lifetime. This is due to the fact that harmonically modulated lifetime measurements are differential measurements (cf. Sect. 6.2).

3.2.3.3 Quasi-Steady-State Approximations

At the time of maximal excess carrier density $t - t_0 = \frac{\omega}{2} + \Delta t$, the derivative of Δn with respect to time must vanish. In order to determine the time shift Δt between maxima of excess carrier density and excess carrier generation rate in the quasi-steady-state limit $\omega^{-1} \gg \tau_{eff}$ (detailed in Sect. 5.1), first order Taylor series expansions of trigonometric functions in Eq. 3.84 in the proximity Δt of $t - t_0 = \frac{T}{2} = \frac{\pi}{\omega}$ are accurate, yielding

$$\frac{d}{dt}\Delta n\left(t = t_0 + \frac{\pi}{\omega} + \Delta t\right) = 0 \overset{\omega^{-1} \gg \tau_{eff}}{\Longleftrightarrow} \Delta t = \tau_{eff}, \tag{3.85}$$

which is the same conclusion as of the above complex treatment of the ordinary time-dependent continuity equation. Beyond, if a quasi-steady-state condition $\omega^{-1} \gg \tau_{eff}$ is satisfied, excess carrier density according to Eq. 3.84 can be shown to correspond to excess carrier generation both multiplied and shifted by effective lifetime throughout the entire modulation cycle. The quasi-steady-state condition allows for

$$\Delta n\,(t) = \frac{G_0 \tau_{eff}}{2}\left(1 - \cos\left(\omega\,(t - t_0)\right) - \omega\tau_{eff} \sin\left(\omega\,(t - t_0)\right)\right). \tag{3.86}$$

Further, in quasi-steady-state the approximations $\cos \tau_{eff}\omega = 1$ and $\sin \tau_{eff}\omega = \tau_{eff}\omega$ are accurate, so with the trigonometric rule $\cos(\alpha \pm \beta) = \cos\alpha\cos\beta \mp \sin\alpha\sin\beta$, excess carrier density becomes

$$\begin{aligned}\Delta n\,(t) &= \frac{G_0 \tau_{eff}}{2}\left(1 - \cos\left(\omega\left(t - t_0 - \tau_{eff}\right)\right)\right) \\ &= G_0 \tau_{eff} \sin^2\left(\omega\left(t - t_0 - \tau_{eff}\right)\right) = \tau_{eff}\, G\left(t - \tau_{eff}\right).\end{aligned} \tag{3.87}$$

3.2.4 *Full Harmonically Modulated Solutions*

It was pointed out in Sect. 3.1.4 that for thick silicon substrates (e.g. ingots), high surface recombination velocities, and shallow excess carrier generation (as typically encountered in many lifetime techniques), dynamic and steady-state expressions of effective lifetime may drastically diverge. Further, in Sect. 3.2.3 it was stated that effective lifetime as encountered in harmonically modulated lifetime techniques neither completely coincides with dynamic nor with steady-state effective lifetime. This ambiguity gave rise to investigate the full solution of boundary value problems associated with the continuity equation (Eq. 3.19)—aiming at a sound interpretation of harmonically modulated effective lifetime in terms of recombination and transport properties. In the following, solutions of relevant boundary value problems based on Green's functions are sketched.

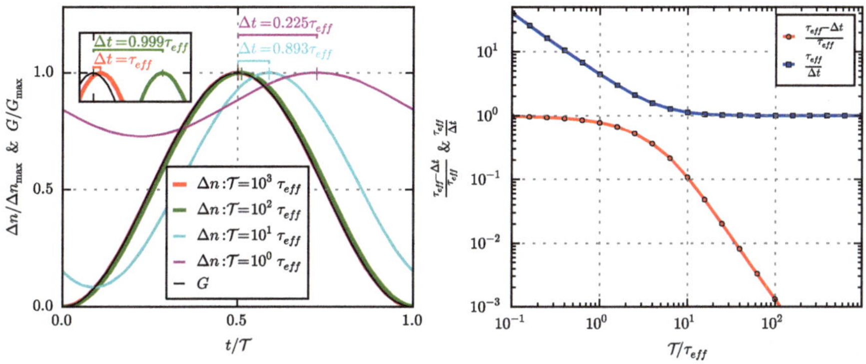

Fig. 3.8 Depiction of the time shift Δt between maxima of excess carrier density Δn and generation rate G for harmonic time modulation at different ratios of periodic time $\mathcal{T}$ and effective lifetime τ_{eff}. *On the left* harmonically time modulated excess carrier density and generation rate according to Eq. 3.84 are plotted, and Δt is highlighted, respectively. An inserted plot focuses on the peak area of the quasi-steady-state curves. *On the right* both the relative deviation and the mismatch between τ_{eff} and Δt are plotted as a function of $\mathcal{T}/\tau_{eff}$

3.2.4.1 Green's Functions of the Heat Equation

The most widespread approach to solve the continuity equation is based on a separation of variables [19–21, 43–46]. Other approaches involve time domain Fourier [47, 48] or Laplace [49, 50] transforms aiming at a reduction to ordinary problems.[24]

Although the above approaches are not generally restricted to homogeneous problems ($G = 0$), most treatments refer to homogeneous problems with finite initial values arising from pulsed or switched-off excess carrier generation. In that sense, the approach of Sanii et al. to frequency-modulated free carrier absorption represents an exception [51, 52]. Therein, a solution based on a Fourier decomposition of the inhomogeneity—which is most appropriate for lock-in techniques—was given.

Alternatively, Green's functions are a popular means to solve heat transport problems [53, 54]. As Eq. 3.19 represents a special case of a heat equation, the Green's function approach has also been adapted to semiconductor problems in the past. With a likely first application to a homogeneous semiconductor problem by Lax and Neustadter [55], the Green's function approach was very recently used by Wang et al. for a transient solution of a homogeneous semi-infinite boundary value problem—aiming at lifetime investigations of silicon ingots [18].

In the following, the integration of Green's functions is exploited in order to solve Eq. 3.19 for a time-dependent inhomogeneity $G(z, t)$ of the form of Eq. 3.54, boundary conditions as specified in Eq. 3.20, and an initial value $\Delta n\,(z, t_0 = 0) = 0$. With the variable transformation

$$\Delta n' = e^{\frac{t}{\tau}} \Delta n, \tag{3.88}$$

[24] Note that this very short digest of literature does not claim to be comprehensive.

the continuity equation as given in Eq. 3.19 is transformed to the standard heat equation

$$\frac{\partial \Delta n'(z,t)}{\partial t} - D_a \frac{\partial^2 \Delta n'(z,t)}{\partial z^2} = G(z,t)\, e^{\frac{t}{\tau}}. \tag{3.89}$$

This transformation allows to adapt the sophisticated methodology of the standard heat equation [53] to the dynamics of excess carriers in silicon.

3.2.4.2 Ingot

The solution of Eq. 3.89 for an initial value $\Delta n'(z, t_0 = 0) = 0$ and a boundary condition[25]

$$D_a \left. \frac{\partial \Delta n'}{\partial z} \right|_{z=0} = S \Delta n'(z=0). \tag{3.90}$$

with an ingot surface recombination velocity S is [54][26]

$$\Delta n(z,t) = \int_0^t dt' \int_0^\infty d\zeta\, \mathcal{G}(z,\zeta,t-t')\, G(\zeta,t')\, e^{\frac{t'-t}{\tau}}, \tag{3.91}$$

with the Green's function

$$\mathcal{G}(z,\zeta,t) = \frac{1}{2\sqrt{\pi D_a t}} \left(\sum_{\pm} e^{-\frac{(z \pm \zeta)^2}{4D_a t}} - \frac{2S}{D_a} \int_0^\infty d\varrho\, e^{-\frac{(z+\zeta+\varrho)^2}{4D_a t} - \frac{S\varrho}{D_a}} \right). \tag{3.92}$$

Recalling the form of excess carrier generation rate according to Eq. 3.54 [27]

$$G(z,t) = j_\gamma(t)\, \alpha_\gamma e^{-\alpha_\gamma z},$$

execution of the integration over ϱ and ζ in Eqs. 3.92 and 3.91 leads to complementary error function expressions that are motivated in Appendix A.2. With such expressions, the solution of the boundary value problem of interest is

[25] The boundary condition Eq. 3.20 is apparently invariant under the transformation $\Delta n \rightarrow \Delta n'$.

[26] Note the retransformation $\Delta n = e^{-\frac{t}{\tau}} \Delta n'$, which is incorporated here.

[27] This treatment is not restricted to harmonic time modulation of excess carrier generation rate.

$$\Delta n\left(z,t\right)=\alpha_{\gamma}\int\limits_{0}^{t}dt'\,j_{\gamma}\left(t'\right)e^{\frac{t'-t}{\tau}}\left(\frac{1}{2}\left(B_{+}+B_{-}\right)+\frac{S}{S-\alpha_{\gamma}D_{a}}\left(B_{S}-B_{+}\right)\right), \tag{3.93}$$

with coefficients

$$B_{\pm}=e^{\pm\alpha_{\gamma}\left(z\pm\alpha_{\gamma}D_{a}\left(t-t'\right)\right)}\operatorname{erfc}\left(\frac{\alpha_{\gamma}D_{a}\left(t-t'\right)\pm\frac{z}{2}}{\sqrt{D_{a}\left(t-t'\right)}}\right) \tag{3.94}$$

$$B_{S}=e^{\frac{S\left(z+S\left(t-t'\right)\right)}{D_{a}}}\operatorname{erfc}\left(\frac{S\left(t-t'\right)+\frac{z}{2}}{\sqrt{D_{a}\left(t-t'\right)}}\right). \tag{3.95}$$

The solution of the continuity equation given in Eq. 3.93 is semi-analytical, as this expression still involves a time integration. Yet, numerical computation of this time integral for arbitrary $j_{\gamma}(t')$ is straightforward and requires little computation effort. Indeed, when compared to a fully numerical finite element approach (in both variables z and t), this semi-analytical solution provides much higher accuracy at virtually negligible computation time. With the semi-analytical solution, the demanding finite element simulation of carrier diffusion (cf. Appendix A.3.2) is overcome. The bottleneck of the finite element approach stems from the combination of two discretization conditions $\Delta z \ll \alpha_{\gamma}^{-1}$ and $\frac{D_a \Delta t'}{\Delta z^2} \ll 1$ for time steps $\Delta t'$ and depth steps Δz, which are particularly limiting for thick substrates and shallow excess carrier generation. In contrast, numerical time integration as required for the semi-analytical solution provided in Eq. 3.93 is fast and straightforward: The factor $e^{\frac{t'-t}{\tau}}$ in Eq. 3.93 justifies a restriction of the integration domain to times $t - t' \geq 0$ with an upper bound given by the condition $e^{\frac{t'-t}{\tau}} \lll 1$. In Sect. 8.4, $t - t' \leq 7\tau$ was chosen. A variable time discretization featuring a reduction of discrete time steps $\Delta t'$ toward $t - t' \to 0$ further boosted accuracy and reduced computation effort.

3.2.4.3 Wafer

For substrates of finite thickness d, the Green's function solution of Eq. 3.89 for boundary conditions specified in Eqs. 3.21 and 3.22 and an initial value $\Delta n'\left(z, t_0 = 0\right) = 0$ shall be briefly sketched along the lines of [54]. Excess carrier density then takes the form

$$\Delta n\left(z,t\right)=\int\limits_{0}^{t}dt'\int\limits_{0}^{d}d\zeta\,\mathcal{G}\left(z,\zeta,t-t'\right)G\left(\zeta,t'\right)e^{\frac{t'-t}{\tau}}, \tag{3.96}$$

which appears identical to the expression for ingots given in Eq. 3.91, except for the finite integration limit $\zeta \leq d$, and a considerably different Green's function

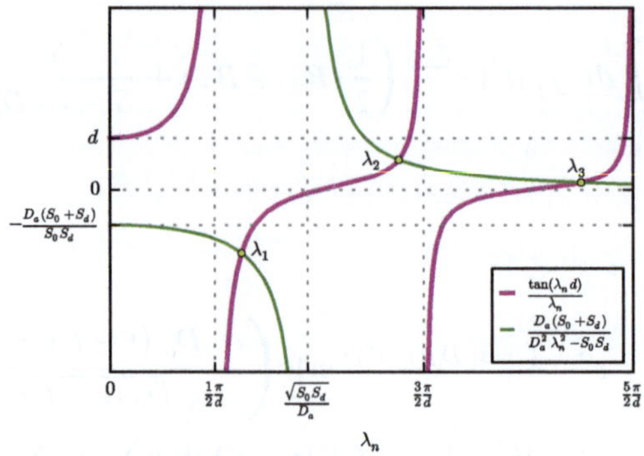

Fig. 3.9 Graphical illustration of solutions λ_n of Eq. 3.98 (for wafers). Any solution λ_n must lie in the interval $[(n-1)\frac{\pi}{d}, n\frac{\pi}{d}]$. The leftmost branch of $\frac{\tan\lambda_n d}{\lambda_n}$ only features an intercept with the rational function $\frac{D_a(S_0+S_d)}{D_a^2\lambda_n^2-S_0S_d}$ for sufficiently small surface recombination velocities, which must satisfy the condition $\frac{\sqrt{S_0S_d}}{D_a} < \frac{\pi}{2d}$

$$\mathcal{G}(z,\zeta,t) = \sum_{n=1}^{\infty} \frac{y_n(z)y_n(\zeta)}{\|y_n\|^2} e^{-\lambda_n^2 D_a t}. \tag{3.97}$$

Here, λ_n is the ensemble of positive solutions of the equation

$$\frac{\tan\lambda_n d}{\lambda_n} = \frac{D_a(S_0+S_d)}{D_a^2\lambda_n^2 - S_0S_d}. \tag{3.98}$$

Solutions of Eq. 3.98 are graphically illustrated in Fig. 3.9. The function y_n is of the form

$$y_n(z) = \cos\lambda_n z + \frac{S_0}{D_a\lambda_n}\sin\lambda_n z, \tag{3.99}$$

with

$$\|y_n\|^2 = \frac{S_d}{2D_a\lambda_n^2}\frac{D_a^2\lambda_n^2+S_0^2}{D_a^2\lambda_n^2+S_d^2} + \frac{S_0}{2D_a\lambda_n^2} + \frac{d}{2}\left(1+\left(\frac{S_0}{D_a\lambda_n}\right)^2\right). \tag{3.100}$$

References

1. S.M. Sze, *Semiconductor Devices: Physics and Technology* (Wiley, Hoboken, 2001)
2. D. E. Kane, R.M. Swanson, Measurement of the emitter saturation current by a contactless photoconductivity decay method, in *Proceedings of the 18th IEEE PVSC* (Las Vegas, 1985), pp. 578–583

3. C. Reichel, F. Granek, J. Benick, O. Schultz-Wittmann, S.W. Glunz, Comparison of emitter saturation current densities determined by quasi-steady-state photoconductance measurements of effective carrier lifetimes at high and low injections, in *Proceedings of the 23rd EU PVSEC* (Valencia, 2008), pp. 1664–1669
4. S. Rein, *Lifetime Spectroscopy: A Method of Defect Characterization in Silicon for Photovoltaic Applications* (Springer, Berlin, 2005)
5. P. Würfel, *Physics of Solar Cells: From Basic Principles to Advanced Concepts* (Wiley-VCH, Weinheim, 2009)
6. A. Richter, S.W. Glunz, F. Werner, J. Schmidt, A. Cuevas, Improved quantitative description of auger recombination in crystalline silicon. Phys. Rev. B **86**, 165202 (2012)
7. P.P. Altermatt, F. Geelhaar, T. Trupke, X. Dai, A. Neisser, E. Daub, Injection dependence of spontaneous radiative recombination in c-Si: experiment, theoretical analysis, and simulation, in *Proceedings of the 5th International Conference on Numerical Simulation and Optoelectronic Devices*, IEEE (Berlin, 2005), pp. 47–48
8. P.P. Altermatt, F. Geelhaar, T. Trupke, X. Dai, A. Neisser, E. Daub, Injection dependence of spontaneous radiative recombination in crystalline silicon: experimental verification and theoretical analysis. Appl. Phys. Lett. **88**, 261901 (2006)
9. J. Dziewior, W. Schmid, Auger coefficients for highly doped and highly excited silicon. Appl. Phys. Lett. **31**, 346–348 (1977)
10. R.A. Sinton, R.M. Swanson, Recombination in highly injected silicon. IEEE Trans. Electron Dev. **34**, 1380–1389 (1987)
11. A. Hangleiter, R. Häcker, Enhancement of band-to-band auger recombination by electron-hole correlations. Phys. Rev. Lett. **65**, 215–218 (1990)
12. R.N. Hall, Germanium rectifier characteristics. Phys. Rev. **83**, 228 (1951)
13. W. Shockley, W.T. Read, Statistics of the recombinations of holes and electrons. Phys. Rev. **87**, 835–842 (1952)
14. T. Markvart, L. Castaner, *Solar Cells: Materials, Manufacture and Operation*. (Elsevier Science, Amsterdam, 2004)
15. M.A. Green, *Solar Cells: Operating Principles, Technology and System Applications*. (University of New South Wales, Sydney, 1992)
16. R.A. Sinton, T. Trupke, Limitations on dynamic excess carrier lifetime calibration methods. Prog. Photovoltaics Res. Appl. **20**, 246–249 (2012)
17. M.A. Green, Analytical expressions for spectral composition of band photoluminescence from silicon wafers and bricks. Appl. Phys. Lett. **99**, 131112 (2011)
18. K. Wang, M.A. Green, H. Kampwerth, Transient photoconductance and photoluminescence from thick silicon wafers and bricks: analytical solutions. Solar Energy Mater. Sol. Cells **111**, 189–192 (2013)
19. J.P. McKelvey, *Solid State and Semiconductor Physics* (Harper & Row, New York, 1966)
20. K.L. Luke, L. Cheng, Analysis of the interaction of a laser pulse with a silicon wafer: determination of bulk lifetime and surface recombination velocity. J. Appl. Phys. **61**, 2282–2293 (1987)
21. A.B. Sproul, M.A. Green, A.W. Stephens, Accurate determination of minority carrier- and lattice scattering-mobility in silicon from photoconductance decay. J. Appl. Phys. **72**, 4161–4171 (1992)
22. A.B. Sproul, Dimensionless solution of the equation describing the effect of surface recombination on carrier decay in semiconductors. J. Appl. Phys. **76**, 2851–2854 (1994)
23. A. Cuevas, D. Macdonald, Measuring and interpreting the lifetime of silicon wafers. Sol. Energy **76**, 255–262 (2004)
24. J.R. Haynes, W.C. Westphal, The drift mobility of electrons in silicon. Phys. Rev. **85**, 680 (1952)
25. J.R. Haynes, W. Shockley, The mobility and life of injected holes and electrons in germanium. Phys. Rev. **81**, 835–843 (1951)
26. J.A. Hornbeck, J.R. Haynes, Trapping of minority carriers in silicon. I. p-type silicon. Phys. Rev. **97**, 311–321 (1955)

27. J.R. Haynes, J.A. Hornbeck, Trapping of minority carriers in silicon. II. n-type silicon. Phys. Rev. **100**, 606–615 (1955)
28. H.Y. Fan, Effect of traps on carrier injection in semiconductors. Phys. Rev. **92**, 1424–1428 (1953)
29. D. Macdonald, A. Cuevas, Trapping of minority carriers in multicrystalline silicon. Appl. Phys. Lett. **74**, 1710–1712 (1999)
30. D. Macdonald, A. Cuevas, Understanding carrier trapping in multicrystalline silicon. Sol. Energy Mater. Sol. Cells **65**, 509–516 (2001)
31. R.A. Sinton, A. Cuevas, M. Stuckings, Quasi-steady-state photoconductance, a new method for solar cell material and device characterization, in *Proceedings of the 25th IEEE PVSC* (Washington, D.C., 1996), pp. 457–460
32. R.A. Sinton, A. Cuevas, Contactless determination of current-voltage characteristics and minority-carrier lifetimes in semiconductors from quasi-steady-state photoconductance data. Appl. Phys. Lett. **69**, 2510–2512 (1996)
33. M. Bail, J. Kentsch, R. Brendel, M. Schulz, Lifetime mapping of Si wafers by an infrared camera (for solar cell production), in *Proceedings of the 28th IEEE PVSC* (Anchorage, 2000), pp. 99–103
34. J. Isenberg, S. Riepe, S.W. Glunz, W. Warta, Imaging method for laterally resolved measurement of minority carrier densities and lifetimes: measurement principle and first applications. J. Appl. Phys. **93**, 4268–4275 (2003)
35. M.C. Schubert, J. Isenberg, W. Warta, Spatially resolved lifetime imaging of silicon wafers by measurement of infrared emission. J. Appl. Phys. **94**, 4139–4143 (2003)
36. R.A. Bardos, T. Trupke, M.C. Schubert, T. Roth, Trapping artifacts in quasi-steady-state photoluminescence and photoconductance lifetime measurements on silicon wafers. Appl. Phys. Lett. **88**, 053504 (2006)
37. T. Trupke, R.A. Bardos, F. Hudert, P.Würfel, J. Zhao, A. Wang, M.A. Green, Effective excess carrier lifetimes exceeding 100 Milliseconds in float zone silicon determined from photoluminescence, in *Proceedings of the 19th EU PVSEC* (Paris, 2004), pp. 758–761
38. T. Trupke, R.A. Bardos, Photoluminescence: a surprisingly sensitive lifetime technique, in *Proceedings of the 31st IEEE PVSC* (Orlando, 2005), pp. 903–906
39. J.A. Giesecke, T. Niewelt, M. Rüdiger, M. Rauer, M.C. Schubert, W. Warta, Broad range injection-dependent minority carrier lifetime from photoluminescence. Sol. Energy Mater. Sol. Cells **102**, 220–224 (2012)
40. W. Spitzer, H.Y. Fan, Infrared absorption in n-type silicon. Phys. Rev. **108**, 268–271 (1957)
41. H. Hara, Y. Nishi, Free carrier absorption in p-type silicon. J. Phys. Soc. Jpn. **21**, 1222 (1966)
42. D.K. Schroder, R.N. Thomas, J.C. Swartz, Free carrier absorption in silicon. IEEE J. Solid-State Circ. **13**, 180–187 (1978)
43. W. Shockley, *Electrons and Holes in Semiconductors* (van Nostrand, New York, 1950)
44. H.Y. Tada, Theoretical analysis of transient solar-cell response and minority carrier lifetime. J. Appl. Phys. **37**, 4595–4596 (1966)
45. B.H. Rose, H.T. Weaver, Determination of effective surface recombination velocity and minority-carrier lifetime in high-efficiency Si solar cells. J. Appl. Phys. **54**, 238–247 (1983)
46. E. Gaubas, J. Vanhellemont, A simple technique for the separation of bulk and surface recombination parameters in silicon. J. Appl. Phys. **80**, 6293–6297 (1996)
47. S.R. Dhariwal, L.S. Kothari, S.C. Jain, Theory of transient photovoltaic effects used for measurement of lifetime of carriers in solar cells. Solid-state Electron. **20**, 297–304 (1977)
48. S.R. Dhariwal, N.K. Vasu, A generalised approach to lifetime measurement in pn junction solar cells. Solid-State Electron. **24**, 915–927 (1981)
49. R.H. Kingston, Switching time in junction diodes and junction transistors. Proc. IRE **42**, 829–834 (1954)
50. M.S. Tyagi, J.F. Nijs, R.J. Van Overstraeten, Effect of surface recombination on the transient decay of excess carriers produced by short wavelength laser pulses. Solid-State Electron. **25**, 411–415 (1982)

51. F. Sanii, R.J. Schwartz, R.F. Pierret, W.M. Au, The measurement of bulk and surface recombination by means of modulated free carrier absorption, in *Proceedings of the 20th IEEE PVSC* (Las Vegas, 1988), pp. 575–580
52. F. Sanii, F.P. Giles, R.J. Schwartz, J.L. Gray, Contactless nondestrcutive measurement of bulk and surface recombination using frequency-modulated free carrier absorption. Solid-State Electron. **35**, 311–317 (1992)
53. H.S. Carslaw, J.C. Jaeger, *Conduction of Heat in Solids*, 2nd edn. (Clarendon Press, Oxford, 1986)
54. A.D. Polyanin, *Handbook of Linear Partial Differential Equations for Engineers and Scientists* (Chapman & Hall/CRC, Boca Raton, 2002)
55. B. Lax, S.F. Neustadter, Transient response of a p-n junction. J. Appl. Phys. **25**, 1148–1154 (1954)

Chapter 4
Luminescence of Silicon

This chapter outlines the theory of spontaneous emission of radiation by non-black bodies such as semiconductors (and particularly indirect semiconductors). The link between electronic properties and luminescent emission is drawn, and elementary approaches to electronic properties from silicon are classified.

4.1 The Generalized Law of Radiation

The interaction between light and matter comprises three fundamental transitions: photon absorption, stimulated photon emission, and spontaneous photon emission. If the electronic system of matter (e.g. of a crystalline solid) is in thermal equilibrium with ambient radiation, transition rates associated with the above transitions must compensate. In the following, rates r_x denote the number of transitions per volume, energy, and time. With the photon energy increment $d\varepsilon_\gamma$, the differential transition rates (transition rates per volume in the energy interval $[\varepsilon_\gamma, \varepsilon_\gamma + d\varepsilon_\gamma]$) are

$$dr_x = r_x d\varepsilon_\gamma. \tag{4.1}$$

4.1.1 Detailed Balance and Derivative Considerations

The differential rates associated with the three fundamental transitions in a semiconductor are the photon absorption rate $dr_\uparrow$, the stimulated photon emission rate $dr_{\downarrow st}$, and the spontaneous photon emission rate $dr_{\downarrow sp}$. As stated above, thermal equilibrium (identical temperatures of distributions of electron system and ambient radiation) implies

$$dr_\uparrow = dr_{\downarrow st} + dr_{\downarrow sp}. \tag{4.2}$$

J. Giesecke, *Quantitative Recombination and Transport Properties in Silicon from Dynamic Luminescence*, Springer Theses, DOI: 10.1007/978-3-319-06157-3_4,
© Springer International Publishing Switzerland 2014

The process of stimulated emission is to be perceived as a negative contribution to photon absorption [1, 2]. Therefore, net absorption rate dr_{abs} corresponds to the balance $dr_{\uparrow} - dr_{\downarrow st}$, and with Eq. 4.2 one thus obtains

$$dr_{abs} = dr_{\uparrow} - dr_{\downarrow st} = dr_{\downarrow sp}. \tag{4.3}$$

The latter equality between net absorption rate and spontaneous emission rate is otherwise referred to as the principle of detailed balance [3]. Expressed as a radiation law for non-black bodies by Kirchhoff [4] long before derivative quantum mechanical concepts such as transitions between discrete energy levels of an electronic system had been known—it provides the basis of the derivation of a generalized law of radiation.

One may define an isotropic absorption coefficient $\alpha_\gamma(\varepsilon_\gamma)$ via the above net absorption rate dr_{abs}. With a density of photons[1] $n_\gamma(\varepsilon_\gamma)d\varepsilon_\gamma$ in the energy interval $[\varepsilon_\gamma, \varepsilon_\gamma + d\varepsilon_\gamma]$, and with the velocity of light $c' = \frac{c_0}{n}$ in a solid with a refractive index with real part n, the absorption coefficient $\alpha_\gamma(\varepsilon_\gamma)$ may be defined as

$$dr_{abs} = \alpha_\gamma(\varepsilon_\gamma)c' n_\gamma(\varepsilon_\gamma)d\varepsilon_\gamma. \tag{4.4}$$

Here, the product $c' n_\gamma(\varepsilon_\gamma)d\varepsilon_\gamma$ can be understood as the superposition of absolute values of all microscopic photon current densities in a volume of interest in the energy interval $[\varepsilon_\gamma, \varepsilon_\gamma + d\varepsilon_\gamma]$. Exploiting the principle of detailed balance (cf. Eq. 4.3), the spontaneous emission rate must likewise read

$$dr_{\downarrow sp} = \alpha_\gamma(\varepsilon_\gamma)c' n_\gamma(\varepsilon_\gamma)d\varepsilon_\gamma. \tag{4.5}$$

Thus, in thermal equilibrium, spontaneous emission rate is proportional to the absorption coefficient. In order to obtain a complete analytical expression for the spontaneous emission rate, an expression for the photon density $n_\gamma(\varepsilon_\gamma)d\varepsilon_\gamma$ is to be found. The following derivation proceeds along the lines of a derivation given by Würfel in 1982 [5]. A very similar reasoning based on detailed balance considerations—and with equivalent conclusions—was outlined by Lasher and Stern already in 1964 [6].

4.1.2 Transition Rates According to Würfel

The conditions of Würfel's calculus are reflected here:

1. Uniform chemical potentials for electrons in the conduction band and holes in the valence band, allowing the densities of such excitations to be described by unique Fermi-Dirac distributions, respectively. This implies a chemical equilibrium with regard to intra-band transitions.

[1] Isotropy of microscopic photon currents accounting for n_γ is explicitly not required here.

2. Thermal equilibrium between the electron/hole system and the crystal lattice. This is equivalent to identical temperatures of the lattice as well as the distributions of free electrons and holes.

Chemical equilibrium with regard to band-to-band transitions is explicitly not required. Therefore, a finite separation $\eta_{\sum} = \varepsilon_{FC} - \varepsilon_{FV}$ of quasi-Fermi energies of electrons and holes is allowed. Under such conditions, the Fermi-Dirac representation

$$f\left(\varepsilon, \mu_c\right) = \frac{1}{e^{\frac{\varepsilon-\mu_c}{k_B T}} + 1} \tag{4.6}$$

is appropriate to describe the number of occupied or unoccupied states in both valence and conduction band. The chemical potential μ_c here corresponds to quasi-Fermi energies of electrons ($\mu_c = \varepsilon_{FC}$) or holes ($\mu_c = \varepsilon_{FV}$), respectively. Würfel [5] described photon density $n_\gamma(\varepsilon_\gamma)d\varepsilon_\gamma$ on the basis of expressions of optical transition rates in terms of Fermi-Dirac distributions.

Absorption: The rate of photon absorption processes leading to the excitation of an electron from the energy interval $[\varepsilon, \varepsilon + d\varepsilon]$ in the valence band to the energy interval $[\varepsilon + \varepsilon_\gamma, \varepsilon + \varepsilon_\gamma + d\varepsilon]$ in the conduction band must be proportional to the density of occupied states in the valence band $D\left(\varepsilon\right) f\left(\varepsilon, \varepsilon_{FV}\right) d\varepsilon$, to the density of unoccupied states in the conduction band $D\left(\varepsilon + \varepsilon_\gamma\right)\left(1 - f\left(\varepsilon + \varepsilon_\gamma, \varepsilon_{FC}\right)\right) d\varepsilon$, and to the density of photons $n_\gamma(\varepsilon_\gamma)d\varepsilon_\gamma$. A proportionality factor $m\left(\varepsilon, \varepsilon_\gamma\right)$ must account for the matrix element or probability per time of the optical transition between the energy intervals $[\varepsilon, \varepsilon + d\varepsilon]$ and $[\varepsilon + \varepsilon_\gamma, \varepsilon + \varepsilon_\gamma + d\varepsilon]$. Also, the factor $m\left(\varepsilon, \varepsilon_\gamma\right)$ ensures the proper dimension of the entire expression as a transition rate per volume.[2] With the latter proportionality factor, the absorption rate may be expressed as the following energy integral:

$$dr_{\uparrow} = d\varepsilon_\gamma n_\gamma(\varepsilon_\gamma) \int d\varepsilon\, m\left(\varepsilon, \varepsilon_\gamma\right) D\left(\varepsilon\right) f\left(\varepsilon, \varepsilon_{FV}\right) D\left(\varepsilon + \varepsilon_\gamma\right)\left(1 - f\left(\varepsilon + \varepsilon_\gamma, \varepsilon_{FC}\right)\right) \tag{4.7}$$

Stimulated Emission: The rate of stimulated photon emission processes leading to the recombination of a conduction band electron in the energy interval $[\varepsilon + \varepsilon_\gamma, \varepsilon + \varepsilon_\gamma + d\varepsilon]$ with a valence band hole in the energy interval $[\varepsilon, \varepsilon + d\varepsilon]$ is proportional to the density of unoccupied states in the valence band (holes) $D\left(\varepsilon\right)\left(1 - f\left(\varepsilon, \varepsilon_{FV}\right)\right) d\varepsilon$, to the density ofoccupied states in the conduction band (electrons) $D\left(\varepsilon + \varepsilon_\gamma\right) f\left(\varepsilon + \varepsilon_\gamma, \varepsilon_{FC}\right) d\varepsilon$, and to the density of photons $n_\gamma(\varepsilon_\gamma)d\varepsilon_\gamma$. With the same proportionality factor $m\left(\varepsilon, \varepsilon_\gamma\right)$ as introduced for the description of absorption rate, the rate of stimulated emission reads

$$dr_{\downarrow st} = d\varepsilon_\gamma n_\gamma(\varepsilon_\gamma) \int d\varepsilon\, m\left(\varepsilon, \varepsilon_\gamma\right) D\left(\varepsilon\right)\left(1 - f\left(\varepsilon, \varepsilon_{FV}\right)\right) D\left(\varepsilon + \varepsilon_\gamma\right) f\left(\varepsilon + \varepsilon_\gamma, \varepsilon_{FC}\right). \tag{4.8}$$

[2] Note that the factor $m\left(\varepsilon, \varepsilon_\gamma\right)$ is identical for all transition rates of interest, involving the absorption rate, the stimulated emission rate, and the spontaneous emission rate.

Spontaneous Emission: As the stimulated emission rate, the rate of spontaneous photon emission processes leading to the recombination of a conduction band electron in the energy interval $[\varepsilon + \varepsilon_\gamma, \varepsilon + \varepsilon_\gamma + d\varepsilon]$ with a valence band hole in the energy interval $[\varepsilon, \varepsilon + d\varepsilon]$ must also be proportional to the density of unoccupied states in the valence band and to the density of occupied states in the conduction band. However, spontaneous emission is distinguished from stimulated emission in that it is proportional to the photon density of states $D_\gamma\left(\varepsilon_\gamma\right) d\varepsilon_\gamma$ rather than to the actual photon density $n_\gamma(\varepsilon_\gamma)d\varepsilon_\gamma$. The density of photon states in a cube with periodic boundary conditions

$$D_\gamma\left(\varepsilon_\gamma\right) d\varepsilon_\gamma = \frac{\varepsilon_\gamma^2}{\pi^2 \hbar^3 c'^3} d\varepsilon_\gamma \tag{4.9}$$

is motivated in Appendix A.4. With the same proportionality factor $m\left(\varepsilon, \varepsilon_\gamma\right)$ as introduced for the description of the above transition rates, the rate of spontaneous emission reads

$$dr_{\downarrow sp} = d\varepsilon_\gamma D_\gamma\left(\varepsilon_\gamma\right) \int d\varepsilon\, m\left(\varepsilon, \varepsilon_\gamma\right) D(\varepsilon)\left(1 - f\left(\varepsilon, \varepsilon_{FV}\right)\right) D\left(\varepsilon + \varepsilon_\gamma\right) f\left(\varepsilon + \varepsilon_\gamma, \varepsilon_{FC}\right). \tag{4.10}$$

4.1.3 Photon Density

With the function

$$g\left(\varepsilon, \varepsilon_\gamma, \varepsilon_{FV}, \varepsilon_{FC}\right) = m\left(\varepsilon, \varepsilon_\gamma\right) D(\varepsilon)\left(1 - f\left(\varepsilon, \varepsilon_{FV}\right)\right) D\left(\varepsilon + \varepsilon_\gamma\right) f\left(\varepsilon + \varepsilon_\gamma, \varepsilon_{FC}\right), \tag{4.11}$$

photon density $n_\gamma(\varepsilon_\gamma)d\varepsilon_\gamma$ can be expressed via the transition rates defined in Eqs. 4.7, 4.8, and 4.10 as

$$n_\gamma\left(\varepsilon_\gamma\right) d\varepsilon_\gamma = d\varepsilon_\gamma \frac{\varepsilon_\gamma^2}{\pi^2 \hbar^3 c'^3} \frac{\int d\varepsilon\, g\left(\varepsilon, \varepsilon_\gamma, \varepsilon_{FV}, \varepsilon_{FC}\right)}{\int d\varepsilon\, g\left(\varepsilon, \varepsilon_\gamma, \varepsilon_{FV}, \varepsilon_{FC}\right)\left(\frac{f(\varepsilon, \varepsilon_{FV})(1 - f(\varepsilon + \varepsilon_\gamma, \varepsilon_{FC}))}{(1 - f(\varepsilon, \varepsilon_{FV})) f(\varepsilon + \varepsilon_\gamma, \varepsilon_{FC})} - 1\right)}. \tag{4.12}$$

Upon insertion of the Fermi-Dirac distribution (as of Eq. 4.6) into the expression in the denominator of Eq. 4.12, this expression substantially simplifies to

$$\frac{f\left(\varepsilon, \varepsilon_{FV}\right)\left(1 - f\left(\varepsilon + \varepsilon_\gamma, \varepsilon_{FC}\right)\right)}{\left(1 - f\left(\varepsilon, \varepsilon_{FV}\right)\right) f\left(\varepsilon + \varepsilon_\gamma, \varepsilon_{FC}\right)} = e^{\frac{\varepsilon_\gamma - \left(\varepsilon_{FC} - \varepsilon_{FV}\right)}{k_B T}} = e^{\frac{\varepsilon_\gamma - \eta_\Sigma}{k_B T}}, \tag{4.13}$$

which is apparently independent of energy ε, such that photon density may read

$$n_\gamma\left(\varepsilon_\gamma\right)d\varepsilon_\gamma = d\varepsilon_\gamma \frac{\varepsilon_\gamma^2}{\pi^2\hbar^3 c'^3}\frac{1}{e^{\frac{\varepsilon_\gamma-\eta_\Sigma}{k_B T}}-1}. \tag{4.14}$$

This expression much resembles Planck's law describing the energy density of a black body in the photon energy interval $[\varepsilon_\gamma, \varepsilon_\gamma + d\varepsilon_\gamma]$ [7], but photons feature a finite chemical potential η_Σ here [5].

4.1.4 The Generalized Planck Law of Radiation

Insertion of the expression found above for photon density $n_\gamma\left(\varepsilon_\gamma\right)d\varepsilon_\gamma$ into Eq. 4.5 yields the generalized law of radiation[3]

$$dr_{\downarrow sp} = d\varepsilon_\gamma \frac{\alpha_\gamma\left(\varepsilon_\gamma\right)\varepsilon_\gamma^2}{\pi^2\hbar^3 c'^2}\frac{1}{e^{\frac{\varepsilon_\gamma-\eta_\Sigma}{k_B T}}-1}. \tag{4.15}$$

Its major distinctions with respect to Planck's law describing black body radiation are

- the consideration of an absorptance less than unity through incorporation of the absorption coefficient $\alpha_\gamma\left(\varepsilon_\gamma\right)$ on the basis of Kirchhoff's law, and
- the consideration of the free energy or chemical potential η_Σ of generated or recombining electron-hole pairs [2]. As photons, which generate electron-hole pairs in a semiconductor, must carry the free energy of such electron-hole pairs, η_Σ could be perceived as a chemical potential of nonthermal radiation [5].

In contrast to black body radiation, nonthermal radiation—such as luminescence in a semiconductor—is characterized by a threshold photon energy in its spectrum due to the band gap [5].

4.1.5 Application to Indirect Semiconductors

The above considerations do not account for the fact that electron-hole pair generation and recombination processes in indirect semiconductors imply generation or annihilation of phonons. An optical transition in an indirect semiconductor could be adequately described in terms of a chemical reaction equation

[3] Note that this expression accounts for isotropic radiation in all directions. Considering radiation only in a solid angle Ω would require accounting for the ratio $\frac{\Omega}{4\pi}$ as another factor on the right side of Eq. 4.15.

$$\gamma \leftrightharpoons e + h \pm \Gamma \tag{4.16}$$

with an equilibrium condition for a minimum of free energy corresponding to

$$\mu_\gamma = \eta_\Sigma \pm \mu_\Gamma. \tag{4.17}$$

Here, Γ denotes a phonon of chemical potential μ_Γ, while μ_γ represents a photonic chemical potential. Generation or annihilation of phonons is accompanied by an additional energy term $\pm\varepsilon_\Gamma$ in the energy balance of the transition described in the reaction Eq. 4.16, and the question is whether this fact has to be accounted for in the detailed balance considerations above, and whether or not the generalized radiation law is altered through additional emission or absorption of phonons.

In 1957, Bowlden found that the relation between emission and absorption characteristics valid for direct transitions would also be valid for indirect transitions [8]. In 1995, Würfel et al. applied the principle of detailed balance to indirect transitions [1]. They additionally accounted for the density of states D_Γ as well as for the occupation probability f_Γ of phonons in the above expressions of transition rates (cf. Eqs. 4.7, 4.8, and 4.10). In conclusion, they showed that the rate of spontaneous recombination features the exact same dependence on the absorption coefficient for both indirect and direct semiconductors. Any further dependencies on phonon energy ε_Γ cancelled out in their calculations, such that the generalized radiation law for indirect semiconductors would read

$$dr_{\downarrow sp} = d\varepsilon_\gamma \frac{\alpha_\gamma\left(\varepsilon_\gamma\right)\varepsilon_\gamma^2}{\pi^2\hbar^3 c'^2}\frac{1}{e^{\frac{\varepsilon_\gamma - \eta_\Sigma \pm \mu_\Gamma}{k_B T}} - 1}. \tag{4.18}$$

The latter expression differs from the expression for direct semiconductors (cf. Eq. 4.15) in that it accounts for the chemical potential of phonons μ_Γ. However, with phonon lifetimes of approximately 10^{-12} s [9] compared to radiative recombination lifetimes ranging between approximately 10^{-2} s at a net dopant concentration of $n_0 \approx 10^{16}\,\mathrm{cm}^{-3}$ and approximately 10^4 s for intrinsic silicon (to be concluded from measurements of the radiative recombination coefficient B [10–12] with $\tau_{rad} = (n_0 B)^{-1}$, cf. 4.2 and Eq. 3.6), the relative change of the occupation number of phonons induced by radiative recombination was argued to be negligible [1, 10]. Therefore, the chemical potential change $|\mu_\Gamma|$ associated with the emission or absorption of a phonon in silicon at room temperature must be much less[4] than the separation of quasi-Fermi energies η_Σ

$$|\mu_\Gamma| \lll \eta_\Sigma. \tag{4.19}$$

This justifies the validity of the generalized radiation law given in Eq. 4.15 for silicon.

[4] Some works set $\mu_\Gamma = 0$ under the specified conditions [1, 10].

4.2 Electronic Properties from Luminescence

Here, the relation between the generalized radiation law and measurable luminescence intensities is outlined. Having drawn the link between electronic properties and excess carrier density in Chap. 3, this section provides the link between excess carrier density and luminescent emission by a semiconductor.

4.2.1 From Spontaneous Emission to Excess Carrier Density

Equation 2.29 describes the relation between the separation of quasi-Fermi energies and the product $n_e n_h = \left(n_e^0 + \Delta n\right)\left(n_h^0 + \Delta n\right)$. As $n_h^0 n_e^0 = n_i^2$, with typical net dopant concentrations in the range of $n_0 = 10^{16}\,\mathrm{cm}^{-3}$ and excess carrier densities $\Delta n \gg \frac{n_i^2}{n_0}$, dark minority carrier concentration can be neglected. With net dopant concentration n_0, the $n_e n_h$-product is then accurately described (for both net acceptors or donors) as

$$n_e n_h = \Delta n \left(n_0 + \Delta n\right). \tag{4.20}$$

With this, the separation of quasi-Fermi energies according to Eq. 2.29 reads

$$\eta_\Sigma = k_B T \ln\left(\frac{1}{n_i^2}\Delta n \left(n_0 + \Delta n\right)\right). \tag{4.21}$$

For all relevant photon energies satisfying $\varepsilon_\gamma - \eta_\Sigma \gg k_B T$, insertion into the generalized radiation law yields

$$dr_{\downarrow sp} = d\varepsilon_\gamma \frac{1}{\pi^2 \hbar^3 c'^2} \frac{1}{n_i^2} \alpha_\gamma\left(\varepsilon_\gamma\right) \varepsilon_\gamma^2 e^{-\frac{\varepsilon_\gamma}{k_B T}} \Delta n \left(n_0 + \Delta n\right). \tag{4.22}$$

Therefore, the spontaneous emission rate of radiation by a semiconductor is to be described as the product of a prefactor comprising fundamental physical and material related constants, a function of photon energy and temperature,[5] and the product of minority and majority carrier densities. With the latter equation, excess carrier density Δn is manifested as the key quantity of a luminescence measurement. Under low-level injection conditions (*lli*) $n_0 \gg \Delta n$, spontaneous emission is even proportional to excess carrier density

$$dr_{\downarrow sp}^{\mathrm{lli}} = d\varepsilon_\gamma \frac{1}{\pi^2 \hbar^3 c'^2} \frac{n_0}{n_i^2} \alpha_\gamma\left(\varepsilon_\gamma\right) \varepsilon_\gamma^2 e^{-\frac{\varepsilon_\gamma}{k_B T}} \Delta n. \tag{4.23}$$

[5] This function contains a Bose factor, it carries the energy dependence due to the photon density of states (cf. Appendix A.4), and it contains the absorption coefficient and the inverse squared intrinsic carrier density.

4.2.2 The Coefficient of Radiative Recombination

With a so-called integral coefficient of radiative recombination B [12], the energy integral over Eq. 4.22 is commonly expressed as

$$\int dr_{\downarrow sp} = \Delta n \left(n_0 + \Delta n\right) \int_0^\infty d\varepsilon_\gamma \frac{dB}{d\varepsilon_\gamma} = B \Delta n \left(n_0 + \Delta n\right). \tag{4.24}$$

With Eq. 4.22, the spectral or differential coefficient of radiative recombination reads

$$\frac{dB}{d\varepsilon_\gamma} = \frac{1}{\pi^2 \hbar^3 c'^2 n_i^2} \alpha_\gamma \left(\varepsilon_\gamma\right) \varepsilon_\gamma^2 e^{-\frac{\varepsilon_\gamma}{k_B T}}, \tag{4.25}$$

and the integral coefficient B is[6]

$$B = \frac{1}{\pi^2 \hbar^3 c'^2 n_i^2} \int_0^\infty d\varepsilon_\gamma \, \alpha_\gamma \left(\varepsilon_\gamma\right) \varepsilon_\gamma^2 e^{-\frac{\varepsilon_\gamma}{k_B T}}. \tag{4.26}$$

The latter expression is somewhat inaccurate with respect to the dependences of the integral coefficient of radiative recombination

$$B = B\left(T, n_0, \Delta n\right). \tag{4.27}$$

For silicon, its temperature dependence stems from the pronounced temperature dependences of the intrinsic carrier concentration $n_i\ (T)$, of the absorption coefficient $\alpha_\gamma\ (T)$, and of the Bose factor $e^{-\frac{\varepsilon_\gamma}{k_B T}}$. However, what is disguised in this representation is the dependence of B on free carrier density. Equations 4.24, 4.25, and 4.26 are only valid for nondegenerate occupation of bands [12]. In 2005, Altermatt et al. pointed out that the radiative recombination coefficient in crystalline silicon is enhanced by the Coulomb attraction between electrons and holes, which is screened at high carrier densities [14, 15]. Throughout this work, the paramerization of the integral coefficient of radiative recombination as proposed by Altermatt [14] (with modifications by Roth and Rosenits [16]) is used. This parametrization is explicitly stated in Appendix A.5. Figure 4.1 shows both, the differential (or spectral [12]) coefficient of radiative recombination $\frac{dB}{d\varepsilon_\gamma}$ as a function of photon energy ε_γ, and the relative integral coefficient of radiative recombination $\frac{B}{B_0}$ as a function of the free carrier sum $2\Delta n + n_0$ and for different temperatures T.

---

[6] Rigorously speaking, the refractive index contained in the velocity of light c' also depends on photon energy ε_γ. However, due to a weak energy dependence within the relevant energy range of the silicon luminescence spectrum, it is deemed sufficient to use its average value within this energy range.

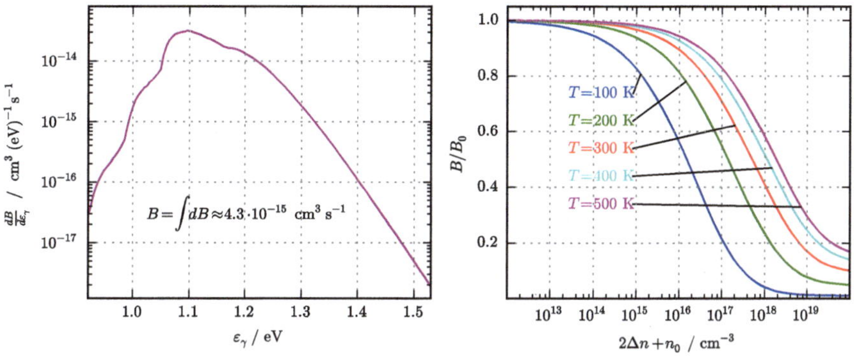

Fig. 4.1 On the *left*, the differential (or spectral [12]) coefficient of radiative recombination $\frac{dB}{d\varepsilon_\gamma}$ is plotted as a function of photon energy ε_γ on the basis of refractive index data and band-to-band 300 K absorption coefficient data tabulated by Green and Keevers [13]. Intrinsic carrier density was set to $n_i = 10^{10}\,\text{cm}^{-3}$ here. On the *right*, the relative integral coefficient of radiative recombination $B/B_0 = B/B\,(T = 0, n_0 = 0, \Delta n = 0)$ is plotted as a function of the sum of free carrier densities $2\Delta n + n_0$ and for different temperatures T (cf. Eq. A.49)

4.2.3 Measured Luminescence Intensity

Spontaneous emission gives rise to measurable photon currents out of a semiconductor substrate. Such photon currents can be detected e.g. via photodetectors. Let Φ denote the luminescence intensity from a semiconductor volume V, measured at an effective solid angle of detection Ω'.[7] Although luminescence *intensities* may be encountered in the form of voltages, currents, or count rates, Φ shall be perceived here as a dimensionless quantity.[8] As such, the photoluminescence *intensity* $d\Phi$ measured in the time interval $[t, t + dt]$ may be expressed as an integral of spontaneous emission rate over photon energy ε_γ and volume V, yielding

$$d\Phi = dt\frac{\Omega'}{4\pi}\int_V d\mathbf{r}\int_0^\infty dr_{\downarrow sp}\left(\varepsilon_\gamma, \mathbf{r}, t\right)\,Q_\Phi\left(\varepsilon_\gamma\right)\,T_\Phi\left(\varepsilon_\gamma\right)\,R_\Phi\left(\varepsilon_\gamma, \mathbf{r}\right). \tag{4.28}$$

Three dimensionless functions Q_Φ, T_Φ, and R_Φ are involved. The function $0 \leq Q_\Phi \leq 1$ accounts for the spectral sensitivity or quantum efficiency of the detector, and the function $0 \leq T_\Phi \leq 1$ accounts for the optical losses along the light path from the semiconductor interface to the detector—involving interface reflection, light absorption along the light path, and reflection or absorption in the detector optics.The

[7] Here, the *effective* solid angle of detection also accounts for refraction of light at the interface between a semiconductor and its surroundings, and for light scattering at nonplanar interfaces.

[8] Note that this perception of the term *intensity* differs from the common physical perception of an energy current density.

function $0 \leq R_\Phi \leq 1$ accounts for self absorption or reabsorption of luminescent radiation within the substrate bulk. As radiative recombination in silicon is in almost any case an unlikely process compared to nonradiative recombination via defects or impurities, reabsorbed photons can be regarded as lost in the balance of photon currents. Therefore, for luminescence detection within a small solid angle around the perpendicular of a silicon substrate with a planar interface, a straightforward expression for R_Φ is given by the Lambert–Beer law

$$R_\Phi\left(\varepsilon_\gamma, z\right) = \begin{cases} e^{-\alpha_\gamma(\varepsilon_\gamma)z}, & \text{or} \\ e^{-\alpha_\gamma(\varepsilon_\gamma)(d-z)}. \end{cases} \tag{4.29}$$

Here, the underlying geometrical convention is that substrate interfaces be distinguished by $z = 0$ and $z = d$ (with thickness d). Either form of Eq. 4.29 applies depending on whether luminescence detection occurs toward the interface at $z = 0$ of a substrate (where carrier injection or illumination commonly takes place) or at $z = d$. For nonplanar interfaces, R_Φ may take a more complicated form, as this implies scattering of non-perpendicularly emitted photons.[9] Assuming lateral invariance of spontaneous light emission from the investigated substrate surface A_Φ,[10] luminescence *intensity* as detected from this (planar) substrate surface becomes[11]

$$d\Phi\left(t\right) = dt\frac{\Omega'}{4\pi}A_\Phi \int_0^\infty d\varepsilon_\gamma \frac{dB}{d\varepsilon_\gamma}Q_\Phi\left(\varepsilon_\gamma\right)T_\Phi\left(\varepsilon_\gamma\right)\int_0^d dz\, e^{-\alpha_\gamma(\varepsilon_\gamma)z}\Delta n\left(z,t\right)\left(n_0 + \Delta n\left(z,t\right)\right). \tag{4.30}$$

The reabsorption function R_Φ brings about an entanglement of dependencies such that the integrals over depth and photon energy do not factorize. For applications relating the information provided by luminescence measurements to recombination and transport properties, it is worthwhile to obtain a straightforward and at the same time most generally valid relation between luminescence *intensity* and excess carrier density. Such a relation is obtained via a function $p_\Phi(z)$, which accounts for the relative depth-dependent probability of luminescence detection. For sufficiently short time intervals $[t_1, t_2]$ in order to ensure negligible variation of excess carrier density, integration over photon energy and time yields the photoluminescence *intensity*

$$\Phi\left(t\right) = b_\Phi \int_0^d dz\, p_\phi\left(z\right)\Delta n\left(z,t\right)\left(n_0 + \Delta n\left(z,t\right)\right), \tag{4.31}$$

[9] For such photons, the light path in the bulk may substantially exceed its projection on the perpendicular.

[10] E.g. the substrate surface corresponding to one pixel of a CCD camera.

[11] As indicated in Eq. 4.29, the argument z of the reabsorption function must be transformed to $d - z$ for luminescence detection from the substrate interface located at $z = d$.

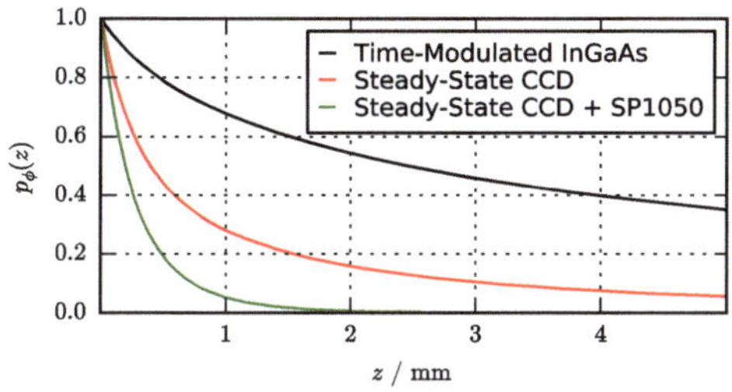

Fig. 4.2 Plot of relative detection probabilities $p_\Phi(z)$ of some luminescence detection systems used in this work [17]. These involve a modulated luminescence system featuring InGaAs detectors (*uppermost curve*), a steady-state PL imaging system based on an infrared-enhanced CCD camera without (*intermediate curve*) and with an additional 1,050 nm optical short pass filter (*bottommost curve*). Here, luminescence emission toward the substrate interface at $z = 0$ is captured. Also, the substrate optical transmission was assumed to be independent of photon energy ε_γ

with a constant

$$b_\Phi = \frac{\Omega'}{4\pi} A_\Phi \int_{t_1}^{t_2} dt \int_0^\infty d\varepsilon_\gamma \frac{dB}{d\varepsilon_\gamma} Q_\Phi\left(\varepsilon_\gamma\right) T_\Phi\left(\varepsilon_\gamma\right). \tag{4.32}$$

This definition of the constant b_ϕ implies a detection probability $p_\Phi = 1$ at the substrate interface oriented toward the detector. Figure 4.2 depicts relative reabsorption probabilities $p_\Phi(z)$ encountered in the luminescence setups used in this work.

4.2.4 Dynamic Luminescence Intensity

Spectrally integrated luminescence *intensity* as detected upon emission from the surface of a silicon device does not a priori carry information about the depth distribution of excess carrier density $\Delta n(z)$. Even though—under certain preconditions—one may access such information from spectrally resolved luminescence measurements [18], we shall focus on the case of spectrally integrated luminescence measurements for now. Under the latter measurement conditions, it may be useful to obtain information about recombination and transport properties without knowledge of the exact depth distribution $\Delta n(z)$. This is most crucial for the ordinary time-dependent treatment of the continuity equation as used in parts of this work, and generally for time-domain approaches, which infer effective excess carrier recombination lifetime regardless of the total magnitude of measured photoluminescence *intensity*.

An often sufficient condition for depth-independent analyses of luminescence measurements in terms of recombination and transport properties is that the function $p_\Phi(z)$ may feature no substantial variation throughout the entire depth $0 < z < d$ of a substrate. If this is the case, reabsorption of photoluminescence is of minor impact.

As can be seen from Fig. 4.2, this condition is approximately satisfied for wafers or solar cells of thickness $d \lesssim 250\,\mu\text{m}$ in the case of the modulated InGaAs detection system. If the above condition is not satisfied, a depth-independent treatment may still be justified if time and depth dependencies of Δn factorize, i.e. if the relative depth profile of excess carrier density is quasi-time-independent. With a depth-averaged excess carrier density

$$\Delta n\,(t) = \frac{1}{d}\int_0^d dz\,\Delta n\,(z,t)\,, \tag{4.33}$$

and a function $\beta\,(z)$ satisfying the condition

$$\int_0^d dz\,\beta\,(z) \overset{!}{=} d, \tag{4.34}$$

the following relation must hold:

$$\Delta n\,(z,t) \overset{!}{\approx} \Delta n\,(t)\,\beta\,(z) \tag{4.35}$$

Integration over substrate depth then yields the photoluminescence *intensity*

$$\Phi\,(t) = c_\Phi \Delta n\,(t)\,(n_0 + \Delta n\,(t))\,. \tag{4.36}$$

with a constant

$$c_\Phi = b_\Phi \int_0^d dz\,\beta\,(z)\,p_\Phi\,(z)\,. \tag{4.37}$$

Equation 4.36 features an apparent inconsistency for high-level injection conditions ($\Delta n \gtrsim n_0$)—manifesting itself upon insertion of $\Delta n\,(z,t)$ as of Eq. 4.35 into Eq. 4.31. It must be noted however, that—provided common dopant concentrations of $n_0 \gtrsim 10^{15}\,\text{cm}^{-3}$—high-level injection is typically not encountered in thin wafers and under open circuit conditions unless recombination rates are sufficiently low in order to satisfy

$$\int_0^d dz\,\beta\,(z) \approx \int_0^d dz\,\beta^2\,(z)\,. \tag{4.38}$$

The use of Eq. 4.36 is justified for many cases of dynamic effective lifetime investigations, which may even include much pronounced depth profiles of excess carrier densities. This is because—even if luminescence reabsorption amounts to a substantial fraction of total emitted luminescence intensity—it only significantly affects the *relative* course of $\Phi\,(t)$ in case of a significant redistribution of $\Delta n\,(z)$ during a dynamic measurement. Under dynamic measurement conditions, only the *relative* course of $\Phi(t)$—rather than its total amplitude—matters.

Concerning the interpretation of dynamic lifetime measurements, caution is indicated in the case of thick samples featuring very high surface recombination velocities—particularly for shallow excess carrier generation [17, 19]. In such cases, the unambiguous distinction between dynamic and steady-state perceptions of effective lifetime is essential (cf. Sects. 3.1.4, 6.1, 8.4).

4.3 Classification of Luminescence Applications

Luminescence applications are manifold, and therefore the following classification of applications neither aspires to be comprehensive nor to adequately represent the chronological order of scientific findings. Rather, it is solely intended to categorize key luminescence approaches to electronic properties.

Equation 4.22 identifies excess carrier density as the cause of spontaneous emission, regardless of the mechanism of carrier excitation. Possible excitation mechanisms are optical electron-hole pair generation (photoluminescence), electrical injection of excess minority carriers in pn-junction devices via application of a forward bias (electroluminescence), or electron-hole pair generation via impingement of fast electrons onto a semiconductor (cathodoluminescence). As the electron-hole pair excitation mechanism is secondary with regard to the generalized radiation law, no distinction shall be made concerning the excitation mechanism in the following classification.

4.3.1 The Ab Initio Class

The generalized law of radiation as given in Eq. 4.15 can be used to determine an implied voltage of a semiconductor substrate from luminescent emission upon steady-state injection of excess carriers. This implied voltage again relates to excess carrier density and to recombination properties—as derived above.

Even though such an approach—as proposed e.g. by Würfel [2]—may succeed in special cases [20, 21], its general applicability is much complicated and constrained by the amount of required a priori assumptions. This becomes clear from Eq. 4.31. While net dopant concentration n_0, the substrate surface of interest A_Φ, the optical properties of substrate and setup accounting for the transmission function $T_\Phi(\varepsilon_\gamma)$, the band-to-band absorption coefficient $\alpha_\gamma(\varepsilon_\gamma)$, the intrinsic carrier density n_i and (with considerable experimental effort) even the total detector quantum efficiency $Q_\Phi(\varepsilon_\gamma)$ might be accessible, it is barely possible to account for the effective solid angle of detection Ω' for nonplanar—and particularly for quasi-Lambertian (cf. Appendix E)—interface morphologies in an analytically straightforward manner.

Further, screening of Coulomb interaction at high free carrier densities [15] may additionally complicate the ab initio approach.

4.3.2 The Ratio Class

Accepting the limitations of the ab initio approach for many applications due to an uncertainty of required a priori assumptions, a broad class of luminescence approaches exploiting luminescence intensity ratios has evolved. Such ratios may relate to distributions of luminescence intensity with respect to both space **r** [22] or photon energy ε_γ [18]—but they are distinguished by the fact that through division of luminescence intensities, most dependencies cancel out, revealing a sufficient and often unambiguous contrast with respect to the properties of interest.

Ancient approaches by Tajima [23] and by Colley and Lightowlers [24] to obtain dopant concentrations from low temperature ratios of phonon-assisted exciton lines and bands could e.g. be attributed to this class. Likewise a more recent spectrally resolved luminescence approach by Würfel et al. [18]—explicitly making use of self-absorption of luminescence—and derivatives thereof [21, 25–27] are to be attributed to this class.

The most prominent and definitely the recently most frequently applied representative of the class of luminescence approaches based on ratios infers differences of implied voltage $\Delta U_{i,j}$ (e.g. across a large area device such as a silicon wafer or a solar cell) from luminescence intensity ratios by exploiting the relation

$$\Phi_i \propto e^{\frac{qU_i}{k_B T}}, \tag{4.39}$$

which yields

$$\Delta U_{i,j} = \frac{k_B T}{q} \ln \frac{\Phi_i}{\Phi_j}. \tag{4.40}$$

This simple and yet most useful relation can be regarded as the essential cause of the boost that luminescence imaging techniques have experienced within the last eight years. With an alleged first luminescence imaging application by Uchida dating back to 1963 [28, 29], the use of CCD detectors by Fuyuki et al. in 2005 [30] and by Trupke et al. in 2006 [31] have triggered a vast amount of derivative photovoltaic applications focusing on the determination of series resistance [32–38], parallel resistance [39, 40], and/or recombination properties [38, 41–43] of silicon photovoltaic devices—to name but some works. Particularly for the resistance techniques, the capability to resolve $\Delta U_{i,j}$ often proved sufficient for a self-contained determination, while for most recombination techniques, an external calibration of *total* implied voltage was necessary.

4.3.3 *The Dynamic Class*

Dynamic approaches to luminescence provide the most immediate access to excess carrier recombination lifetime, as the decay of luminescence intensity is easily related to the decay of excess carrier density (cf. Eq. 4.36). The immediate access to carrier lifetime leads to a minimum of required a priori assumptions about material properties and a virtually negligible impact of self-absorption, interface morphology, and optical interface properties due to the irrelevance of total photoluminescence amplitude. Yet, these advantages are obtained at the cost of drawbacks such as carrier trapping (cf. Sect. 3.1.6) and complications arising from a different perception of effective carrier lifetime [17, 19] (to be addressed in Sect. 6.1) and from differential measurement designs [44–48] (to be addressed in Sect. 6.2).

4.3.3.1 Transient Luminescence Decay

With radiative recombination dominating the total recombination rate, transient photoluminescence decay has been prevalent e.g. for exciton or minority carrier lifetime analyses in direct band gap compound semiconductors [49–51], but it had also been exploited for lifetime investigations in indirect band gap semiconductors such as non-crystalline [52] or crystalline silicon [53, 54] in the past. A very recent theoretical study investigates the impact of bulk and surface recombination effects on transient photoluminescence decay in very thick silicon substrates (e.g. ingots) [55].

Recent photovoltaic applications related to transient photoluminescence decay have extensively focused on spatially resolved measurements. Herlufsen et al. [56, 57] adapted an imaging approach of effective lifetime based on the combined local transient and steady-state analysis of photoluminescence intensity in an electronically shuttered InGaAs CMOS camera, which had originally been proposed by Ramspeck et al. [58, 59] for transient free carrier absorption imaging. Kiliani et al. implemented a fully transient effective lifetime analysis based on a sophisticated modulation of the detector signal via a rotating shutter wheel [60, 61] or via an image intensifier unit [62], thereby enabling fast simultaneous shuttering of an entire CCD chip.

4.3.3.2 Harmonically Modulated Luminescence

A brief timeline of harmonically modulated approaches to recombination and transport properties of semiconductors has been sketched in the introduction, with modulated luminescence approaches by Tolstoi and Feofilov [63], by Guidotti et al. [64, 65], by Mandelis et al. [66],[12] by Trupke and Bardos [53, 54, 69], and by Brüggemann and Reynolds [70].

[12] with derivative phase imaging applications by Melnikov et al. [67] and by Sun et al. [68].

Most modulated luminescence lifetime analyses can be classified as either phase-sensitive (commonly lock-in-based) approaches [64, 66, 70], or as self-consistent analyses of time-dependent signals (e.g. photocurrents) [63, 69, 71].

This work elaborates the relation between self-consistent lifetime analyses of time-dependent photocurrents and the above-mentioned phase-sensitive group of lifetime approaches. Further, constraints of the phase-sensitive group of approaches are identified (cf. Sect. 6.2), and a pathway to overcome such constraints of the very powerful and most sensitive lock-in approach is outlined [48, 72].

References

1. P. Würfel, S. Finkbeiner, E. Daub, Generalized Planck's radiation law for luminescence via indirect transitions. Appl. Phys. A **60**, 67–70 (1995)
2. P. Würfel, *Physics of Solar Cells: From Basic Principles to Advanced Concepts* (Wiley-VCH, Weinheim, 2009)
3. W. van Roosbroeck, W. Shockley, Photon-radiative recombination of electrons and holes in germanium. Phys. Rev. **94**, 1558–1560 (1954)
4. G. Kirchhoff, Über das Verhältnis zwischen dem Emissionsvermögen und dem Absorptionsvermögen der Körper für Wärme und Licht. Annalen der Physik **185**(2), 275–301 (1860)
5. P. Würfel, The chemical potential of radiation. J. Phys. C **15**, 3967–3985 (1982)
6. G. Lasher, F. Stern, Spontaneous and stimulated recombination radiation in semiconductors. Phys. Rev. **133**, A553–A563 (1964)
7. M. Planck, *Theorie der Wärmestrahlung*, (Barth, Leipzig, 1906)
8. H. Bowlden, Detailed balance in nonvertical transitions. J. Phys. Chem. Solids **3**, 115–117 (1957)
9. P.A. Temple, C.E. Hathaway, Multiphonon Raman spectrum of silicon. Phys. Rev. B **7**, 3685–3697 (1973)
10. E. Daub, Photolumineszenz von Silizium, Ph.D. thesis, University of Karlsruhe (1995)
11. H. Schlangenotto, H. Maeder, W. Gerlach, Temperature dependence of the radiative recombination coefficient in silicon. Phys. Stat. Sol. A **21**, 357–367 (1974)
12. T. Trupke, M.A. Green, P. Würfel, P.P. Altermatt, A. Wang, J. Zhao, R. Corkish, Temperature dependence of the radiative recombination coefficient of intrinsic crystalline silicon. J. Appl. Phys. **94**, 4930–4937 (2003)
13. M.A. Green, M.J. Keevers, Optical properties of intrinsic silicon at 300K. Prog. Photovoltaics Res. Appl. **3**, 189–192 (1995)
14. P.P. Altermatt, F. Geelhaar, T. Trupke, X. Dai, A. Neisser, E. Daub, Injection dependence of spontaneous radiative recombination in c-Si: experiment, theoretical analysis, and simulation, in *Proceedings of the 5th International Conference on Numerical Simulation and Optoelectronic Devices*, IEEE (Berlin, 2005), pp. 47–48
15. P.P. Altermatt, F. Geelhaar, T. Trupke, X. Dai, A. Neisser, E. Daub, Injection dependence of spontaneous radiative recombination in crystalline silicon: experimental verification and theoretical analysis. Appl. Phys. Lett. **88**, 261901 (2006)
16. P. Rosenits, Electrical characterisation of crystalline silicon thin-film material. Ph.D. thesis, Universität Konstanz (2010).
17. J.A. Giesecke, R.A. Sinton, M.C. Schubert, S. Riepe, W. Warta, Determination of bulk lifetime and surface recombination velocity of silicon ingots from dynamic photoluminescence. IEEE J. Photovoltaics **3**, 1311–1318 (2013)
18. P. Würfel, T. Trupke, T. Puzzer, E. Schffer, W. Warta, S.W. Glunz, Diffusion lengths of silicon solar cells from luminescence images. J. Appl. Phys. **101**, 123110 (2007)

19. R.A. Sinton, T. Trupke, Limitations on dynamic excess carrier lifetime calibration methods. Prog. Photovoltaics: Res. Appl. **20**, 246–249 (2012)
20. J.A. Giesecke, M. Kasemann, W. Warta, Determination of local minority carrier diffusion lengths in crystalline silicon from luminescence images. J. Appl. Phys. **106**, 014907 (2009)
21. J.A. Giesecke, M. Kasemann, M.C. Schubert, P. Würfel, W. Warta, Separation of local bulk and surface recombination in crystalline silicon from luminescence reabsorption. Prog. Photovoltaics: Res. Appl. **18**, 10–19 (2010)
22. P. Gundel, F.D. Heinz, M.C. Schubert, J.A. Giesecke, W. Warta, Quantitative carrier lifetime measurement with micron resolution. J. Appl. Phys. **108**, 033705 (2010)
23. M. Tajima, Determination of boron and phosphorus concentration in silicon by photoluminescence analysis. Appl. Phys. Lett. **32**, 719–721 (1978)
24. P.M. Colley, E.C. Lightowlers, Calibration of the photoluminescence technique for measuring B, P and Al concentrations in Si in the range 10^{12} to 10^{15} cm^{-3} using fourier transform spectroscopy. Semicond. Sci. Technol. **2**, 157 (1987)
25. B. Mitchell, T. Trupke, J.W. Weber, J. Nyhus, Bulk minority carrier lifetimes and doping of silicon bricks from photoluminescence intensity ratios. J. Appl. Phys. **109**, 083111 (2011)
26. B. Mitchell, J.W. Weber, D. Walter, D. Macdonald, T. Trupke, On the method of photoluminescence spectral intensity ratio imaging of silicon bricks: advances and limitations. J. Appl. Phys. **112**, 063116 (2012)
27. M.P. Peloso, J.S. Lew, T. Trupke, M. Peters, R. Utama, A.G. Aberle, Evaluating the electrical properties of silicon wafer solar cells using hyperspectral imaging of luminescence. Appl. Phys. Lett. **99**, 221915 (2011)
28. I. Uchida, Observation of the recombination radiation from silicon pn junction. Jpn. J. Appl. Phys. **2**, 561–564 (1963)
29. D. Hinken, C. Schinke, S. Herlufsen, A. Schmidt, K. Bothe, R. Brendel, Experimental setup for camera-based measurements of electrically and optically stimulated luminescence of silicon solar cells and wafers. Rev. Sci. Instr. **82**, 033706 (2011)
30. T. Fuyuki, H. Kondo, T. Yamazaki, Y. Takahashi, Y. Uraoka, Photographic surveying of minority carrier diffusion length in polycrystalline silicon solar cells by electroluminescence. Appl. Phys. Lett. **86**, 262108 (2005)
31. T. Trupke, R.A. Bardos, M.C. Schubert, W. Warta, Photoluminescence imaging of silicon wafers. Appl. Phys. Lett. **89**, 044107 (2006)
32. T. Trupke, E. Pink, R.A. Bardos, M.D. Abbott, Spatially resolved series resistance of silicon solar cells obtained from luminescence imaging. Appl. Phys. Lett. **90**, 093506 (2007)
33. D. Hinken, K. Ramspeck, K. Bothe, B. Fischer, R. Brendel, Series resistance imaging of solar cells by voltage dependent electroluminescence. Appl. Phys. Lett. **91**, 182104 (2007)
34. H. Kampwerth, T. Trupke, J.W. Weber, Y. Augarten, Advanced luminescence based effective series resistance imaging of silicon solar cells. Appl. Phys. Lett. **93**, 202102 (2008)
35. B. Michl, M. Kasemann, J. Giesecke, M. Glatthaar, A. Schütt, J. Carstensen, H. Föll, S. Rein, W. Warta, H. Nagel, in Application of luminescence imaging based series resistance measurement methods in an industrial environment, in *Proceedings of the 23rd EU PVSEC* (Valencia, 2008), pp. 1176–1181
36. J. Haunschild, M. Glatthaar, M. Kasemann, S. Rein, E.R. Weber, Fast series resistance imaging for silicon solar cells using electroluminescence. Phys. Stat. Sol. (RRL) **3**, 227–229 (2009)
37. D. Kiliani, A. Herguth, G. Hahn, V. Rutka, M. Junk, in Fitting of lateral resistances in silicon solar cells to electroluminescence images, in *Proceedings of the 24th EU PVSEC* (Hamburg, 2009), pp. 2088–2090
38. M. Glatthaar, J. Haunschild, M. Kasemann, J. Giesecke, W. Warta, S. Rein, Spatially resolved determination of dark saturation current and series resistance of silicon solar cells. Phys. Stat. Sol. (RRL) **4**, 13–15 (2010)
39. M. Kasemann, D. Grote, B. Walter, W. Kwapil, T. Trupke, Y. Augarten, R.A. Bardos, E. Pink, M.D. Abbott, W. Warta, Luminescence imaging for the detection of shunts on silicon solar cells. Prog. Photovoltaics: Res. Appl. **16**, 297–305 (2008)

40. Y. Augarten, T. Trupke, M. Lenio, J. Bauer, O. Breitenstein, J. Weber, R. Bardos, Luminescence shunt imaging: qualitative and quantitative shunt images using photoluminescence imaging, in *Proceedings of the 24th EU PVSEC* (Hamburg, 2009), pp. 27–31
41. S. Herlufsen, J. Schmidt, D. Hinken, K. Bothe, R. Brendel, Photoconductance-calibrated photoluminescence lifetime imaging of crystalline silicon. Phys. Stat. Sol. (RRL) **2**, 245–247 (2008)
42. D. Hinken, K. Bothe, K. Ramspeck, S. Herlufsen, R. Brendel, Determination of the effective diffusion length of silicon solar cells from photoluminescence. J. Appl. Phys. **105**, 104516 (2009)
43. J.A. Giesecke, M.C. Schubert, B. Michl, F. Schindler, W. Warta, Minority carrier lifetime imaging of silicon wafers calibrated by quasi-steady-state photoluminescence. Solar Energy Mater. Solar Cells **95**, 1011–1018 (2011)
44. R. Brendel, Note on the interpretation of injection-level-dependent surface recombination velocities. Appl. Phys. A **60**, 523–524 (1995)
45. A.G. Aberle, J. Schmidt, R. Brendel, On the data analysis of light-biased photoconductance decay measurements. J. Appl. Phys. **79**, 1491–1496 (1996)
46. F.M. Schuurmans, A. Schönecker, A.R. Burgers, W.C. Sinke, Simplified evaluation method for light-biased effective lifetime measurements. Appl. Phys. Lett. **71**, 1795–1797 (1997)
47. J. Schmidt, Measurement of differential and actual recombination parameters on crystalline silicon wafers. IEEE Trans. Electron Dev. **46**, 2018–2025 (1999)
48. J.A. Giesecke, S.W. Glunz, W. Warta, Understanding and resolving the discrepancy between differential and actual minority carrier lifetime. J. Appl. Phys. **113**, 073706 (2013)
49. A. Ehrhardt, W. Wettling, A. Bett, Transient photoluminescence decay study of minority carrier lifetime in GaAs heteroface solar cell structures. Appl. Phys. A **53**, 123–129 (1991)
50. J. Cui, H. Wang, F. Gan, X. Huang, Z. Cai, Q. Li, Z. Yu, Picosecond transient photoluminescence spectra of ZnSe–ZnS strained-layer superlattices grown on GaAs (001) by molecular beam epitaxy. Appl. Phys. Lett. **61**, 1540–1542 (1992)
51. B. Monemar, P.P. Paskov, J.P. Bergman, G. Pozina, A.A. Toropov, T.V. Shubina, T. Malinauskas, A. Usui, Transient photoluminescence of shallow donor bound excitons in GaN. Phys. Rev. B **82**, 235202 (2010)
52. S. Finkbeiner, J. Weber, M. Rosenbauer, M. Stutzmann, Transient photoluminescence decay in porous silicon and siloxene. J. Lumin. **57**, 231–234 (1993)
53. T. Trupke, R.A. Bardos, F. Hudert, P.Würfel, J. Zhao, A. Wang, M.A. Green, Effective excess carrier lifetimes exceeding 100 milliseconds in float zone silicon determined from photoluminescence, in *Proceedings of the 19th EU PVSEC* (Paris, 2004), pp. 758–761
54. T. Trupke, R.A. Bardos, in Photoluminescence: a surprisingly sensitive lifetime technique, in *Proceedings of the 31st IEEE PVSC* (Orlando, 2005), pp. 903–906
55. K. Wang, M.A. Green, H. Kampwerth, Transient photoconductance and photoluminescence from thick silicon wafers and bricks: analytical solutions. Solar Energy Mater. Solar Cells **111**, 189–192 (2013)
56. S. Herlufsen, K. Ramspeck, D. Hinken, A. Schmidt, J. Müller, K. Bothe, J. Schmidt, R. Brendel, Photoluminescence lifetime imaging for the characterisation of silicon wafers. Phys. Stat. Sol. (RRL) **5**, 25–27 (2011)
57. S. Herlufsen, K. Bothe, J. Schmidt, R. Brendel, S. Siegmund, Dynamic photoluminescence lifetime imaging of multicrystalline silicon bricks. Solar Energy Mater. Solar Cells **106**, 42–46 (2012)
58. K. Ramspeck, S. Reissenweber, J. Schmidt, K. Bothe, R. Brendel, Dynamic carrier lifetime imaging of silicon wafers using an infrared-camera-based approach. Appl. Phys. Lett. **93**, 102104 (2008)
59. K. Ramspeck, K. Bothe, J. Schmidt, R. Brendel, Combined dynamic and steady-state infrared camera based carrier lifetime imaging of silicon wafers. J. Appl. Phys. **106**, 114506 (2009)
60. D. Kiliani, G. Micard, B. Raabe, G. Hahn, Time resolved photoluminescence imaging for carrier lifetime mapping of silicon wafers, in *Proceedings of the 25th EU PVSEC* (Valencia, 2010), pp. 1363–1366

61. D. Kiliani, G. Micard, B. Steuer, B. Raabe, A. Herguth, G. Hahn, Minority charge carrier lifetime mapping of crystalline silicon wafers by time-resolved photoluminescence imaging. J. Appl. Phys. **110**, 054508 (2011)
62. D. Kiliani, A. Herguth, G. Micard, J. Ebser, G. Hahn, Time-resolved photoluminescence imaging with electronic shuttering using an image intensifier unit. Solar Energy Mater. Solar Cells **106**, 55–59 (2012)
63. N.A. Tolstoi, P.P. Feofilov, Novel method to investigate the processes of rise and decay of luminescence (translation). Bull. Russ. Acad. Sci. Phys. **13**, 211–217 (1949)
64. D. Guidotti, J.S. Batchelder, J.A. Van Vechten, A. Finkel, Nondestructive depth profiling of carrier lifetimes in full silicon wafers. Appl. Phys. Lett. **48**, 68–70 (1986)
65. D. Guidotti, J.S. Batchelder, A. Finkel, P.D. Gerber, J.A. Van Vechten, Modeling of luminescence phase delay for nondestructive characterization of Si wafers. J. Appl. Phys. **66**, 2542–2553 (1989)
66. A. Mandelis, J. Batista, D. Shaughnessy, Infrared photocarrier radiometry of semiconductors: physical principles, quantitative depth profilometry, and scanning imaging of deep subsurface electronic defects. Phys. Rev. B **67**, 205208 (2003)
67. A. Melnikov, A. Mandelis, J. Tolev, P. Chen, S. Huq, Infrared lock-in carrierography (photocarrier radiometric imaging) of Si solar cells. J. Appl. Phys. **107**, 114513 (2010)
68. Q. Sun, A. Melnikov, A. Mandelis, Quantitative self-calibrating lock-in carrierographic lifetime imaging of silicon wafers. Appl. Phys. Lett. **101**, 242107 (2012)
69. T. Trupke, R.A. Bardos, M.D. Abbott, Self-consistent calibration of photoluminescence and photoconductance lifetime measurements. Appl. Phys. Lett. **87**, 184102 (2005)
70. R. Brüggemann, S. Reynolds, Modulated photoluminescence studies for lifetime determination in amorphous-silicon passivated crystalline-silicon wafers. J. Non-Cryst. Solids **352**, 1888–1891 (2006)
71. A.R. Engler, C.J. Kevane, Direct reading minority carrier lifetime measuring apparatus. Rev. Sci. Instr. **28**, 548–551 (1957)
72. J.A. Giesecke, S.W. Glunz, W. Warta, Determination of actual carrier lifetime from differential measurements. Energy Procedia **38**, 59–65 (2013)

Chapter 5
Harmonically Modulated Lifetime

This chapter covers principles of harmonically modulated carrier lifetime techniques. Relative timescales of modulation are addressed, and different concepts of a quasi-steady-state condition are discussed. Beyond, self-consistency of the continuity equation is introduced as a powerful concept to obtain injection-dependent effective lifetime from harmonically modulated measurement designs. The self-consistent approach is classified and its essential characteristics are outlined.

5.1 Quasi-Steady-State Measurement Conditions

Several assets of dynamic lifetime investigations of semiconductors have been mentioned in previous chapters: The most immediate access to carrier lifetime via time domain techniques eliminates amplitude artifacts and requires virtually no spectral assumptions. In some special cases, no a priori assumptions are required at all for a lifetime determination.

Among dynamic carrier lifetime techniques, harmonically modulated techniques are essentially characterized by an explicitly variable timescale of the modulation of excess carrier generation rate. In the range between *fully transient* analyses of excess carrier decay upon abrupt and discrete switching of excess carrier generation on the one hand, and true steady-state measurement conditions featuring no temporal variation of excess carrier generation on the other hand, harmonically modulated lifetime techniques essentially enable any intermediate measurement conditions between these extremes.

5.1.1 Definition of Dynamic Regimes

It is useful to classify harmonically modulated measurements in terms of the relative timescales of excess carrier generation (modulation time $\mathcal{T}$) and recombination (effective carrier lifetime τ_{eff}).

J. Giesecke, *Quantitative Recombination and Transport Properties in Silicon from Dynamic Luminescence*, Springer Theses, DOI: 10.1007/978-3-319-06157-3_5,

© Springer International Publishing Switzerland 2014

The Transient Limit: Truly transient measurement conditions are encountered if the timescale $\mathcal{T}$ of a significant variation of excess carrier generation is negligible compared to any relevant effective lifetime. If effective lifetime is comprised out of a superposition of more than one significant modes of excess carrier decay, the transient limit is distinguished by the condition

$$\tau_j \pi \gg \mathcal{T}, \tag{5.1}$$

while τ_j represents the decay time of the highest significant mode of excess carrier decay according to Eq. 3.31, i.e. the shortest relevant decay time involved.
True Steady State: The other extreme concerning the timescale $\mathcal{T}$ of a significant variation of excess carrier generation is the discrete case of a true steady state

$$\frac{dG}{dt} = 0. \tag{5.2}$$

Notably, the true steady state is **not** equivalent to the limit $\mathcal{T} \rightarrow \infty$. In Sect. 6.1 measurement conditions yielding substantial deviations between effective lifetime measurement results obtained in a harmonically modulated measurement configuration satisfying the condition $\mathcal{T} \rightarrow \infty$ versus under true steady-state conditions are detailed. Although in the limit $\mathcal{T} \rightarrow \infty$, signal amplitudes and excess carrier distributions are expected to be indistinguishable from true steady state, the dynamic lifetime interpretation may still substantially differ from a steady-state lifetime interpretation. Therefore, the limit $\mathcal{T} \rightarrow \infty$ is referred to as part of the so-called *quasi-steady-state* regime, here.
The Quasi-Steady-State Regime: A quasi-steady-state condition may be generally distinguished by a relaxation time of a system under investigation much less than the timescale of a significant temporal variation of a deliberate impact on the system (e.g. a driving force). A rigorous quasi-steady-state condition for lifetime investigations of semiconductors would be

$$\tau_1 \pi \ll \mathcal{T}. \tag{5.3}$$

Here, τ_1 represents the decay time of the most slowly decaying mode of excess carrier density (cf. e.g. Eq. 3.31).
The Quasi-Transient Regime: A measurement shall be referred to as quasi-transient if any significant decay mode satisfies the condition

$$\tau_j \pi \gtrsim \mathcal{T}. \tag{5.4}$$

Harmonically modulated measurement conditions which satisfy neither a quasi-steady-state condition nor a quasi-transient condition shall be classified as *intermediate*. Figure 3.8 illustrates cases of harmonically modulated experiments in the quasi-steady-state, intermediate, and quasi-transient regime of a mono-exponential excess carrier decay (i.e. featuring only one mode of excess carrier decay).

5.1.2 The Benefit of Quasi-Steady-State

The key benefits of a quasi-steady-state conditions can be classified in two categories: The first category of major benefits relates to analytical simplifications brought about by the quasi-steady-state condition.

I The interpretation of the phase shift between excess carrier density and generation rate as effective lifetime is only one example of a first order approximation of high accuracy in quasi-steady-state.
II The derivation of a simple and yet accurate expression of harmonically modulated effective lifetime as a superposition of decay times of the decay modes of excess carrier density in Sect. 6.1.2 is solely enabled by a quasi-steady-state condition.
III The spatially resolved effective lifetime calibration of photoluminescence images to be elaborated in Sect. 8.3 builds upon a quasi-steady-state condition.

The second category of benefits relates to an accurate approximation of steady-state carrier diffusion conditions.

i In contrast to transient excess carrier decay, a quasi-steady-state condition enforces defined injection conditions.
ii Quasi-steady-state implies a straightforward temporal synchronization of excess carrier density and generation rate. If a rigorous quasi-steady-state condition is satisfied, excess carrier density is nearly identical at times of equal generation rate $G_{up} = G_{down}$ during a modulation cycle, and $\Delta n\,(z)$ virtually coincides with the steady-state solution corresponding to this excess carrier generation rate, respectively.
iii A timescale of a significant variation of excess carrier generation rate much greater than the decay time of the fundamental mode of excess carrier density implies negligible diffusion dynamics. Therefore, a significant redistribution of the depth profile of excess carrier density during a modulation cycle can only be caused by the injection dependence of recombination (and transport) properties (dependencies of $\Delta n\,(z, t)$ factorize, cf. Eq. 4.35). Notably, this does not hold for quasi-transient or intermediate measurement conditions, which allow a substantial temporal imbalance of $\Delta n\,(z)$ due to the rapid decay of higher order decay modes of excess carrier density.
iv If the relative depth profile of $\Delta n\,(z)$ remains constant throughout a modulation cycle (as reasoned above), there is no impact of self-absorption of luminescence in the substrate bulk on lifetime.

It must be noted that the quasi-steady-state regime may feature disadvantages compared to faster modulation as well. The most prominent potential cause of artifacts to be addressed in this context is carrier trapping, as described in Sect. 3.1.6. Further, effective lifetime $\tau_{eff,h}$ as obtained from the time shift between excess carrier density and generation rate in harmonically modulated experiments (even in the limit $\mathcal{T} \rightarrow \infty$) is not to be generally confused with steady-state effective lifetime $\tau_{eff,s}$. This is most crucial for high surface recombination velocities, as elaborated in Sect. 6.1.

5.1.3 Quasi-Steady-State Perceptions

Quasi-steady-state conditions have long been used, and they are by far not restricted to harmonically modulated experiments. In fact, the most prevalent quasi-steady-state photoconductance lifetime technique (QSSPC) as developed by Sinton et al. in the mid 1990s [1, 2] apparently exploits a quasi-steady-state condition. Due to the prevalence of QSSPC in silicon photovoltaics, Sinton's perception of the quasi-steady-state condition has much influenced how this condition is commonly understood. However, there is a distinct difference between Sinton's perception and the one sketched above and upheld throughout this work:

QSSPC (cf. Appendix B.1) features a temporal variation of the optical irradiation intensity of a Xe flash lamp on a timescale much greater than the decay time of the fundamental mode of excess carrier density. This quasi-steady-state condition essentially ensures a straightforward temporal synchronization of excess carrier generation rate and a corresponding excess carrier density. A secondary benefit of this quasi-steady-state condition is that higher order decay modes—if present—contribute to total recombination rate in an unambiguous and virtually time-independent manner. In contrast, under transient measurement conditions, the rapid decay of higher order decay modes brings about dynamics that can be a priori excluded in quasi-steady-state. Yet, the essential difference with respect to the quasi-steady-state perception upheld here is that QSSPC only needs a temporal variation of excess carrier generation rate in order to access injection-dependent lifetime. The amplitude-sensitive determination of carrier lifetime via calibration of the inductance of a resonant circuit and via carrier mobility models does not require any temporal variation, whereas this variation is vital for stand-alone lifetime techniques based on harmonical modulation of excess carrier generation. This distinction is clarified by rearranging the time-dependent continuity equation (Eq. 3.75) to an effective lifetime expression, as e.g. proposed by Nagel et al. [3]

$$\tau_{eff} = \frac{\Delta n\,(t)}{G(t) - \frac{d\Delta n(t)}{dt}}. \tag{5.5}$$

In any case, quasi-steady-state implies

$$G(t) \gg \left| \frac{d\Delta n\,(t)}{dt} \right| \tag{5.6}$$

here, but while QSSPC would function even for $\frac{d\Delta n(t)}{dt} = 0$, this is not true for a harmonically modulated lifetime technique, which actually infers lifetime from the ratio between G and $\frac{d\Delta n}{dt}$—as will be shown in the following section.

5.2 Self-Consistency of the Continuity Equation

Under quasi-steady-state measurement conditions, the ordinary time-dependent treatment of the continuity equation as outlined in Sect. 3.2.3 allows the straightforward determination of an effective lifetime as the time shift between excess carrier density and generation rate. Self-consistency of the continuity equation provides an alternative—albeit to some extent equivalent—approach to effective lifetime, which is even more broadly applicable. The broader range of application results from the fact that the self-consistent approach is not restricted to the validity of a quasi -steady-state condition, and that—in contrast to the phase approach—it resolves actual rather than differential lifetime (cf. Sect. 6.2).

5.2.1 The Principle of Self-Consistency

With a self-consistent and injection-dependent effective carrier lifetime $\tau_\sigma(\Delta n)$, the time-dependent continuity equation in the form of Eq. 3.75 shall be evoked as

$$\tau_\sigma(\Delta n(t)) = \frac{\Delta n(t)}{G(t) - \frac{d\Delta n(t)}{dt}}. \tag{5.7}$$

Further, let us recall time-dependent luminescence intensity according to Eq. 4.36

$$\Phi(t) = c_\Phi \Delta n(t)(n_0 + \Delta n(t)). \tag{5.8}$$

If the scaling factor c_Φ is known, Eq. 5.8 yields $\Delta n(t)$ as its positive solution

$$\Delta n(t) = \frac{-c_\Phi n_0 + \sqrt{(c_\Phi n_0)^2 + 4c_\Phi \Phi(t)}}{2c_\Phi}. \tag{5.9}$$

However, c_Φ is usually unknown, and its ab initio determination is far from straightforward for reasons outlined in Sect. 4.3.1.

The self-consistent analysis of a harmonically modulated experiment provides a means to determine the scaling factor c_Φ, and to thereby also determine effective lifetime $\tau_\sigma(\Delta n)$. Insertion of the positive solution $\Delta n(t)$ of Eq. 5.8 into the continuity equation yields a time trajectory of lifetime expressions $\tau_\sigma(t)|_{c_\Phi}$. If the scaling factor c_Φ is not properly chosen, lifetime expressions corresponding to the ascending and descending branches of harmonically modulated excess carrier generation rate must deviate from each other. This so-called lifetime *hysteresis* [5]

$$\Delta\tau_\sigma(\Delta n_0)|_{c_\Phi} = \tau_\sigma\left(\Delta n_0\left(t_\uparrow\right)\right)|_{c_\Phi} - \tau_\sigma\left(\Delta n_0\left(t_\downarrow\right)\right)|_{c_\Phi} \tag{5.10}$$

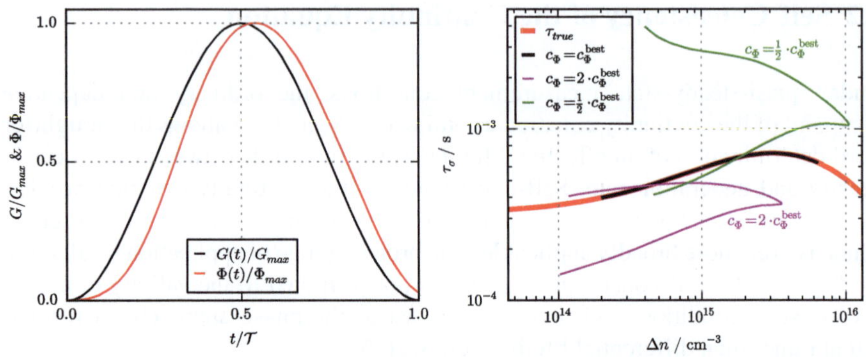

Fig. 5.1 On the *left*, relative luminescence intensity $\Phi(t)$ and excess carrier generation rate $G(t)$ of a modeled (noise-free) harmonically modulated photoluminescence experiment (in an intermediate dynamic regime) are depicted. On the *right*, self-consistent effective lifetime interpretations of the photoluminescence measurement (as modeled on the *left*) are plotted. The self-consistent analysis is demonstrated to only reproduce true lifetime, if the scaling factor c_Φ is chosen appropriately. Only then is *hysteresis* $\Delta\tau_\sigma(\Delta n_0)|_{c_\Phi}$ eliminated. Model assumptions were an excess carrier generation rate peaking at a depth-averaged value of $G_{max} = 10^{19}\,\text{cm}^{-3}\text{s}^{-1}$, a net dopant concentration of $n_0 = 1.5 \times 10^{16}\,\text{cm}^{-3}$, and a modulation time $\mathcal{T} = 12.5 \times 10^{-3}$ s. True lifetime as used for the modeled experiment is also plotted on the *right*. In contrast to a real experiment, this model experiment is based on the assumption of an injection-independent coefficient of radiative recombination [4]

is expected to be an injective function of the scaling factor c_Φ. In an idealized (noise-free) case, the appropriate scaling factor is then determined by requiring

$$\Delta\tau_\sigma(\Delta n_0)|_{c_\Phi} \stackrel{!}{=} 0. \tag{5.11}$$

A numerical algorithm implemented in the lifetime measurement and analysis software developed in the course of this work (cf. Sect. 7.1) performs a minimization (rather than an elimination) of *hysteresis* in practical cases of experimental data superimposed by noise. From a rigorous point of view, the scaling factor c_Φ is proportional to the integral coefficient of radiative recombination, which is a function of free carrier density [4]. This dependency can be iteratively taken into account. Figure 5.1 shows both, relative luminescence intensity $\Phi(t)$ and excess carrier generation rate $G(t)$ of a modeled harmonically modulated photoluminescence experiment representing an intermediate dynamic regime. The intermediate dynamic regime is deemed most suitable for a high-contrast visualization of *hysteresis*, and along the way it substantiates the applicability of the self-consistent principle beyond the quasi-steady-state regime. What is also depicted in Fig. 5.1 is a self-consistent interpretation of the modeled photoluminescence measurement in terms of effective lifetime $\tau_\sigma(\Delta n(t))$. Self-consistency is only achieved, if the scaling factor c_Φ is chosen appropriately.

Note that the principle of self-consistency of the continuity equation is not restricted to the analysis of effective lifetime [6, 7]. In fact, Trupke and Bardos initially proposed a method to determine excess carrier generation rate in calibrated photoconductance measurements (with effective lifetime determined by another means) [6].

5.2.2 Self-Consistency and Phase Information

Focusing on the signals of a harmonically modulated lifetime measurement as e.g. depicted on the left of Fig. 5.1, one may find an intuitive access to the principle of self-consistency: relative luminescence intensity is smaller than relative excess carrier generation rate in the ascending branch of a modulation cycle, while it is greater in the descending branch. Also, the derivative of excess carrier density (or luminescence intensity) with respect to time is apparently positive in the ascending branch while being negative in the descending branch. Let us consider two times of equal excess carrier density $\Delta n\left(t_\uparrow\right) = \Delta n\left(t_\downarrow\right)$ in the ascending ($\uparrow$) and descending ($\downarrow$) branch of a modulation cycle, respectively. According to Eq. 5.7 self-consistency is achieved if a scaling factor c_Φ for luminescence intensity is found such that the difference between the derivatives of excess carrier density with respect to time exactly compensates the difference between generation rates at times $t_\uparrow$ and $t_\downarrow$:

$$G\left(t_\uparrow\right) - G\left(t_\downarrow\right) \stackrel{!}{=} \left.\frac{d\Delta n}{dt}\right|_{t_\uparrow} - \left.\frac{d\Delta n}{dt}\right|_{t_\downarrow} \tag{5.12}$$

This is the initial point of a proof of analogy of harmonically modulated lifetime $\tau_{eff,h}$ as the quasi-steady-state time shift between excess carrier density and generation rate on the one hand and self-consistent carrier lifetime τ_σ on the other hand—to be outlined in the following.

5.2.2.1 Analogy of Self-Consistent and Harmonically Modulated Lifetime

According to the quasi-steady-state solution of the time-dependent continuity equation as motivated in Eq. 3.87, with an arbitrary time $\Delta > 0$ excess carrier densities at times $t_\downarrow = \frac{T}{2} + \tau_{eff,h} + \Delta$ and $t_\uparrow = \frac{T}{2} + \tau_{eff,h} - \Delta$ must coincide. Insertion of both, quasi-steady-state excess carrier density as of Eq. 3.87 and generation rate as of Eq. 3.83 into the self-consistency condition Eq. 5.12 (while exploiting $\sin\left(\frac{\pi}{2} + \alpha\right) = \cos\alpha$ and derivative rules) yields

$$\cos^2\left(\frac{\pi}{T}\left(\tau_{eff,h} + \Delta\right)\right) - \cos^2\left(\frac{\pi}{T}\left(\tau_{eff,h} - \Delta\right)\right) = \frac{4\pi\tau_{eff,h}}{T}\sin\frac{\pi\Delta}{T}\cos\frac{\pi\Delta}{T}. \tag{5.13}$$

The left side of this equation can be rearranged via the trigonometric rules $\cos(\alpha \pm \beta) = \cos\alpha\cos\beta \mp \sin\alpha\sin\beta$ and $\cos^2\alpha = \frac{1}{2}(1 + \cos 2\alpha)$, such that

$$-\sin\frac{2\pi\tau_{eff,h}}{T}\sin\frac{2\pi\Delta}{T}=\frac{4\pi\tau_{eff,h}}{T}\sin\frac{\pi\Delta}{T}\cos\frac{\pi\Delta}{T}. \tag{5.14}$$

Aspiring an expression for self-consistent lifetime, one may insert excess carrier density as of Eq. 3.87 and generation rate as of Eq. 3.83 into the continuity equation as provided in Eq. 5.7:[1]

$$\tau_\sigma\left(t_\downarrow\right)=\frac{\tau_{eff,h}\cos^2\frac{\pi\Delta}{T}}{\cos^2\frac{\pi}{T}\left(\tau_{eff,h}+\Delta\right)-\frac{2\pi\tau_{eff,h}}{T}\sin\frac{\pi\Delta}{T}\cos\frac{\pi\Delta}{T}} \tag{5.15}$$

Again exploiting both, the above-mentioned trigonometric rules and the self-consistency condition as of Eq. 5.14, the denominator on the right side of Eq. 5.15 simplifies such that

$$\tau_\sigma\left(t_\downarrow\right)=\frac{\tau_{eff,h}\cos^2\frac{\pi\Delta}{T}}{\frac{1}{2}\left(1+\cos\frac{2\pi\tau_{eff,h}}{T}\cos\frac{2\pi\Delta}{T}\right)}. \tag{5.16}$$

The quasi-steady-state condition allows accurate approximation of $\cos\frac{2\pi\tau_{eff,h}}{T}=1$, which finally leads to the identity

$$\tau_\sigma\left(t_\downarrow\right)=\tau_{eff,h}. \tag{5.17}$$

As the time Δ was arbitrarily chosen, this identity must hold for any $t_\downarrow$, yielding[2]

$$\tau_\sigma=\tau_{eff,h}. \tag{5.18}$$

Therefore, the self-consistent principle essentially yields effective lifetime τ_σ as the phase-shift between excess carrier density and generation rate—herein referred to as harmonically modulated effective lifetime $\tau_{eff,h}$.

5.2.2.2 Deviations Between Self-Consistent and Harmonically Modulated Lifetime

The identity expressed in Eq. 5.18 is restricted to both,

- a quasi-steady-state condition, and
- the condition of injection-independent effective lifetime.

Deviations may occur if any of these conditions is violated. Note that both conditions were premises of the above proof: While the premise of quasi-steady-state was much exploited throughout this proof (e.g. the excess carrier density inserted into

---

[1] Here, the time $t_\downarrow$ is chosen for evaluation of the continuity equation, but $t_\uparrow$ could be chosen alike.
[2] Notably, this is valid only for injection-independent effective lifetime.

the self-consistency condition was a quasi-steady-state approximation), injection independence of effective lifetime was implicitly assumed through the arbitrary choice of a time Δ in combination with one constant $\tau_{eff,h}$ for any Δ.

Self-consistent lifetime is an actual carrier lifetime, whereas the time shift $\tau_{eff,h}$ is found to be a differential lifetime. Substantial deviations between τ_σ and $\tau_{eff,h}$ therefore not only occur in the intermediate to quasi-transient dynamic range, but also in the case of injection-dependent recombination properties. A full derivation is given in Sect. 6.2.

5.2.2.3 Note on the History of Self-Consistent Hysteresis Elimination

Historically, early phase-sensitive lifetime investigations were actually conducted by means of a self-consistent *hysteresis* elimination of time trajectories associated with excess carrier density and generation rate [8–10]. Notably, there are systematic analogies between those historic works and the concept of self-consistency as introduced by Trupke and Bardos in [6] and [5].

5.2.3 Essential Properties of Self-Consistent Lifetime

A definition of the range of validity of a self-consistent lifetime analysis is followed by a compilation of some relevant implications of the dynamic nature of self-consistent lifetime—which are valid for all harmonically modulated dynamic lifetime techniques. Further, it is discussed how systematic errors of assumptions about net dopant concentration and excess carrier generation rate propagate to self-consistent lifetime errors.

5.2.3.1 Range of Validity

Several limitations of harmonically modulated effective lifetime and self-consistent lifetime have been addressed in this chapter. Here, a compilation of the range of validity of self-consistent lifetime τ_σ and a distinction with respect to harmonically modulated lifetime $\tau_{eff,h}$ shall be outlined:

- **Effective Lifetime**: Both, self-consistent lifetime τ_σ and harmonically modulated lifetime $\tau_{eff,h}$ represent *effective* lifetimes. The distinct definition of an *effective* lifetime requires a rule on how different modes of excess carrier decay are aggregated in an *effective* lifetime. Such a rule is derived in Sect. 6.1.2 (cf. Eq. 6.24), and it is identical for both $\tau_{eff,h}$ and τ_σ. Accordingly, deviations from steady-state effective lifetime $\tau_{eff,s}$ may occur for high surface recombination velocities.

- **Quasi-Steady-State**: Self-consistent lifetime τ_σ is **not** restricted to a quasi-steady-state condition, whereas the phase shift interpretation in terms of an effective lifetime $\tau_{eff,h}$ is (cf. Fig. 3.8).
- **Differential Lifetime**: The principle of self-consistency enables an actual lifetime interpretation rather than a differential one. As outlined in Sect. 6.2, this is a clear distinction from harmonically modulated lifetime $\tau_{eff,h}$, which yields a differential lifetime interpretation.
- **Trapping**: Self-consistent lifetime τ_σ and harmonically modulated lifetime $\tau_{eff,h}$ are likewise exposed to a trapping artifact—as discussed in Sect. 3.1.6.

5.2.3.2 The Dynamic Nature of Self-Consistent Lifetime

As demonstrated in Sect. 5.2.2, the self-consistent lifetime analysis is distinctly dynamic. As such, artifacts as e.g. due to the nonuniformity of the depth profile of excess carrier density leave the determination of self-consistent lifetime virtually unaffected.

A highly nonuniform depth profile of both excess carrier generation rate and excess carrier density leads to a substantial underestimation of the assigned excess carrier density of a self-consistent lifetime analysis due to arithmetic depth averaging.[3] Excess carrier density at the location of a significant dynamic signal generation (caused e.g. by radiative recombination) then substantially exceeds its depth average. Yet, this only has a corrigible impact [11] on the injection level assigned to self-consistent lifetime rather than on self-consistent lifetime itself.

Lateral nonuniformity of excess carrier density (e.g. due to lateral lifetime variations) may affect a lumped self-consistent lifetime (and likewise harmonically modulated lifetime) in a different manner than it affects steady-state effective lifetime. Generally, relatively high lifetimes feature a higher weight in a lumped dynamic (e.g. self-consistent) lifetime analysis than in a lumped steady-state (amplitude-sensitive) analysis. This is thoroughly elaborated in Sect. 8.3. A lumped self-consistent lifetime analysis integrating over a substrate volume featuring a nonuniform distribution of excess carrier density cannot be assigned a distinct value of excess carrier density. Rather, the lumped lifetime analysis yields an *intensity weighted* average of lifetimes corresponding to a distinct excess carrier generation rate.

It has afore been argued that reabsorption (or self-absorption) of radiative recombination would be virtually without any impact on dynamic lifetime analyses (as e.g. self-consistent lifetime). This essential property is due to the dynamic nature of the self-consistent lifetime analysis. Reabsorption primarily affects amplitudes, which do not matter here. Higher order reabsorption artifacts due to a redistribution

[3] Complications may generally arise in case of a combined occurence of a nonuniform depth profile of excess carrier density and a highly injection-dependent bulk lifetime. Any lifetime measurement technique would in that case yield a superposition of different lifetimes. Yet, this is a general constraint in lifetime metrology rather than a particular caution related to self-consistent or dynamic lifetime analyses.

of the depth profile of excess carrier density during a modulation cycle may additionally be avoided by enforcing a quasi-steady-state condition—leaving the injection-dependence of recombination properties as the only loophole for a redistribution of $\Delta n(z)$ to occur.

5.2.3.3 The Impact of Net Dopant Concentration

In contrast to a straightforward lifetime interpretation from the phase shift between excess carrier density (photoluminescence intensity) and generation rate, a self-consistent lifetime analysis of a harmonically modulated luminescence experiment takes net dopant concentration n_0 as an additional argument.[4] Therefore, at deviations from low-level injection conditions, systematic errors of an assumed net dopant concentration may propagate to self-consistent lifetime.

Low-Level Injection: Self-consistent lifetime can be shown to be independent of net dopant concentration n_0 under low-level injection conditions $\Delta n \ll n_0$. Under such conditions, the relation between photoluminescence intensity and excess carrier density as of Eq. 5.8 simplifies to

$$\Phi(t) = c_\Phi n_0 \Delta n(t) . \tag{5.19}$$

As detailed in Sect. 5.2.1, the self-consistent lifetime analysis adapts the scaling factor c_Φ of Eq. 5.8 such that the lifetime *hysteresis* (cf. Eq. 5.7) between the ascending and decending branches of a modulation cycle is eliminated. Yet, in the special case of $\Phi \propto \Delta n$, it is justified to perceive the product $c_\Phi n_0$ as an *effective* scaling factor. A potential mismatch factor δ between an assumed net dopant concentration n_0 and its true value is compensated by an inverse mismatch factor δ^{-1} between the scaling factor c_Φ resulting from a self-consistent lifetime analysis and its physical definition according to Eq. 4.37. Therefore, under low-level injection conditions the self-consistent lifetime interpretation does not at all suffer from a systematic error of the assumed value of net dopant concentration n_0, as also illustrated in Fig. 5.2.

Departure from Low-Level Injection: If low-level injection conditions are violated, the quadratic term in Eq. 5.8 becomes non-negligible. Consequently, the self-consistent lifetime analysis of a modulated luminescence experiment based on erroneous assumptions about net dopant concentration implies a discrepancy between model and reality, which the self-consistent evaluation algorithm cannot compensate. Equation 5.9 clearly illustrates that in the nonlinear case $\Phi \not\propto \Delta n$, the mismatch factor δ between an assumed net dopant concentration n_0 and itstrue value cannot

---

[4] Net dopant concentration does **not** necessarily affect any self-consistent lifetime analysis. Rather, it is specific to intermediate- to high-level injection luminescence analyses, as luminescence intensity is proportional to the product of minority and majority carrier density (cf. Eq. 5.8). For instance, a modulated photoconductance lifetime analysis would be independent of n_0—even under high-level injection conditions.

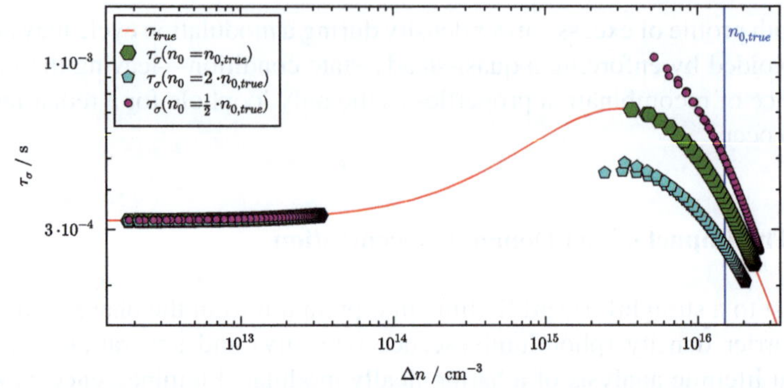

Fig. 5.2 Illustration of the variable impact of erroneous assumptions about net dopant concentration n_0 on a self-consistent lifetime analysis of a luminescence experiment. Two model experiments of harmonically modulated photoluminescence were self-consistently analyzed with a correct value of $n_{0,true}$, and with deviating values $2 \cdot n_{0,true}$ and $\frac{1}{2} \cdot n_{0,true}$, respectively. While under low-level injection conditions ($\Delta n \ll n_0$), erroneous assumptions about n_0 are of negligible impact, there is a pronounced impact upon departure from low-level injection. Model parameters: the topmost 8 % of modeled signals $G(t)$ and $\Phi(t)$ were used for self-consistent *hysteresis* elimination. Modulation time: $\mathcal{T} = 10^{-1}$ s

be compensated by an inverse mismatch factor δ^{-1} on the scaling factor c_Φ. Such potential failure of a self-consistent lifetime analysis is demonstrated in Fig. 5.2.

The dependence of self-consistent lifetime on net dopant concentration at deviations from low-level injection conditions is exploited for an analysis of net dopant concentration via harmonically modulated photoluminescence, as detailed in Sect. 9.1 and in [12]. Also note that, in the course of the present work, several alternative harmonically modulated lifetime approaches were put forth, which do not depend on net dopant concentration (cf. [13–15], as also detailed in Sect. 8.1).

5.2.3.4 The Impact of Generation Rate

A systematic error of the assumed depth-averaged excess carrier generation rate G may result from proportional systematic errors of the impinging photon current density $j_{\gamma,i}$, the transmissivity of the irradiated substrate interface $1 - R_\gamma$, and the inverse substrate thickness d^{-1} (cf. Eq. 3.58).

Recalling the self-consistency condition given in Eq. 5.12, the magnitude of the derivative of excess carrier density with respect to time $\frac{d\Delta n}{dt}$ is to be adapted to the magnitude of excess carrier generation rate G such that Eq. 5.12 is satisfied. Therefore, a systematic error of G propagates to a **proportional** systematic error of $\frac{d\Delta n}{dt}$ and consequently of Δn alike. Notably, due to the dynamic nature of the self-consistent lifetime analysis, this error does **not** propagate to self-consistent lifetime τ_σ, which is only sensitive to phase information and relative (rather than total) signal amplitudes.

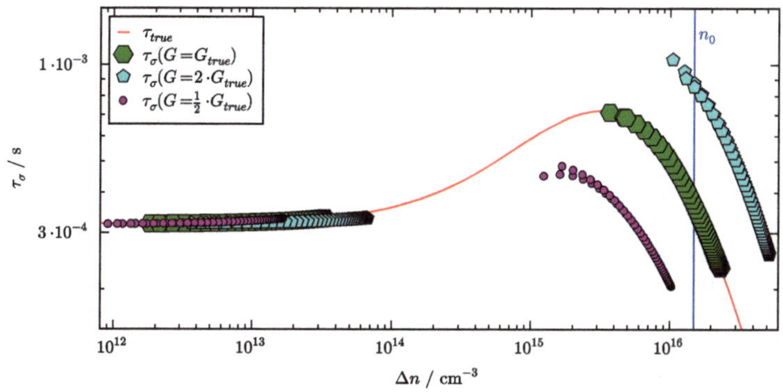

Fig. 5.3 Illustration of the variable impact of erroneous assumptions about depth-averaged excess carrier generation rate G on a self-consistent lifetime analysis of a luminescence experiment. The exact same model experiments as evaluated with varying net dopant concentration in Fig. 5.2 were used for a self-consistent evaluation with varying generation rate. Under low-level injection conditions ($\Delta n \ll n_0$), erroneous assumptions about G only cause a proportional mismatch of the assigned excess carrier density. In contrast under high-level injection conditions, erroneous assumptions about G affect both self-consistent lifetime τ_σ and the excess carrier density assigned to it

It follows from Eq. 5.12 that the essential impact of erroneous assumptions about excess carrier generation rate G on self-consistent lifetime τ_σ (Δn) is the proportional shift of the assigned excess carrier density Δn. This shift occurs regardless of the injection level. Thus, if lifetime features a pronounced injection dependence, the shift of the assinged Δn level may also bring about a misinterpretation of self-consistent lifetime τ_σ.

Beyond, under high-level injection conditions, the shift of the level of excess carrier density Δn assigned to a self-consistent lifetime τ_σ may be accompanied by a systematic error of τ_σ, as can be seen in Fig. 5.3. This effect is to be understood as a higher order artifact of a differential lifetime measurement (cf. Sect. 6.2): The self-consistent analysis represents a combined analysis of both, the phase shift as well as relative curve shapes of measured signals $\Phi(t)$ and $G(t)$. Under quasi-steady-state conditions, the relative curve shape of $\Phi(t)$ may only deviate from the relative curve shape of $G(t)$ for two reasons:

1. Injection dependence of carrier lifetime
2. High-level injection nonlinearity of $\Phi(\Delta n)$.

In the case of erroneous assumptions about generation rate G, these two causes cannot be adequately separated by the self-consistent analysis due to an erroneous assignment of excess carrier density. As the phase shift between $\Phi(t)$ (or likewise $\Delta n(t)$) and $G(t)$ is highly sensitive to the injection dependence of lifetime (detailed in Sect. 6.2), this may lead to lifetime artifacts of a self-consistent lifetime analysis under high-level injection—although a self-consistent lifetime analysis is under normal circumstances unaffected by differential lifetime artifacts.

References

1. R.A. Sinton, A. Cuevas, M. Stuckings, Quasi-steady-state photoconductance, a new method for solar cell material and device characterization, in *Proceedings of the 25th IEEE PVSC*, (Washington, D.C., 1996), pp. 457–460
2. R.A. Sinton, A. Cuevas, Contactless determination of current-voltage characteristics and minority-carrier lifetimes in semiconductors from quasi-steady-state photoconductance data. Appl. Phys. Lett. **69**, 2510–2512 (1996)
3. H. Nagel, C. Berge, A.G. Aberle, Generalized analysis of quasi-steady-state and quasi-transient measurements of carrier lifetimes in semiconductors. J. Appl. Phys. **86**, 6218–6221 (1999)
4. P.P. Altermatt, F. Geelhaar, T. Trupke, X. Dai, A. Neisser, E. Daub, Injection dependence of spontaneous radiative recombination in crystalline silicon: experimental verification and theoretical analysis. Appl. Phys. Lett. **88**, 261901 (2006)
5. T. Trupke, R.A. Bardos, M.D. Abbott, Self-consistent calibration of photoluminescence and photoconductance lifetime measurements. Appl. Phys. Lett. **87**, 184102 (2005)
6. T. Trupke, R.A. Bardos, Self-consistent determination of the generation rate from photoconductance measurements. Appl. Phys. Lett. **85**, 3611–3613 (2004)
7. W.S. Liang, K.J. Weber, Y.L. Ren, Impact of laterally non-uniform carrier lifetime on photoconductance-based lifetime measurements with self-consistent calibration. Prog. Photovoltaics Res. Appl. **21**, 1640–1644 (2013)
8. N.A. Tolstoi, P.P. Feofilov, Novel method to investigate the processes of rise and decay of luminescence (translation). Bull. Russ. Acad. Sci. Phys. **13**, 211–217 (1949)
9. A.R. Engler, C.J. Kevane, Direct reading minority carrier lifetime measuring apparatus. Rev. Sci. Instr. **28**, 548–551 (1957)
10. E. Harnik, A. Many, N.B. Grover, Phase shift method of carrier lifetime measurements in semiconductors. Rev. Sci. Instr. **29**, 889–891 (1958)
11. S. Bowden, R.A. Sinton, Determining lifetime in silicon blocks and wafers with accurate expressions for carrier density. J. Appl. Phys. **102**, 124501 (2007)
12. J.A. Giesecke, M.C. Schubert, W. Warta, Measurement of net dopant concentration via dynamic photoluminescence. J. Appl. Phys. **112**, 063704 (2012)
13. J.A. Giesecke, M.C. Schubert, D. Walter, W. Warta, Minority carrier lifetime in silicon wafers from quasi-steady-state photoluminescence. Appl. Phys. Lett. **97**, 092109 (2010)
14. J.A. Giesecke, M.C. Schubert, W. Warta, Self-sufficient minority carrier lifetime in silicon from quasi-steady-state photoluminescence. Phys. Stat. Sol. A **209**, 2286–2290 (2012)
15. J.A. Giesecke, S.W. Glunz, W. Warta, Understanding and resolving the discrepancy between differential and actual minority carrier lifetime. J. Appl. Phys. **113**, 073706 (2013)

Chapter 6
Constraints of Dynamic Carrier Lifetime Techniques

Dynamic carrier lifetime techniques feature inherent advantages over steady-state techniques. Many of these advantages are related to the immediate access to time domain, which allows a straightforward interpretation without models and a priori assumptions. However, a decay time measurement is not necessarily a carrier lifetime measurement, and caution is required in cases of locally confined (nonuniform) excess carrier generation and recombination. This chapter details key constraints and limitations of dynamic, and particularly harmonically modulated, carrier lifetime techniques—elaborated in the course of this work.

6.1 Nonuniform Generation and Recombination

Measured excess carrier recombination lifetimes are usually *effective* lifetimes. In Sect. 3.1.4 different perceptions of effective carrier lifetime were introduced, comprising a steady-state perception yielding the ratio between cumulative excess carrier density and generation rate on the one hand, and a dynamic perception yielding the relaxation time of the fundamental decay mode of excess carrier density on the other hand. Steady-state and dynamic lifetime expressions may drastically differ—particularly in the case of highly localized and nonuniform excess carrier recombination (e.g. at interfaces).

Focusing on the case of excess carrier generation much confined to a highly recombinative substrate interface, Sinton and Trupke [1] recently addressed the problem of diverging effective lifetimes obtained from fully transient versus steady-state measurement designs. This section elaborates how the above considerations by Sinton

J. Giesecke, *Quantitative Recombination and Transport Properties in Silicon from Dynamic Luminescence*, Springer Theses, DOI: 10.1007/978-3-319-06157-3_6,
© Springer International Publishing Switzerland 2014

and Trupke relate to the mismatch between steady-state effective lifetime $\tau_{eff,s}$ and harmonically modulated effective lifetime $\tau_{eff,h}$.[1, 2] It is detailed

1. how the mismatch between dynamic and steady-state effective lifetime expressions relates to effective lifetime $\tau_{eff,h}$ as obtained from a harmonically modulated measurement design (cf. Sect. 6.1.1), and
2. how the weighting of excess carrier decay modes differs between quasi-steady-state harmonically modulated and steady-state measurement designs (cf. Sect. 6.1.2).

6.1.1 Harmonical Modulation: Dynamic or Steady-State

In order to be able to interpret the result of a dynamic lifetime measurement design with regard to steady-state device operation, the relation between a specific dynamic effective lifetime and steady-state effective lifetime must be clear.

Harmonically modulated lifetime techniques are a class of dynamic decay time measurement techniques. This fact suggests that dynamic effective lifetime $\tau_{eff,d}$ relates to harmonically modulated effective lifetime $\tau_{eff,h}$ at least to some extent. However, harmonically modulated lifetime techniques differ from fully transient techniques in that the timescale of the variation of generation rate may vary with respect to lifetime. A harmonically modulated measurement design allows quasi-transient, quasi-steady-state, or intermediate measurement conditions (cf. sect. 5.1). The extent to which harmonically modulated lifetime $\tau_{eff,h}$ is related to dynamic or steady-state effective lifetimes also depends on the ratio between modulation time $\mathcal{T}$ and the relaxation time of the fundamental decay mode of excess carrier density $\tau_{eff,d}$.

Quasi-Transient Operation: A quasi-transient harmonically modulated lifetime measurement design $\mathcal{T} \sim \tau_{eff,d}$ brings about substantial changes of excess carrier generation rate during the slowest decay time $\tau_{eff,d}$ of the semiconductor under investigation. Therefore, higher order decay modes, which would very quickly die out in a fully transient measurement design (featuring an abrupt and discrete switching of excess carrier generation), are only partly or weakly sustained. This brings about a substantially weaker contribution of such higher order decay modes to the total excess carrier recombination rate than under steady-state conditions, and consequently $\tau_{eff,h}$ is then largely affected by dynamic effective lifetime $\tau_{eff,d}$, albeit not identical. The faster time modulation proceeds, the closer harmonically modulated lifetime $\tau_{eff,h}$ and dynamic effective lifetime $\tau_{eff,d}$ approach.

[1] For nonuniform excess carrier recombination and quasi-steady-state operation conditions, no distinction between harmonically modulated lifetime $\tau_{eff,h}$ as derived from phase analyses and self-consistent lifetime τ_{σ} is needed, as these expressions coincide here (cf. Sect. 5.2). Note that this does not hold for injection-dependent recombination properties (cf. Sect. 6.2).

[2] The account given in this section is adapted from the corresponding original publication [2].

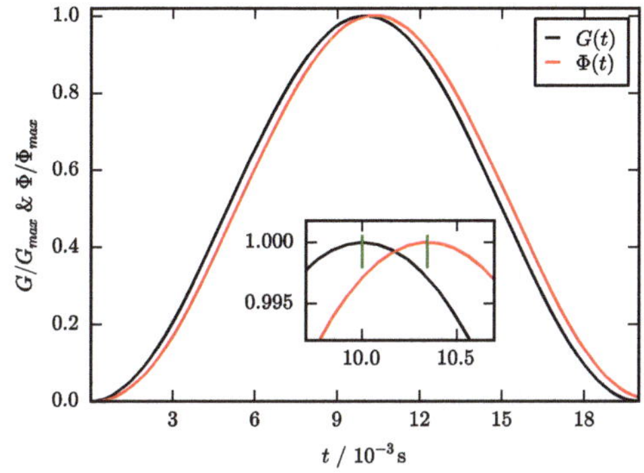

Fig. 6.1 Plot of both, harmonically modulated cumulative excess carrier generation rate $G(t)$ and PL intensity $\Phi(t)$ of a silicon ingot as it would be measured with the harmonically modulated luminescence setup at Fraunhofer ISE for the given electronic properties (cf. Eq. 4.31). Curves are normalized to a maximum of unity, respectively. PL intensity $\Phi(t)$ results from the solution of the continuity equation provided in Eq. 3.93, also accounting for the setup-specific luminescence reabsorption probability $p_\Phi(z)$. Model assumptions are: $S = 10^5\,\text{cm s}^{-1}$, $D_a = 27\,\text{cm}^2\,\text{s}^{-1}$, $\tau = 10^{-3}$ s, $\alpha_\gamma = 417\,\text{cm}^{-1}$, and $\mathcal{T} = 0.02$ s. As can be seen in the detailed plot of maximal signals, harmonically modulated lifetime (time shift between $G(t)$ and $\Phi(t)$) amounts to $\tau_{eff,h} = 343\,\mu$s, whereas steady-state lifetime is only $\tau_{eff,s} = 15.3\,\mu$s, yielding a mismatch greater than 22. Adapted after [3]

Quasi-Steady-State Operation: If a quasi-steady-state condition $\mathcal{T} \gg \tau_{eff,d}$ is satisfied, excess carrier generation rate only negligibly changes during the slowest decay time $\tau_{eff,d}$ of the semiconductor under investigation. Therefore, quickly decaying higher order decay modes are steadily sustained and consequently these modes steadily contribute to the total excess carrier recombination rate. In this perception, one might expect harmonically modulated lifetime $\tau_{eff,h}$ to coincide with steady-state lifetime $\tau_{eff,s}$—even in the extreme case of excess carrier generation much confined to highly recombinative interfaces, which was particularly addressed by Sinton and Trupke [1]. However, the above reasoning of $\tau_{eff,h} = \tau_{eff,s}$ may be inaccurate, and in some cases[3] the mismatch between $\tau_{eff,h}$ and $\tau_{eff,s}$ may amount to several orders of magnitude. The reason for this is that the harmonically modulated effective lifetime concept weights the decay modes of excess carrier density in a different manner than the steady-state concept—even if these modes feature the same distribution of mode weights in both steady-state and dynamic measurement configurations.[4] An ab initio analytical derivation of this effect will be provided in the following section. Fig. 6.1 provides an example of a very large mismatch between $\tau_{eff,h}$ and $\tau_{eff,s}$.

[3] e.g. thick substrates with both, high surface recombination velocity and bulk lifetime, cf. Sect. 8.4

[4] This is the case, if a rigorous quasi-steady-state condition is satisfied.

Table 6.1 Comparison between effective lifetime $\tau_{eff,h} = \tau_\sigma$ as measured via self-consistent quasi-steady-state harmonically modulated photoluminescence (modulation time $\mathcal{T} = 12.5 \times 10^{-3}$ s) and *expected* steady-state effective lifetime $\tau_{eff,s}$ (according to the steady-state solution of the continuity equation, cf. Sect. 3.2.2) on a p-type 1 Ω cm wafer ($d = 253\,\mu$m thick) with one optimally passivated interface (featuring a 30 nm thick AlO_x coating) and another highly recombinative interface (mechanically polished [4], cf. Sect. 9.2). Parameters for the calculation of $\tau_{eff,s}$: $\alpha_\gamma = 417\,\text{cm}^{-1}$ ($\lambda_\gamma = 880$ nm monochromatic irradiation), $\tau = 10^{-3}$ s, $D_a = 27\,\text{cm}^2\,\text{s}^{-1}$. Experimental results τ_σ correspond to a transmitted photon current density of $j_{\gamma,0} = 1.75 \times 10^{17}\,\text{cm}^{-2}\,\text{s}^{-1}$ (impinging photon current density peaking at an equivalent of one sun (cf. Appendix A.6) at an interface transmissivity of $1 - R_\gamma = 0.7$), but were injection-independent. As can be seen, the theory of Eq. 6.25 is experimentally confirmed here

Configuration	$\tau_{eff,h} = \tau_\sigma/\mu s$	$\tau_{eff,s}/\mu s$	$\tau_{eff,h}/\tau_{eff,s}$	
			Experiment	Theory
$S_0 \to \infty$, $S_d \to 0$	8.0 ± 0.4	2.0	4.0 ± 0.2	4.1
$S_0 \to 0$, $S_d \to \infty$	9.8 ± 0.2	11.5	0.85 ± 0.02	0.85

6.1.2 Harmonically Modulated Weighting of Decay Modes

As outlined in Sect. 6.1.1, the consideration of transient effects associated with dynamic effective lifetime $\tau_{eff,d}$ may to some extent explain the discrepancy between modulated lifetime $\tau_{eff,h}$ and steady-state effective lifetime $\tau_{eff,s}$ if a quasi-steady-state condition is not satisfied. It may even serve as a qualitative and phenomenological explanation for the discrepancy between modulated lifetime $\tau_{eff,h}$ and steady-state effective lifetime $\tau_{eff,s}$ in some special cases satisfying a quasi-steady-state condition. Yet, there are configurations of nonuniform recombination properties yielding a modulated lifetime $\tau_{eff,h}$ which can neither quantitatively nor qualitatively be explained by solely considering dynamic effective lifetime $\tau_{eff,d}$. For instance, an explanation solely considering dynamic effective lifetime is not capable of explaining the substantial mismatch between $\tau_{eff,h}$ and $\tau_{eff,s}$ encountered for very high S_0 in the presence of a most rigorous quasi-steady-state condition ($\mathcal{T} \to \infty$).

6.1.2.1 On the Apparent Paradox of $\tau_{eff,h} < \tau_{eff,s}$

For very large interface recombination velocities S_d on wafers, substantial deviations $\tau_{eff,h} < \tau_{eff,s}$ are found. This has been encountered experimentally (cf. Table 6.1), and substantiated via both semi-analytical (cf. Sect. 3.2.4) and finite-element (cf. Appendix A.3.2) solutions of the continuity equation.

The key to fully understand the mismatch between quasi-steady-state harmonically modulated and steady-state effective lifetime expressions in the presence of highly localized excess carrier recombination lies in the consideration of the fact that these two measurement designs feature a fundamentally differing weighting of excess carrier decay modes. The following line of reasoning enables a full quantitative understanding of the mismatch between quasi-steady-state harmonically modulated lifetime $\tau_{eff,h}$ and steady-state effective lifetime $\tau_{eff,s}$ through expression of effective lifetimes in terms of mode weights and decay times of the modes of excess carrier decay. Based on a quasi-steady-state harmonically modulated measurement design, an expression of steady-state effective lifetime is derived from the ratio of depth-integrated excess carrier density and generation rate, whereas an expression of harmonically modulated effective lifetime is obtained from the time shift between excess carrier density and generation rate.

6.1.2.2 Steady-State Versus Harmonically Modulated Decay Mode Weighting

Along the lines of Eqs. 3.54 and 3.57, let us assume a sinusoidally modulated excess carrier generation rate featuring a mono-exponential depth decay due to monochromatic irradiation

$$G(z, t) = \alpha_\gamma j_{\gamma,0} e^{-\alpha_\gamma z} \sin^2\left(\frac{\pi t}{T}\right). \tag{6.1}$$

This expression can be inserted into the Green's function solution of the continuity equation given in Eq. 3.96. Rearrangement of the (linear) integrals of that expression yields

$$\Delta n(z, t) = j_{\gamma,0} \sum_{n=1}^{\infty} \frac{y_n(z)}{\|y_n\|^2} \int_0^d d\zeta\, y_n(\zeta)\, \alpha_\gamma e^{-\alpha_\gamma \zeta} \int_0^t dt' \sin^2\left(\frac{\pi t'}{T}\right) e^{-\kappa_n (t-t')}, \tag{6.2}$$

with the inverse decay time corresponding to the n-th mode of excess carrier decay

$$\kappa_n = \frac{1}{\tau} + \lambda_n^2 D_a, \tag{6.3}$$

and with the n-th positive solution λ_n of the transcendental Eq. 3.98, which must (cf. Fig. 3.9) satisfy

$$(n-1)\frac{\pi}{d} \leq \lambda_n \leq n\frac{\pi}{d}. \tag{6.4}$$

6.1.2.3 Steady-State Decay Mode Weighting

Recalling the definition of steady-state effective lifetime as of Eq. 3.24, $\tau_{eff,s}$ is the ratio of depth-integrated excess carrier density and generation rate. With Eq. 6.2, the depth integral of excess carrier density can be expressed as

$$\int_0^d dz\, \Delta n\,(z,t) = j_{\gamma,0} \sum_{n=1}^{\infty} \beta_n \int_0^t dt'\, \sin^2\left(\frac{\pi t'}{T}\right) e^{-\kappa_n (t-t')}, \tag{6.5}$$

with the depth-integrated weight of the n-th mode of excess carrier decay[5]

$$\beta_n = \frac{1}{\|y_n\|^2} \int_0^d dz\, y_n\,(z) \int_0^d d\zeta\, y_n\,(\zeta)\, \alpha_\gamma e^{-\alpha_\gamma \zeta}, \tag{6.6}$$

which satisfies the normalization condition

$$\sum_{n=1}^{\infty} \beta_n = 1. \tag{6.7}$$

The depth-integrated excess carrier generation rate as of Eq. 3.54 is

$$\int_0^d dz\, G\,(z,t) = j_\gamma\,(t) \left(1 - e^{-\alpha_\gamma d}\right). \tag{6.8}$$

Without loss of generality[6], we can assume $\alpha_\gamma d \gg 1$ such that depth-integrated excess carrier generation rate simply corresponds to the transmitted photon current density $j_\gamma\,(t)$. Therefore, according to the definition of steady-state effective lifetime, it then reads

$$\tau_{eff,s} = \frac{j_{\gamma,0}}{j_\gamma\,(t)} \sum_{n=1}^{\infty} \beta_n \int_0^t dt'\, \sin^2\left(\frac{\pi t'}{T}\right) e^{-\kappa_n (t-t')}. \tag{6.9}$$

Considering a quasi-steady-state condition

$$T \gg \pi \kappa_1^{-1}, \tag{6.10}$$

the sine term in the latter integral varies only negligibly for all *relevant* integration times $t - t'$ which are sufficiently short in order to satisfy the condition

$$e^{-\kappa_1 (t-t')} \not\lll 1. \tag{6.11}$$

[5] Note that $\|y_n\|^2$ has the dimension of a length, thereby rendering β_n dimensionless.

[6] For non-negligible transmission through the substrate bulk, cumulative excess carrier density would be reduced by the same factor as cumulative generation rate. Also note that Eq. 6.8 neglects internal reflection of irradiation light, which is however irrelevant if $\alpha_\gamma d \gg 1$.

Therefore, if the above quasi-steady-state condition is satisfied, the above sine term must satisfy

$$\left|\frac{\sin^2\left(\frac{\pi t}{T}\right) - \sin^2\left(\frac{\pi t'}{T}\right)}{\sin^2\left(\frac{\pi t}{T}\right)}\right| \lll 1 \qquad (6.12)$$

for all relevant integration times $t - t'$. This justifies to take the sine term out of the integral, yielding steady-state effective lifetime

$$\tau_{eff,s} = \sum_{n=1}^{\infty} \beta_n \int_0^t dt'\, e^{-\kappa_n(t-t')}, \qquad (6.13)$$

which is easily integrated

$$\tau_{eff,s} = \sum_{n=1}^{\infty} \frac{\beta_n}{\kappa_n}\left(1 - e^{-\kappa_n t}\right). \qquad (6.14)$$

Again applying the quasi-steady-state condition (6.10), $\kappa_1 t \gg 1$ holds for all relevant times t, yielding the straightforward steady-state effective lifetime expression

$$\tau_{eff,s} = \sum_{n=1}^{\infty} \frac{\beta_n}{\kappa_n}. \qquad (6.15)$$

Hence, steady-state effective lifetime corresponds to a linear superposition of decay times κ_n^{-1} weighted by the depth-integrated weights β_n of the modes of excess carrier decay.

6.1.2.4 Harmonically Modulated Decay Mode Weighting

Harmonically modulated effective lifetime represents the time shift betweeen excess carrier density and generation rate. Under quasi-steady-state operation conditions (as assumed here) and for injection-independent recombination properties[7], harmonically modulated effective lifetime and self-consistent effective lifetime expressions coincide (cf. Sect. 5.2). Therefore, both phase-sensitive (e.g. lock-in) lifetime analyses as well as self-consistent harmonically modulated lifetime analyses are likewise affected by what is to be derived in the following.

[7] The impact of the injection dependence of recombination properties on harmonically modulated lifetime measurements and other light-biased decay time measurements is thoroughly elaborated in Sect. 6.2.

Here—just as in the above derivation for steady-state effective lifetime—an expression of quasi-steady-state harmonically modulated effective lifetime

$$\tau_{eff,h} = \tau_{eff,h}\left(\kappa_n, \beta_n\right) \tag{6.16}$$

as the time shift between depth-integrated excess carrier density and generation rate is aspired as an exclusive function of the inverse decay times κ_n and the depth-integrated weights β_n of the modes of excess carrier decay.

In awareness of the fact that excess carrier generation as of Eq. 6.1 peaks at the time $t = \frac{\mathcal{T}}{2}$, the time of a maximum of depth-integrated excess carrier density

$$t^* = \frac{\mathcal{T}}{2} + \tau_{eff,h} \tag{6.17}$$

is sought. With the expression of depth-integrated excess carrier density as provided in Eq. 6.5, the derivation is sparked by the maximum condition

$$\sum_{n=1}^{\infty} \beta_n \frac{d}{dt}\left(e^{-\kappa_n t} \int_0^t dt' \sin^2\left(\frac{\pi t'}{\mathcal{T}}\right) e^{\kappa_n t'} \right)\Bigg|_{t=t^*} = 0. \tag{6.18}$$

Expression of the function $\sin^2\left(\frac{\pi t'}{\mathcal{T}}\right)$ in terms of complex exponential functions

$$\sin^2\left(\frac{\pi t'}{\mathcal{T}}\right) = \frac{1}{4}\left(2 - e^{\frac{2\pi i t'}{\mathcal{T}}} - e^{-\frac{2\pi i t'}{\mathcal{T}}}\right) \tag{6.19}$$

allows integration over t' such that the above maximum condition (Eq. 6.18) reads

$$\sum_{n=1}^{\infty} \beta_n \frac{d}{dt}\left(\frac{2}{\kappa_n}\left(1 - e^{-\kappa_n t}\right) - \frac{e^{\frac{2\pi i t}{\mathcal{T}}} - e^{-\kappa_n t}}{\frac{2\pi i}{\mathcal{T}} + \kappa_n} - \frac{e^{-\frac{2\pi i t}{\mathcal{T}}} - e^{-\kappa_n t}}{-\frac{2\pi i}{\mathcal{T}} + \kappa_n} \right)\Bigg|_{t=t^*} = 0. \tag{6.20}$$

This derivation aims at the proof of a mismatch between steady-state effective lifetime and quasi-steady-state harmonically modulated effective lifetime, regardless of the modulation time $\mathcal{T}$. Therefore, the expression to be derived here must hold even for the most rigorous quasi-steady-state condition. With the quasi-steady-state condition (6.10) $\mathcal{T} \gg \pi\kappa_1^{-1}$, the derivative $\frac{d}{dt}e^{-\kappa_n t}$ is negligible compared to any competing term at times $t > \frac{\mathcal{T}}{2}$, so the maximum condition (6.20) substantially simplifies to

$$\sum_{n=1}^{\infty} \beta_n \frac{d}{dt}\left(\frac{e^{\frac{2\pi i t}{\mathcal{T}}}}{\frac{2\pi i}{\mathcal{T}} + \kappa_n} + \frac{e^{-\frac{2\pi i t}{\mathcal{T}}}}{-\frac{2\pi i}{\mathcal{T}} + \kappa_n} \right)\Bigg|_{t=t^*} = 0. \tag{6.21}$$

Execution of the derivative with respect to time yields the maximum condition

$$\sum_{n=1}^{\infty} \frac{\beta_n}{\kappa_n^2 + \frac{4\pi^2}{\mathcal{T}^2}} \left(\frac{8\pi^2}{\mathcal{T}^2} \cos\left(\frac{2\pi t^*}{\mathcal{T}}\right) - \kappa_n \frac{4\pi}{\mathcal{T}} \sin\left(\frac{2\pi t^*}{\mathcal{T}}\right)\right) = 0. \quad (6.22)$$

Assuming $\tau_{eff,h}$ to correspond to a weighted superposition of decay times κ_n^{-1}, it must be confined within the interval $0 < \tau_{eff,h} \leq \kappa_1^{-1}$. With κ_1^{-1} as a definite upper bound of $\tau_{eff,h}$, $\mathcal{T} \gg \pi\tau_{eff,h}$ always holds if the quasi-steady-state condition $\mathcal{T} \gg \pi\kappa_n^{-1}$ is satisfied. Therefore, linear Taylor series expansions of cosine and sine expressions of Eq. 6.22 in the temporal vicinity of $t = \frac{\mathcal{T}}{2}$ are highly accurate at times $t^* = \frac{\mathcal{T}}{2} + \tau_{eff,h}$. Further, the quasi-steady-state condition also allows to simplify the denominator of the leftmost fraction in Eq. 6.22 to κ_n^2, yielding yet another substantial simplification of the maximum condition, which becomes

$$\sum_{n=1}^{\infty} \frac{\beta_n}{\kappa_n^2} \left(1 - \kappa_n \tau_{eff,h}\right) = 0 \quad (6.23)$$

upon insertion of $t^* = \frac{\mathcal{T}}{2} + \tau_{eff,h}$. This is equivalent to the expression of harmonically modulated effective lifetime

$$\tau_{eff,h} = \frac{\sum_{n=1}^{\infty} \frac{\beta_n}{\kappa_n^2}}{\sum_{n=1}^{\infty} \frac{\beta_n}{\kappa_n}} \quad (6.24)$$

as an exclusive function of the inverse decay times κ_n and the depth-integrated weights β_n of the modes of excess carrier decay. Notably, and in contrast to the steady-state effective lifetime expression derived in Eq. 6.15, this expression corresponds to an explicitly nonlinear superposition of decay times κ_n^{-1}, providing the key to quantitatively resolve the mismatch between quasi-steady-state harmonnically modulated lifetime and steady-state effective lifetime.

$$\frac{\tau_{eff,h}}{\tau_{eff,s}} = \frac{\sum_{n=1}^{\infty} \frac{\beta_n}{\kappa_n^2}}{\left(\sum_{n=1}^{\infty} \frac{\beta_n}{\kappa_n}\right)^2} \quad (6.25)$$

6.1.2.5 Resolution of the Apparent Paradox of $\tau_{eff,h} < \tau_{eff,s}$

Comparing Eqs. 6.15 and 6.24, harmonically modulated and steady-state effective lifetime expressions coincide if there is only one significant mode of excess carrier decay, i.e. if $\beta_1 = 1$ and $\beta_n = 0 \;\; \forall \;\; n \neq 1$. This would be the exclusive condition for $\tau_{eff,h} = \tau_{eff,s}$ if the weights β_n were restricted to positive numbers. However,

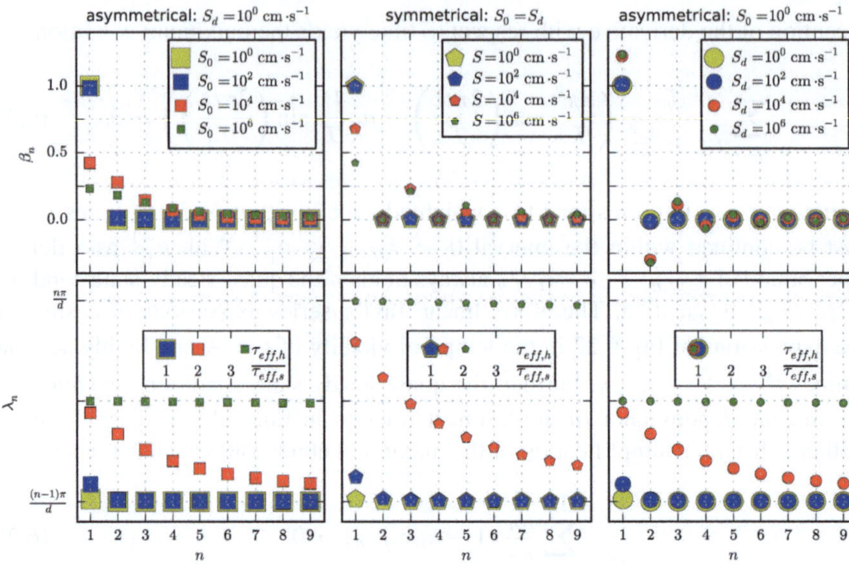

Fig. 6.2 Depiction of the weights β_n of lower order modes of excess carrier decay (cf. Eq. 6.6), of the corresponding solutions λ_n (with $\kappa_n = \tau^{-1} + \lambda_n^2 D_a$) of the transcendental Eq. 3.98, and of the corresponding quasi-steady-state mismatch $\tau_{eff,h}/\tau_{eff,s}$ (*inserted plots*) for representative configurations of surface recombination velocities S_0 and S_d. Model parameters: $d = 200\,\mu$m, $\alpha_\gamma = 417\,\text{cm}^{-1}$, $D_a = 27\,\text{cm}^2\,\text{s}^{-1}$, $\tau = 10^{-3}$ s. Partially negative weights β_n enable $\tau_{eff,h} < \tau_{eff,s}$ in some configurations—particularly for high S_d. Notably—and as also follows from Eq. 3.98—λ_n and κ_n are invariant upon exchange of S_0 and S_d. An adapted figure is to be found in [2]

this is not the case, as negative weights $\beta_n < 0$ are possible. The existence of negative weights $\beta_n < 0$ enables realistic configurations of electronic properties yielding $\tau_{eff,h} < \tau_{eff,s}$. The most prominent example is a silicon wafer featuring interface recombination velocities $S_0 \to 0$ and $S_d \to \infty$. Beyond, this fact enables coinciding values of harmonically modulated and steady-state effective lifetime due to compensation of lifetime contributions of positive and negative weights β_n, respectively.

Figure 6.2 shows the weights β_n of lower order modes of excess carrier decay, corresponding solutions λ_n of the transcendental Eq. 3.98, and the corresponding quasi-steady-state mismatch $\tau_{eff,h}/\tau_{eff,s}$ for representative configurations of surface recombination velocities S_0 and S_d, respectively. Further, Fig. 6.3 provides a systematic overview of the quasi-steady-state mismatch $\tau_{eff,h}/\tau_{eff,s}$ as a function of surface recombination velocities of $d = 200\,\mu$m thick wafers featuring a high bulk lifetime of $\tau = 10^{-3}$ s. Table 6.1 compares self-consistent effective lifetimes $\tau_\sigma = \tau_{eff,h}$ encountered in quasi-steady-state harmonically modulated photoluminescence measurements on an asymmetrically passivated float zone wafer (with $S_0 \to 0, \infty$ and $S_d \to \infty, 0$) with steady-state effective lifetime $\tau_{eff,s}$. The experimental result $\tau_{eff,h}/\tau_{eff,s}$ well confirms the theory of Eq. 6.25.

In Fig. 6.4 the course of the mismatch between harmonically modulated and steady-state effective lifetimes is plotted as a function of wafer thickness. Approach-

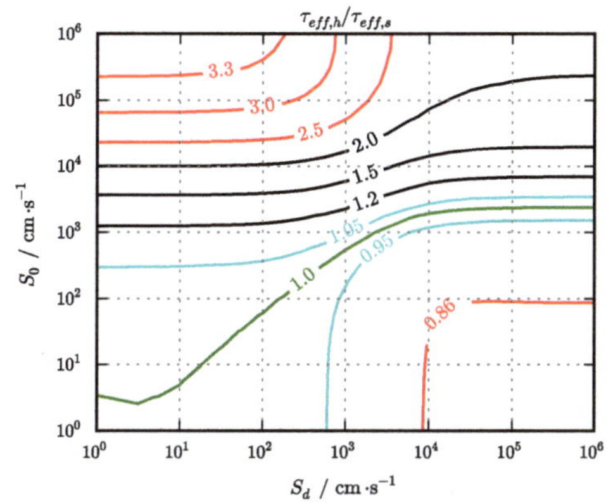

Fig. 6.3 Systematic overview of the quasi-steady-state mismatch $\tau_{eff,h}/\tau_{eff,s}$ as a function of surface recombination velocities S_0 (*illuminated interface*) and S_d (*opposite interface*) for a 1 Ω cm p-type silicon wafer of thickness $d = 200\,\mu$m. Model parameters are identical to Fig. 6.2. An adapted figure is to be found in [2]

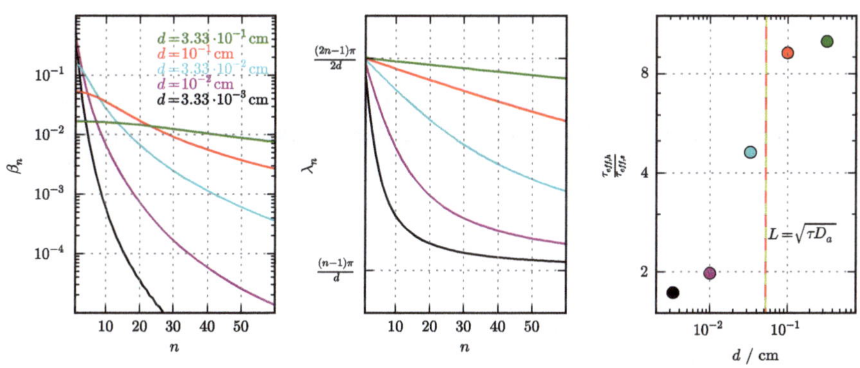

Fig. 6.4 Depiction of the depth averaged mode weights β_n, of the solutions λ_n of the transcendenal equation (Eq. 3.98), and of the resulting mismatch between harmonically modulated lifetime $\tau_{eff,h}$ and steady-state effective lifetime $\tau_{eff,s}$ as a function of wafer thickness d. The thicker a wafer substrate, the smaller are λ_n and the broader is the distribution β_n in the case of dominant surface recombination. This collection of figures is intended to illustrate how the theory of quasi-steady-state decay mode weighting relates to the very large mismatch $\tau_{eff,h}/\tau_{eff,s}$ typically encountered on very thick samples featuring high surface recombination velocity. Model parameters: $\alpha_\gamma = 417\,\text{cm}^{-1}$, $D_a = 27\,\text{cm}^2\,\text{s}^{-1}$, $S_0 = 10^5\,\text{cm}\,\text{s}^{-1}$, $S_d = 10^0\,\text{cm}\,\text{s}^{-1}$, $\tau = 10^{-4}\,\text{s}$

ing constant values at both, wafer thicknesses d much greater and much less than minority carrier diffusion length $L = \sqrt{\tau D_a}$, it is demonstrated that the differences in weighting of decay modes may also explain the very largemismatch $\tau_{eff,h}/\tau_{eff,s}$ typically encountered on very thick samples featuring very high surface recombination velocities (as e.g. demonstrated for ingots in Fig. 6.1).

6.1.3 On the Interpretation of Modulated Lifetime

One may conclude from the considerations and derivations given in the previous sections that the deviations between harmonically modulated effective lifetime $\tau_{eff,h}$ and steady-state effective lifetime $\tau_{eff,s}$ in cases of highly recombinative interfaces may partly be caused by a transient effect of a rapid extinction of higher order modes of excess carrier decay—particularly in the case of a quasi-transient or intermediate modulated measurement design. However, for the occasionally encountered substantial mismatch between $\tau_{eff,h}$ and $\tau_{eff,s}$ under quasi-steady-state measurement conditions, the extinction of higher order decay modes is not the actual cause. Rather, the latter mismatch—which is notably independent of modulation time $\mathcal{T}$ if a quasi-steady-state condition is satisfied—can fully be explained by the differences between the harmonically modulated and steady-state weighting of decay modes, respectively.

The line of reasoning of Sinton and Trupke in [1] suggests a *per se* overestimation of steady-state effective lifetime by any dynamic measurement technique. Such perception gave rise to what is herein referred to as the *apparent paradox of* $\tau_{eff,h} < \tau_{eff,s}$. The resolution of this paradox (cf. Sect. 6.1.2) proves that the reasoning as of [1] is not fully comprehensive with regard to modulated lifetime techniques—particularly under quasi-steady-state conditions.

Also, previous attempts to phenomenologically explain the mismatch between steady-state and harmonically modulated lifetime expressions via a temporal imbalance of the depth profile of excess carrier density and an accompanying redistribution of mode weights β_n as suggested in [5] appear to feature some inaccuracy. In the limit $\mathcal{T} \to \infty$, the temporal imbalance of the depth profile of excess carrier density in the course of a harmonic modulation becomes infinitely small, whereas the mismatch between $\tau_{eff,h}$ and $\tau_{eff,s}$ may sustain at a constant level. The explanation attempt via a temporal imbalance of $\Delta n\left(z\right)$ can only explain deviations between $\tau_{eff,h}$ and $\tau_{eff,s}$ caused by the quasi-transient extinction of higher order modes of excess carrier decay. This effect may indeed occur (and contribute to a mismatch) in a harmonically modulated measurement design if a quasi-steady-state condition is violated. Yet it is negligible compared to the effect elaborated in Sect. 6.1.2 under quasi-steady-state conditions.

If a rigorous quasi-steady-state condition is violated, the temporal imbalance of the depth profile $\Delta n\left(z\right)$ can be qualitatively illustrated by an approximation based on the solution of the continuity equation. With excess carrier generation rate as of Eq. 3.54, Eq. 6.2 may be expressed as

$$\Delta n\left(z,t\right) = \sum_{n=1}^{\infty} \frac{y_n\left(z\right)}{\|y_n\|^2} \int_0^d d\zeta\, y_n\left(\zeta\right) \alpha_\gamma e^{-\alpha_\gamma \zeta} \int_0^t dt'\, j_\gamma\left(t'\right) e^{-\kappa_n\left(t-t'\right)}. \quad (6.26)$$

For minor deviations from rigorous quasi-steady-state, the first order Taylor expansion of photon current density

$$j_\gamma\left(t'\right) = j_\gamma\left(t - \left(t - t'\right)\right) = j_\gamma\left(t\right) - \left.\frac{dj_\gamma}{dt}\right|_t \left(t - t'\right) + o\left(\left(t - t'\right)^2\right) \quad (6.27)$$

is reasonable for sufficiently short maximally *relevant* times $t - t'$. Thus, with a frequency

$$\kappa^* = j_\gamma^{-1}\left(t\right) \left.\frac{dj_\gamma}{dt}\right|_t, \quad (6.28)$$

$j_\gamma\left(t'\right)$ reasonably coincides with the expression

$$j_\gamma\left(t'\right) \approx j_\gamma\left(t\right) e^{-\kappa^*\left(t-t'\right)}. \quad (6.29)$$

Insertion into the solution of the continuity equation yields

$$\Delta n\left(z, t\right) \approx \sum_{n=1}^{\infty} \frac{y_n\left(z\right)}{\|y_n\|^2} \int_0^d d\zeta\, y_n\left(\zeta\right) \alpha_\gamma e^{-\alpha_\gamma \zeta} j_\gamma\left(t\right) \int_0^t dt'\, e^{-\left(\kappa_n+\kappa^*\right)\left(t-t'\right)}. \quad (6.30)$$

If a quasi-steady-state condition is violated, κ^* may become non-negligible compared to inverse decay time κ_1 of the fundamental mode of excess carrier decay. The qualitative implications of this would be

1. An apparent **enhancement** of κ_1 in the *ascending* branch of a modulation cycle, which is equivalent to an apparent **lowering** of the effective decay time of the fundamental mode of excess carrier decay.
2. An apparent **lowering** of κ_1 in the *descending* branch of a modulation cycle, which is equivalent to an apparent **enhancement** of the effective decay time of the fundamental mode of excess carrier decay.

If modulation proceeds sufficiently fast in order to violate $\kappa_1 \gg \kappa^*$, this may lead to a substantial deviation of the depth profile of excess carrier density between the ascending and the descending branch of a modulation cycle. This effect is most pronounced in high lifetime regions of thick substrates (e.g. ingots) distant from highly recombinative interfaces, which are much dominated by the slowly decaying lower order modes of excess carrier decay, as depicted in Fig. 6.5. Yet, it must be noted that this temporal imbalance of the depth profile $\Delta n\left(z\right)$ is a transient effect, which does not essentially give rise to the mismatch between quasi-steady-state harmonically modulated lifetime $\tau_{eff,h}$ and steady-state effective lifetime $\tau_{eff,s}$.

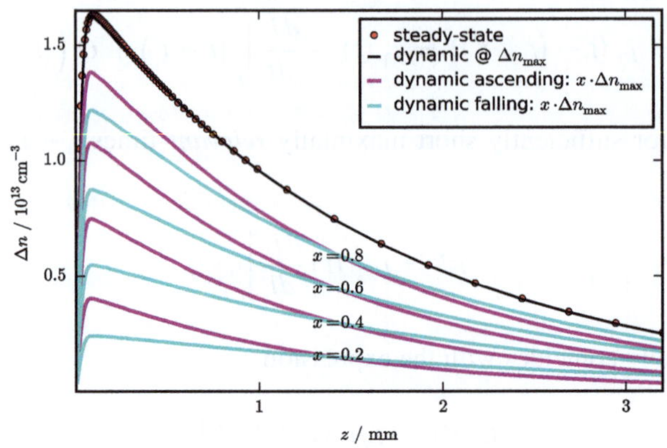

Fig. 6.5 Depiction of the temporal imbalance of the depth profile of excess carrier density of a silicon ingot during a harmonic modulation cycle. Depth profiles of excess carrier density according to Eq. 3.91 are plotted at times t_x of equal depth-integrated excess carrier density (with x representing the fraction of maximal excess carrier density during a modulation cycle). Model assumptions: $S = 10^5\,\text{cm}\,\text{s}^{-1}$, $D_a = 27\,\text{cm}^2\,\text{s}^{-1}$, $\tau = 10^{-3}\,\text{s}$, $\alpha_\gamma = 417\,\text{cm}^{-1}$, $\mathcal{T} = 0.02\,\text{s}$, and $j_{\gamma,0} = 1.78 \times 10^{17}\,\text{cm}^{-2}\,\text{s}^{-1}$. The steady-state solution according to Eq. 3.72 is also plotted. Notably, this temporal imbalance of $\Delta n(z)$ is a transient effect, which does not essentially give rise to the mismatch between quasi-steady-state harmonically modulated lifetime $\tau_{eff,h}$ and steady-state effective lifetime $\tau_{eff,s}$. Adapted after [3]

6.2 Differential Lifetime

In this section, previous findings about the discrepancy between differential and actual recombination properties are generalized and provided with a quantitative theoretical framework.[8] The said discrepancy is motivated as a function of characteristic irradiation conditions of light-biased dynamic lifetime measurements. Both light-biased differential dynamic lifetime measurements as well as measurements of effective carrier lifetime as inferred from the time shift between harmonically modulated excess carrier density and generation rate are shown to represent a most affected limiting case. Dynamic carrier lifetime measurement techniques are classified in terms of whether or not they suffer from the addressed discrepancy. Further, novel pathways in lifetime metrology, resulting from the **theory of light-biased decay time** to be derived in the following, are outlined. A generally valid approach to obtain actual recombination properties from the affected methods is sketched in Sect. 8.1.

6.2.1 Previous Findings on Differential Lifetime

In 1995, Brendel, Aberle, and Schmidt identified systematic deviations of surface recombination velocity measurements via light-biased photoconductance decay from

[8] The account given in this section is adapted from the corresponding original publications [6, 7].

actual (true) surface recombination velocities [8, 9]. Also, they managed to attribute such deviations to the fact that light-biased decay time measurements yielded differential rather than actual recombination properties. In conclusion, the integration of differential surface recombination velocities measured over the entire injection range up to a point of interest was deemed inevitable for a correct determination of actual injection-dependent surface recombination velocity from differential decay time measurements [9].

Analogous observations to the findings of Brendel, Aberle, and Schmidt were made in the course of this work via harmonically modulated luminescence lifetime investigations—notably without any additional bias-light. Aiming at overcoming the requirement of net dopant concentration n_0 for the self-consistent lifetime analysis upon departure from low-level injection conditions (cf. Sect. 5.2.3), the time shift between modulated excess carrier generation rate and luminescence intensity was initially interpreted as effective lifetime—as one might conclude from a quasi-steady-state treatment of the time-dependent continuity equation (cf. Eq. 3.87). Yet, measured time shifts proved to only coincide with effective lifetime in the rarely encountered special case of a vanishing injection dependence of lifetime. There was twofold evidence for the latter observation:

1. **Experimental Evidence**: The self-consistent lifetime analysis only coincided with the simplistic *time shift* lifetime analysis, if self-consistent lifetime featured no injection dependence.
2. **Numerical Modeling**: The time shift between the finite-element solution (cf. Appendix A.3.1) of the time-dependent continuity equation $\Delta n(t)$ and generation rate $G(t)$ only happened to coincide with the underlying true (actual) carrier lifetime, if the latter was injection-independent.

The time shift between maxima of excess carrier density and generation rate proved to be greater than effective carrier lifetime in the case of a positive derivative of lifetime with respect to excess carrier density $\frac{d\tau_{eff}}{d\Delta n} > 0$, whereas it proved to be less than effective carrier lifetime in the case of $\frac{d\tau_{eff}}{d\Delta n} < 0$. This qualitative analogy with respect to the findings of Brendel, Aberle, and Schmidt provided the initial point of a quantitative theoretical investigation, to be outlined in the following.

6.2.2 Theory of Light-Biased Decay Time

The assumption of the time-dependent continuity equation (Eq. 3.75) with an effective lifetime τ_{eff} implies no loss of generality with regard to the conclusions drawn herein.[9] Two problems shall be addressed in the following:

[9] Aberle et al. previously pointed out that their derivation for surface recombination velocities was transferrable to bulk recombination [9]. The same applies to effective lifetime, being representative of the sum of bulk and surface recombination rates— as already addressed by both Schuurmans et al. [10] and by Schmidt [11].

1. It is elaborated, whether the structural modification imposed on the continuity equation through the injection dependence of effective lifetime τ_{eff} affects the determination of lifetime from decay time measurements.
2. The impact of bias light on the determination of lifetime from decay time measurements is systematically elaborated.

6.2.2.1 Structural Modification of the Continuity Equation

Consider the case of an abrupt switch-off of a steady-state excess carrier generation rate with $G(t \leq 0) = G_0 = \frac{\Delta n_0}{\tau_{eff,0}}$ and $G(t > 0) = 0$. With a first order Taylor expansion of injection-dependent effective lifetime

$$\tau_{eff} = \tau_{eff,0} + (\Delta n - \Delta n_0) \left. \frac{d\tau_{eff}}{d\Delta n} \right|_{\Delta n_0} + O\left((\Delta n - \Delta n_0)^2\right) \tag{6.31}$$

for very short times $t > 0$, the linearized continuity equation takes the form

$$\frac{d\Delta n}{dt} + u\Delta n = v, \tag{6.32}$$

with coefficients

$$u = \frac{1 - G_0 \frac{d\tau_{eff}}{d\Delta n}\big|_{\Delta n_0}}{\tau_{eff,0}} \tag{6.33}$$

$$v = -G_0^2 \left. \frac{d\tau_{eff}}{d\Delta n} \right|_{\Delta n_0}. \tag{6.34}$$

The solution of Eq. 6.32 for very short times $t > 0$ is

$$\Delta n(t) = \left(\Delta n_0 - \frac{v}{u}\right) e^{-ut} + \frac{v}{u}, \tag{6.35}$$

and evaluation of a measurable effective decay time τ_m for positive times $t \to 0$ then yields

$$\tau_m(t \to 0) = -\left. \frac{\Delta n}{\frac{d\Delta n}{dt}} \right|_{t \to 0} = \tau_{eff,0}. \tag{6.36}$$

Thus, one may conclude that the mismatch between differential decay time τ_m and actual effective lifetime τ_{eff} is not solely caused by the structural modification of the continuity equation through the injection dependence of lifetime. As will be derived in the following, a mismatch occurs only if the injection dependence of carrier lifetime is accompanied by a light bias.

6.2.2.2 Impact of Bias Light

Let us now consider the case of an abrupt change of a steady-state excess carrier generation rate with $G(t \leq 0) = G_0 = \frac{\Delta n_0}{\tau_{eff,0}}$ and $G(t > 0) = \xi G_0$ with a bias light parameter $0 \leq \xi < 1$. According to the time-dependent continuity equation (Eq. 3.75), for positive times $t \rightarrow 0$ the relation

$$\left.\frac{d\Delta n}{dt}\right|_{t\rightarrow 0} = -G_0(1-\xi) \tag{6.37}$$

holds. In the calculation of the measurable effective decay time $\tau_m(t \rightarrow 0)$,[10] the fact that excess carrier density decays to an equilibrium value $\Delta n(t \rightarrow \infty) \neq 0$ has to be taken into account such that

$$\tau_m = \frac{\Delta n_0 - \Delta n(t \rightarrow \infty)}{G_0(1-\xi)}. \tag{6.38}$$

The latter equation reveals the differential nature of the measurable decay time τ_m, which relates a change of excess carrier generation rate to a change of excess carrier density. Here, the injection dependence of τ_{eff} comes into play, because the equilibrium value $\Delta n(t \rightarrow \infty)$ depends on the derivative of effective lifetime with respect to excess carrier density. As $\left.\frac{d\Delta n}{dt}\right|_{t\rightarrow\infty}$ must vanish, insertion of the first order Taylor expansion of effective lifetime (cf. Eq. 6.31) into the continuity equation (Eq. 3.75) yields

$$\Delta n(t \rightarrow \infty) = \xi G_0 \left(\tau_{eff,0} + (\Delta n(t \rightarrow \infty) - \Delta n_0)\left.\frac{d\tau_{eff}}{d\Delta n}\right|_{\Delta n_0}\right). \tag{6.39}$$

Explicitly solving this equation for $\Delta n(t \rightarrow \infty)$, one obtains

$$\Delta n(t \rightarrow \infty) = \xi \Delta n_0 \frac{1 - G_0 \left.\frac{d\tau_{eff}}{d\Delta n}\right|_{\Delta n_0}}{1 - \xi G_0 \left.\frac{d\tau_{eff}}{d\Delta n}\right|_{\Delta n_0}}. \tag{6.40}$$

The general validity of the latter equation in the limit $\xi \rightarrow 1$ (regardless of the curvature of $\tau_{eff}(\Delta n)$) is shown in Appendix A.7. Insertion into Eq. 6.38 then leads to the final expression

$$\tau_m = \frac{\tau_{eff,0}}{1 - \xi G_0 \left.\frac{d\tau_{eff}}{d\Delta n}\right|_{\Delta n_0}}. \tag{6.41}$$

Note that due to the linear approximation of effective lifetime in Eqs. 6.39 and 6.41 is also an approximation. However, in the limit $\xi \rightarrow 1$ it is exact, regardless of the

[10] $\tau_m(t \rightarrow 0)$ shall be denoted as τ_m for reasons of simplicity.

curvature of $\tau_{eff}\,(\Delta n)$ (cf. Appendix A.7). For finite differences between ξ and unity, the accuracy of Eq. 6.41 much depends on the linearity of $\tau_{eff}\,(\Delta n)$ in the proximity of Δn_0. In most cases, it is deemed to be of reasonable accuracy for values $\xi > 0.9$, which are typically encountered in the case of so-called *small signal* measurements such as light-biased microwave photoconductance decay.

6.2.3 Implications of the Theory of Light-Biased Decay Time

Equation 6.41 is essential for the theory of light-biased decay time. It provides a quantitative measure of the discrepancy between measurable decay time τ_m and actual lifetime $\tau_{eff,0}$ depending on both, the bias light parameter ξ and the injection dependence $\frac{d\tau_{eff}}{d\Delta n}\big|_{\Delta n_0}$ of lifetime. Previous quantitative statements about the discrepancy's dependence on $\frac{d\tau_{eff}}{d\Delta n}\big|_{\Delta n_0}$ are confirmed [11], and an accurate rule on how to relate measured decay times τ_m to actual effective lifetime $\tau_{eff,0}$ is provided.

6.2.3.1 Discussion of Limiting Cases

Injection-Independent Case: For $\frac{d\tau_{eff}}{d\Delta n}\big|_{\Delta n_0} = 0$, according to Eq. 6.41 the mismatch between measured decay time τ_m and actual effective lifetime $\tau_{eff,0}$ vanishes, regardless of a light bias (cf. circular symbols in Fig. 6.6).
Case of $\xi = 0$: This is the case of a decay time measurement without any bias light. A transient injection-dependent measurement of decay time [12–14]

$$\tau_m\,(\Delta n) = -\frac{\Delta n}{\frac{d\Delta n}{dt}} \tag{6.42}$$

after an abrupt and—most importantly—*complete* switch-off of generation rate yields actual effective lifetime $\tau_m = \tau_{eff,0}$, regardless of the injection dependence of lifetime (cf. Fig. 6.6). This particularly applies to the widespread lifetime analysis on the basis of transient or quasi-transient photoconductance [15, 16].
Limit $\xi \to 1$: This is the case of an infinitesimal temporal change of excess carrier generation rate, which also applies to harmonically modulated lifetime measurement designs. For $\xi \to 1$, Eq. 6.41 clearly indicates a maximal impact of the injection dependence of lifetime on the mismatch between measured decay time τ_m and actual effective lifetime $\tau_{eff,0}$—as also visualized in Fig. 6.6.

6.2.3.2 Relation to Harmonically Modulated Measurement Designs

The measurement of the time shift between quasi-steady-state harmonically modulated excess carrier generation rate $G(t)$ and excess carrier density $\Delta n(t)$ is in the

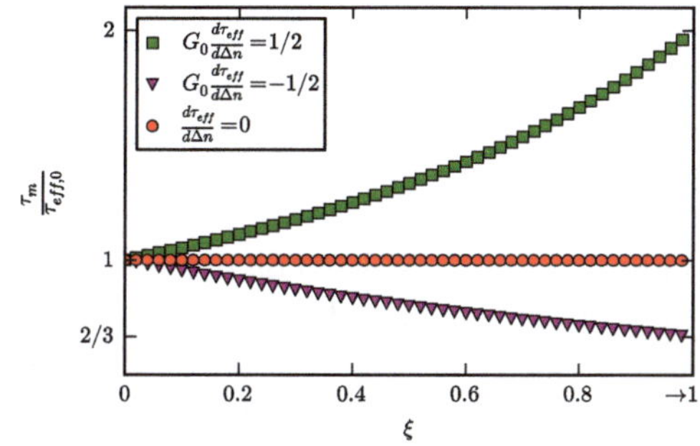

Fig. 6.6 Illustration of Eq. 6.41, depicting the ratio between measurable decay time τ_m and actual effective lifetime $\tau_{eff,0}$ as a function of the bias light parameter $\xi < 1$, with $G\,(t \leq 0) = G_0 = \frac{\Delta n_0}{\tau_{eff,0}}$ and $G\,(t > 0) = \xi G_0$. Three situations are plotted, representing different injection dependencies of lifetime, respectively. Adapted after [6]

end a measurement of decay time τ_m according to Eq. 6.41 in the limit $\xi \rightarrow 1$. This is shown in the following:

The quasi-steady-state solution of the continuity equation (Eq. 3.87) reveals a time shift between excess carrier density and generation rate equal to effective carrier lifetime τ_{eff}, if the latter is injection-independent. However, for the general case of injection-dependent lifetime, this is not true. This can be shown by drawing on Eq. 6.32, which is the linearization of the continuity equation, incorporating the injection dependence of effective lifetime. Other than for the previously used step-type excess carrier generation, let us now assume an inhomogeneity of the form

$$G\,(t) = G_0 + G_1 \sin^2 \frac{\pi t}{T}, \tag{6.43}$$

featuring a sufficiently faint amplitude G_1 in order to ensure accuracy of the linearization which Eq. 6.32 is based upon.[11] Therefore, the total inhomogeneity is in this case $v + G\,(t)$, and the corresponding solution of Eq. 6.32 must be a superposition of the solutions to the inhomogeneities $v + G_0$ and $G_1 \sin^2 \frac{\pi t}{T}$. With an initial value $\Delta n\,(t = 0) = G_0 \tau_{eff,0}$, the solution thus reads

$$\Delta n\,(t) = \tau_{eff,0} \left(G_0 + G_1 \sin^2 \left(\frac{\pi}{T} \left(t - u^{-1} \right) \right) \right), \tag{6.44}$$

thereby confirming the validity of Eq. 6.41 in the limit $\xi \rightarrow 1$ for phase-sensitive harmonically modulated lifetime measurements.[12]

A quasi-steady-state harmonically modulated phase-sensitive lifetime approach [17–22] yields a time shift $\tau_{eff,h} = \lim_{\xi \rightarrow 1} \tau_m = u^{-1}$ corresponing to differential rather

---

[11] For infinitesimal G_1, Eq. 6.32 would be exact.

[12] Note that the constant bias generation rate G_0 does not give rise to the mismatch between τ_m and $\tau_{eff,0}$, here. It solely enables the use of the linearized Eq. 6.32 at a finite generation rate.

than actual lifetime, irrespective of the existence of an additional steady-state bias light. In fact, a continuously modulated irradiation is the only true representative of the infinitesimal limit $\xi \rightarrow 1$, whereas a light-biased decay time measurement as described in [9] is always restricted to finite differences between ξ and unity due to noise. The interpretation of harmonic modulation as a light-biased time sequence of infinitesimal changes of excess carrier generation rate provides a descriptive understanding of the identity of decay time τ_m and time shift u^{-1}. An infinitesimal change of excess carrier generation rate gives rise to relaxation of excess carrier density at a decay time u^{-1}. The superposition of such relaxation processes must consequently bring about a phase lag u^{-1} between modulated excess carrier density and generation rate.

6.2.3.3 On a Previous Quantification of a Differential Lifetime Effect

In 1999, Schmidt [11] investigated the mismatch between differential lifetime as inferred from modulated free carrier absorption [17, 18] and actual lifetime as inferred from quasi-steady-state photoconductance [15, 23] in an injection regime dominated by Auger recombination: he experimentally found a mismatch of $\tau_{eff,0} = 3\tau_m$, which excellently matched his theoretical prediction. Notably, Eq. 6.41 in the limit $\xi \rightarrow 1$ yields the same result: harmonically modulated lock-in free carrier absorption represents a differential lifetime measurement in this very limit, regardless whether an additional steady-state light bias is applied. With a simplified Auger lifetime parametrization $\tau_{eff} = \frac{a}{\Delta n^2}$, it follows that $G_0 = \frac{\Delta n_0}{\tau_{eff,0}} = \frac{\Delta n_0^3}{a}$ and $\frac{d\tau_{eff}}{d\Delta n}\Big|_{\Delta n_0} = -\frac{2a}{\Delta n_0^3}$. Insertion of these expressions into Eq. 6.41 yields $\tau_{eff,0} = 3\tau_m$.

6.2.3.4 Resulting Novel Pathways in Lifetime Metrology

Equation 6.38 explicitly reveals the differential nature of the decay time τ_m as a *change* of excess carrier density per *change* of generation rate. The perspective of Eq. 6.38 provides a descriptive insight into the origin of the discrepancy between differential and actual lifetime, which is also illustrated in Fig. 6.7. With steady-state excess carrier density $\Delta n = \tau_{eff} G_0$ and an injection-dependent lifetime $\frac{d\tau_{eff}}{d\Delta n} \neq 0$, the relative change of excess carrier density cannot equal the relative change of generation rate. For $\frac{d\tau_{eff}}{d\Delta n} > 0$, the relative change of excess carrier density must be greater than that of generation rate, whereas for $\frac{d\tau_{eff}}{d\Delta n} < 0$ the relative change of excess carrier density must be less than that of generation rate.

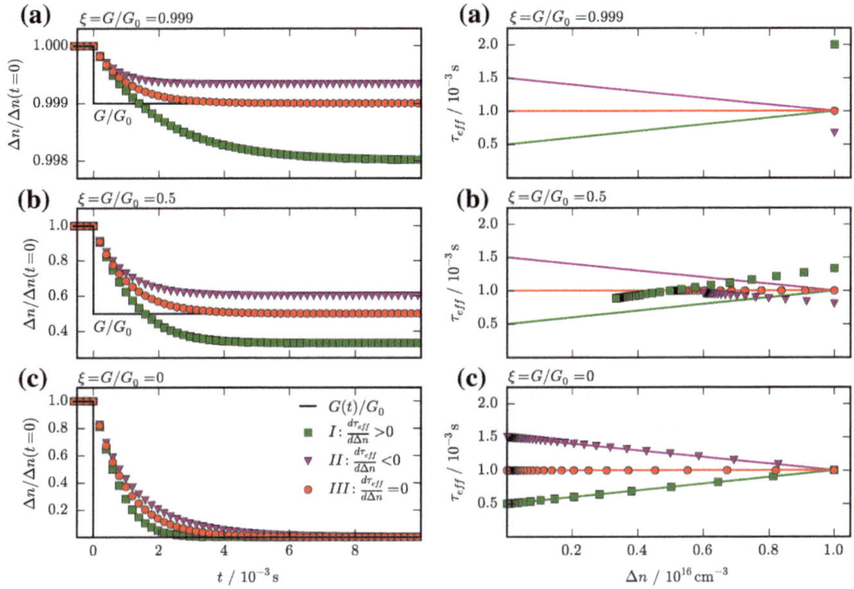

Fig. 6.7 On the *left*, the relative course of excess carrier density in time t after an abrupt change of generation rate at $t = 0$ is depicted for three lifetime scenarios plotted as *solid lines* on the *right*, respectively. $\Delta n\,(t)$ was modeled according to the finite-element approach outlined in Appendix A.3.1. The left plots reveal the difference between the relative changes of excess carrier density and generation rate for $\frac{d\tau_{eff}}{d\Delta n} \neq 0$ and $\xi > 0$. The symbols on the right represent decay time $\tau_m\,(\Delta n\,(t))$ according to Eq. 6.47. Here, the mismatch between initial decay time and actual lifetime is found to be identical to the mismatch between the relative changes of excess carrier density and generation rate. This substantiates the differential nature of a decay time measurement, and confirms every aspect of the essential Eq. 6.41 of the theory of light-biased decay time. Adapted after [6]

According to Eq. 6.38, the measurable decay time in the limit $\xi \rightarrow 1$ becomes

$$\lim_{\xi \to 1} \tau_m = \frac{d\Delta n}{dG}. \tag{6.45}$$

Analogous to derivations by Aberle et al. [9] and Schuurmans et al. [10], injection-dependent actual lifetime is thus

$$\tau_{eff}\left(\Delta n_0 = \tau_{eff} G_0\right) = \lim_{\xi \to 1} \frac{1}{G_0} \int_0^{G_0} dG\, \tau_m\,(G)\,. \tag{6.46}$$

Historically, it was concluded from Eq. 6.46 that measurement and integration of the decay time $\tau_m\,(G)$ over all generation rates $G < G_0$ would be inevitable for a correct determination of actual recombination properties from differential ones. In fact, such measurements are time-consuming and additionally complicated by limitations at $G \rightarrow 0$ due to any instruments' sensitivity limit [9]. In contrast, Eq. 6.41 reveals a pathway to correctly determine actual lifetime from a combination of at least two measurements $\tau_m(G_i)$ to be conducted at adjacent excess carrier generation rates G_i. In Sect. 8.1.3, a concept to obtain actual injection-dependent effective lifetime $\tau_{eff}(\Delta n)$ from phase-sensitive harmonically modulated luminescence measurements is detailed—making use of the theory of light-biased decay time. In this implementation, referred to as *d2a* (*differential-to-actual*), the extensive numerics exploited by previous phase-sensitive approaches—to be detailed in Sect. 8.1.1 [20] and in Sect. 8.1.2 [22]—are replaced by the straightforward analytical relation of Eq. 6.41.

Beyond the implementation of the *d2a* technique for harmonically modulated luminescence measurements [7], this technique likewise provides a pathway for powerful phase-sensitive continuously modulated lock-in lifetime approaches [17–19, 24, 25] to be able to access actual lifetime τ_{eff} rather than differential lifetime $\tau_m = u^{-1}$ in the future.

6.2.4 Numerical Validation

In order to quantitiatively validate the conclusions of Eq. 6.41, the solution of the time-dependent continuity equation (Eq. 3.75) for an injection-dependent lifetime $\tau(\Delta n)$ was modeled via a finite-element approach, as outlined in Appendix A.3.1. The main advantage of the finite element representation lies in its capability to compute an accurate solution $\Delta n(t)$ for virtually any $G(t)$ and $\tau_{eff}(\Delta n)$, if the discretization conditions specified in Appendix A.3.1 are met.

The models presented here in order to validate the theory of light-biased decay time involve both,

(A) an abrupt (Heaviside-like) change of excess carrier generation rate, and
(B) a harmonical modulation of excess carrier generation rate.

In both cases, the initial/maximal excess carrier generation rate was set to $G_0 = 10^{19}\,\text{cm}^{-3}$. Further, three distinct (and linear) lifetime scenarios were systematically investigated, comprising

(I) a positive slope of $\tau_{eff}(\Delta n)$ distinguished by $G_0 \frac{d\tau_{eff}}{d\Delta n} = \frac{1}{2}$,
(II) a negative slope of $\tau_{eff}(\Delta n)$ distinguished by $G_0 \frac{d\tau_{eff}}{d\Delta n} = -\frac{1}{2}$, and
(III) an injection-independent lifetime ($\frac{d\tau_{eff}}{d\Delta n} = 0$).

The three lifetime scenarios were chosen such that both, effective lifetime $\tau_{eff,0} = 10^{-3}\,\text{s}$ as well as quasi-steady-state excess carrier density $\tau_{eff,0}G_0 = 10^{16}\,\text{cm}^{-3}$ would coincide at the generation rate G_0 for any scenario. Aiming at a straightforward quantitative validation of Eq. 6.41, lifetime curves $\tau_{eff}(\Delta n)$ were chosen to be linear functions. A separate validation for realistic nonlinear lifetime curves is shown in Fig. 6.10.

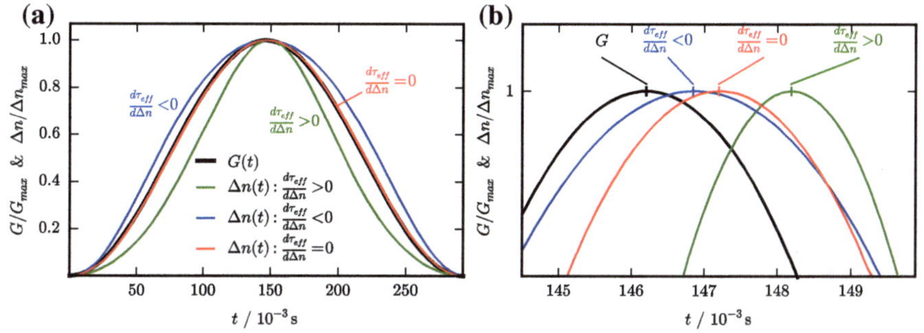

Fig. 6.8 On the *left* (**a**), relative curve shapes of harmonically modulated excess carrier generation rate G and excess carrier density Δn (modeled via a finite-element approach (cf. Appendix A.3) according to the lifetime scenarios shown in Fig. 6.7) are depicted. On the right (**b**), a detailed plot of the peak area of a modulation cycle is shown, clearly illustrating how time shifts between maxima of $\Delta n\ (t)$ and $G\ (t)$ deviate in spite of identical peak lifetimes $\tau_{eff,0} = 10^{-3}$ s. With time shifts coinciding with decay time τ_m in the limit $\xi \rightarrow 1$, the validity of the essential Eq. 6.41 is hereby confirmed. Adapted after [6]

6.2.4.1 Abrupt Change of Generation Rate

For an abruptly changing excess carrier generation rate, $\Delta n\ (t)$ was modeled for three different values of the bias light parameter ξ, involving *(a)* $\xi = 0.999$ as a value close to the limit $\xi \rightarrow 1$, *(b)* $\xi = 0.5$ as an intermediate value, and *(c)* $\xi = 0$ for the case without bias light. Injection-dependent decay time was in any case modeled as

$$\tau_m\ (\Delta n\ (t)) = -\frac{\Delta n\ (t) - \Delta n\ (t \rightarrow \infty)}{\frac{d\Delta n}{dt}}. \tag{6.47}$$

Figure 6.7 depicts excess carrier density as a function of time for all lifetime scenarios and for all three values of ξ. In the case of injection-dependent lifetime and $\xi > 0$, the relative change of excess carrier density differs from the relative change of generation rate. This fact is quantitatively illustrated in Fig. 6.7, revealing the origin of the discrepancy between actual and differential lifetime.

In the limit $\xi \rightarrow 1$, a value of $G_0 \frac{d\tau_{eff}}{d\Delta n} = \pm\frac{1}{2}$ leads to a factor between relative changes of excess carrier density and generation rate of $2(+)$ and $\frac{2}{3}(-)$. In the intermediate case of $\xi = 0.5$, these factors are $\frac{4}{3}(+)$ and $\frac{4}{5}(-)$, and for $\xi = 0$, they are both unity. Note the quantitative analogy to the mismatch between decay time and actual lifetime, which is also shown in Fig. 6.7. Every aspect of Eq. 6.41 is quantitatively substantiated by the simulation results shown in Fig. 6.7.

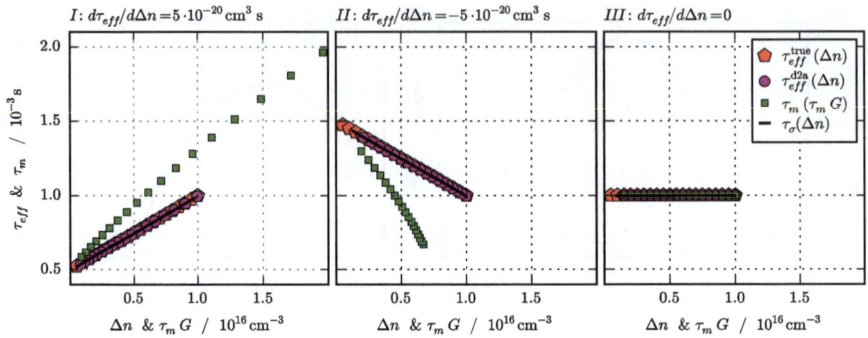

Fig. 6.9 Injection-dependent time shifts $\tau_m\ (\tau_m G)$ between harmonically modulated generation rate and excess carrier density were compared to true injection-dependent lifetime for the three linear scenarios of $\tau_{eff}\ (\Delta n)$ discussed here. All results of τ_m clearly confirm Eq. 6.41 in the limit $\xi \rightarrow 1$. Additionally, two injection-dependent lifetime analyses are performed here: a self-consistent lifetime analysis $\tau_\sigma\ (\Delta n)$ (cf. Sect. 5.2), and a *differential-to-actual* lifetime analysis $\tau_{\mathrm{d2a}}\ (\Delta n)$ (cf. Sect. 8.1.3) based on Eq. 6.41. Both analyses accurately yield true injection-dependent lifetime in any of the investigated lifetime scenarios. Therefore, integration over the entire injection range is not necessary in order to correctly determine injection-dependent lifetime $\tau_{eff}\ (\Delta n)$ from measurements of decay time $\tau_m\ (\tau_m G)$. Adapted after [6]

6.2.4.2 Harmonic Modulation of Generation Rate

Figure 6.8 depicts both, a sinusoidal modulation cycle of optical irradiation of a silicon substrate as well as the modeled quasi-steady-state excess carrier density of the substrate for the aforementioned three linear lifetime scenarios, which are also depicted in Fig. 6.7. From the plot of a total modulation cycle (*a*), it can be seen that the injection dependence has a crucial impact on the relative curve shapes of excess carrier density $\Delta n\ (t)$. A positive derivative of lifetime with respect to excess carrier density enhances the absolute value of the peak curvature of $\Delta n\ (t)$, whereas a negative derivative lowers the absolute value of the peak curvature. These curvatures already bear information about the mismatch between differential lifetime and actual lifetime, which is the ratio between relative changes of excess carrier density and generation rate (cf. Eq. 6.38). In the detailed plot (*b*) of curve peaks on the right of Fig. 6.8, time shifts between maxima of excess carrier density and generation rate are highlighted. Notably, these time shifts exactly coincide with predictions of decay time in the limit $\xi \rightarrow 1$ according to Eq. 6.41. Although all three lifetime scenarios feature the same peak lifetime of $\tau_{eff,0}\left(\Delta n_0 = 10^{16}\,\mathrm{cm}^{-3}\right) = 10^{-3}$ s, the time shifts deviate from actual lifetime just as decay times τ_m according to Eq. 6.41.

In Fig. 6.9, time shifts between generation rate and modelled excess carrier density were determined not only at maximal signals but also at lower generation rates.[13] Comparison of such injection-dependent time shifts to actual (true) injection-dependent lifetime again confirms Eq. 6.41: with a constant value of $\frac{d\tau_{eff}}{d\Delta n}$ the ratio

[13] The exact algorithm of the determination of injection-dependent time shifts from harmonically modulated measurements is detailed in Sect. 8.1.2 and in [22].

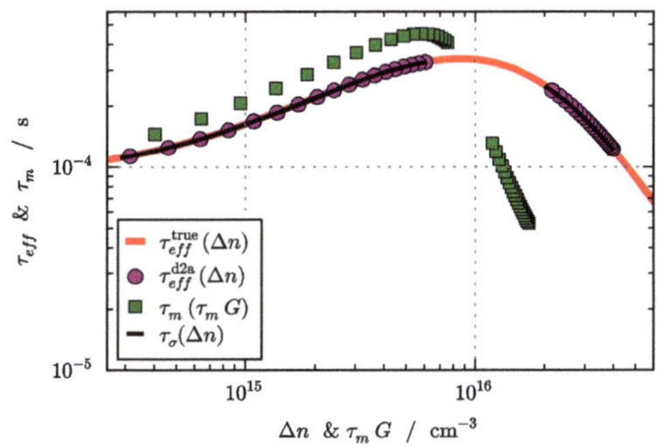

Fig. 6.10 Demonstration of the applicability of both the self-consistent and the *d2a* (*differential-to-actual*) lifetime analyses to a nonlinear lifetime curve $\tau_{eff}(\Delta n)$: Self-consistent (τ_σ) and *d2a* (τ_{d2a}) lifetime analyses were performed on simulated dynamic photoluminescence measurements on the basis of the true carrier lifetime curve shown in the plot. Two measurements were simulated at injection conditions such that they covered the *Shockley-Read-Hall* ascent and the *Auger* descent of carrier lifetime, respectively. It can be seen that time shifts $\tau_m(\tau_m G)$ substantially differ from actual carrier lifetime. Despite the nonlinearity of $\tau_{eff}^{true}(\Delta n)$, both lifetime analyses accurately yield true injection-dependent effective carrier lifetime. Adapted after [6]

between decay time τ_m (time shift) and actual lifetime τ_{eff} approaches unity when approaching very small generation rates $G \rightarrow 0$. Also, two dynamic lifetime analyses of the modeled data shown in Fig. 6.8 were performed: a self-consistent solution of the continuity equation ([26], detailed in Sect. 5.2), and a phase-sensitive *d2a* lifetime analysis based on Eq. 6.41 (detailed in Sect. 8.1.3). Both analyses yield actual lifetimes for any of the modeled scenarios, demonstrating both, the general validity of the self-consistent approach for injection-dependent lifetime measurements, and the possibility to correctly determine actual lifetime from differential lifetime measurements via the *d2a* technique—notably **without integration** over the entire injection range.

The simulation results presented so far were deliberately based on linear lifetime curves $\tau_{eff}(\Delta n)$ in order to instructively point out their quantitative agreement with Eq. 6.41. Yet, the determination of actual injection-dependent carrier lifetime from decay time measurements via Eq. 6.41 is by no means restricted to linear lifetime curves, rather it is generally valid (cf. Appendix A.7). This is demonstrated in Fig. 6.10, where the same analyses as of Fig. 6.9 are applied to modeled dynamic photoluminescence measurements in the injection ranges of both the *Shockley-Read-Hall* ascent and the *Auger* descent of a typical injection-dependent lifetime curve. Despite the fact that this lifetime curve features a pronounced curvature in the targeted injection ranges, both the self-consistent approach as well as the *d2a* analysis yield actual effective carrier lifetime.

References

1. R.A. Sinton, T. Trupke, Limitations on dynamic excess carrier lifetime calibration methods. Prog. Photovoltaics: Res. Appl. **20**, 246–249 (2012)
2. J.A. Giesecke, W. Warta, Understanding carrier lifetime measurements at nonuniform recombination. Appl. Phys. Lett. **104**, 082103 (2014)
3. J.A. Giesecke, R.A. Sinton, M.C. Schubert, S. Riepe, W. Warta, Determination of bulk lifetime and surface recombination velocity of silicon ingots from dynamic photoluminescence. IEEE J. Photovoltaics **3**, 1311–1318 (2013)
4. A.B. Sproul, M.A. Green, A.W. Stephens, Accurate determination of minority carrier-and lattice scattering-mobility in silicon from photoconductance decay. J. Appl. Phys. **72**, 4161–4171 (1992)
5. J.A. Giesecke, F. Schindler, M. Bühler, M.C. Schubert, W. Warta, Accurate determination of minority carrier mobility in silicon from quasi-steady-state photoluminescence. J. Appl. Phys. **113**, 213705 (2013)
6. J.A. Giesecke, S.W. Glunz, W. Warta, Understanding and resolving the discrepancy between differential and actual minority carrier lifetime. J. Appl. Phys. **113**, 073706 (2013)
7. J.A. Giesecke, S.W. Glunz, W. Warta, Determination of actual carrier lifetime from differential measurements. Energy Procedia **38**, 59–65 (2013)
8. R. Brendel, Note on the interpretation of injection-level-dependent surface recombination velocities. Appl. Phys. A **60**, 523–524 (1995)
9. A.G. Aberle, J. Schmidt, R. Brendel, On the data analysis of light-biased photoconductance decay measurements. J. Appl. Phys. **79**, 1491–1496 (1996)
10. F.M. Schuurmans, A. Schönecker, A.R. Burgers, W.C. Sinke, Simplified evaluation method for light-biased effective lifetime measurements. Appl. Phys. Lett. **71**, 1795–1797 (1997)
11. J. Schmidt, Measurement of differential and actual recombination parameters on crystalline silicon wafers. IEEE Trans. Electron Dev. **46**, 2018–2025 (1999)
12. H.W. Curtis, R.L. Verkuil, A high signal-to-noise oscillator for contactless measurement of photoinduced carrier lifetimes, in *Lifetime Factors in Silicon: A Symposium*, ed. by R.D. Westbrook (Astm International, Philadelphia, 1980), pp. 210–224
13. D.E. Kane, R.M. Swanson, Measurement of the emitter saturation current by a contactless photoconductivity decay method, in *Proceedings of the 18th IEEE PVSC* (Las Vegas, 1988), pp. 578–583 (1985)
14. D. Klein, F. Wuensch, M. Kunst, The determination of charge-carrier lifetime in silicon. Phys. Stat. Sol B **245**, 1865–1876 (2008)
15. R.A. Sinton, A. Cuevas, Contactless determination of current-voltage characteristics and minority-carrier lifetimes in semiconductors from quasi-steady-state photoconductance data. Appl. Phys. Lett. **69**, 2510–2512 (1996)
16. H. Nagel, C. Berge, A.G. Aberle, Generalized analysis of quasi-steady-state and quasi-transient measurements of carrier lifetimes in semiconductors. J. Appl. Phys. **86**, 6218–6221 (1999)
17. F. Sanii, F.P. Giles, R.J. Schwartz, J.L. Gray, Contactless nondestrcutive measurement of bulk and surface recombination using frequency-modulated free carrier absorption. Solid-State Electron. **35**, 311–317 (1992)
18. S.W. Glunz, W. Warta, High-resolution lifetime mapping using modulated free-carrier absorption. J. Appl. Phys. **77**, 3243–3247 (1994)
19. R. Brüggemann, S. Reynolds, Modulated photoluminescence studies for lifetime determination in amorphous-silicon passivated crystalline-silicon wafers. J. Non-Cryst. Solids **352**, 1888–1891 (2006)
20. J.A. Giesecke, M.C. Schubert, D. Walter, W. Warta, Minority carrier lifetime in silicon wafers from quasi-steady-state photoluminescence. Appl. Phys. Lett. **97**, 092109 (2010)
21. J.A. Giesecke, D. Walter, F. Kopp, P. Rosenits, M. C. Schubert, W. Warta, Simultaneous determination of carrier lifetime and net dopant concentration of silicon wafers from photoluminescence, in *Proceedings of the 35th IEEE PVSC* (Honolulu, 2010), pp. 847–851

22. J.A. Giesecke, M.C. Schubert, W. Warta, Self-sufficient minority carrier lifetime in silicon from quasi-steady-state photoluminescence. Phys. Stat. Sol. A **209**, 2286–2290 (2012)
23. R.A. Sinton, A. Cuevas, M. Stuckings, Quasi-steady-state photoconductance, a new method for solar cell material and device characterization, in *Proceedings of the 25th IEEE PVSC* (Washington, D.C., 1996), pp. 457–460
24. A. Mandelis, J. Batista, D. Shaughnessy, Infrared photocarrier radiometry of semiconductors: physical principles, quantitative depth profilometry, and scanning imaging of deep subsurface electronic defects. Phys. Rev. B **67**, 205208 (2003)
25. Q. Sun, A. Melnikov, A. Mandelis, Quantitative self-calibrating lock-in carrierographic lifetime imaging of silicon wafers. Appl. Phys. Lett. **101**, 242107 (2012)
26. T. Trupke, R.A. Bardos, M.D. Abbott, Self-consistent calibration of photoluminescence and photoconductance lifetime measurements. Appl. Phys. Lett. **87**, 184102 (2005)

Chapter 7
Evolution of the Experimental Setup

This chapter gives an account of the evolution of the harmonically modulated luminescence experimental setup in the course of this work. An account of the state-of-the-art experimental setup is followed by the description of key innovations of the experimental setup.

7.1 Harmonically Modulated Luminescence Setup

In the following, relevant information on the harmonically modulated photoluminescence setup is provided.[1] A schematic illustration of the present harmonically modulated photoluminescence setup at Fraunhofer ISE is depicted in Fig. 7.1.

7.1.1 Data Acquisition and Software

The lifetime measurement & analysis software *QSSL Lifetime*[2] initiated in the course of this work is connected to the modulated luminescence experimental setup via data aquisition devices (two-way analog-digital converters) allowing for both, a real time control of the output power of a monochromatic light source through a modulated output voltage, and a simultaneous time-dependent acquisition of signals proportional to both photoluminescence intensity and irradiance.

7.1.1.1 The Software *QSSL Lifetime*

Here, the essential features of *QSSL Lifetime* shall briefly be summarized.

---

[1] Modulated electroluminescence is covered in Sect. 7.4.

[2] *QSSL Lifetime* is written in the *Python* programming language.

J. Giesecke, *Quantitative Recombination and Transport Properties in Silicon from Dynamic Luminescence*, Springer Theses, DOI: 10.1007/978-3-319-06157-3_7,
© Springer International Publishing Switzerland 2014

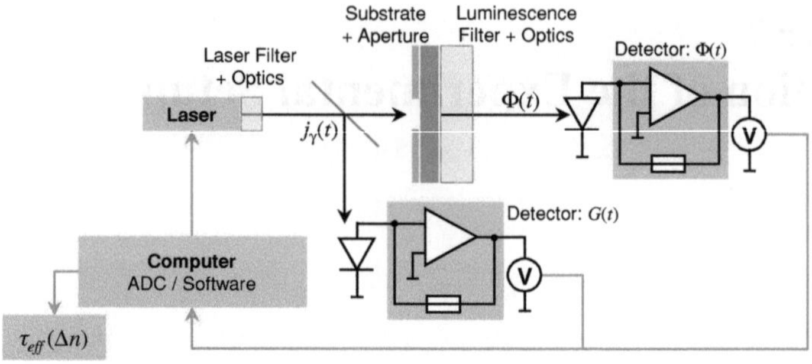

Fig. 7.1 Scheme of the modulated photoluminescence setup used in this work. A software-controlled harmonic waveform defines a sinusoidal optical power output of a monochromatic light source (laser). The laser light is filtered and shaped such that a homogeneous lateral irradiation profile impinges on the substrate area of interest, which may optionally be covered by an aperture. The radiative recombination $\Phi(t)$ of the substrate is measured via a photodetector connected to a current amplifier. Another photodetector (+ current amplifier) measures a signal proportional to the photon current impinging on the substrate, which can be scaled to a depth-averaged excess carrier generation rate $G(t)$. Both signals (voltages) are simultaneously digitized via an ADC unit (analog-digital conversion), before they are synchronized and analyzed in terms of injection-dependent effective lifetime $\tau_{eff}\,(\Delta n)$. The luminescence detector (+ filters and optics) may likewise be placed on the irradiated substrate side, e.g. if the substrate features an opaque interface such as a full area metalization

Data Structure: The dynamic lifetime analysis of an array of raw data containing a time vector t_i as well as both, luminescence intensity $\Phi(t_i)$ and (relative) generation rate $G(t_i)$ represents the core of *QSSL Lifetime*.[3] The analysis proceeds either self-consistently or according to the phase-sensitive *d2a* method, as described in Sect. 8.1.3. Raw data, relevant parameters, and the resulting injection-dependent lifetime $\tau_{eff}\,(\Delta n)$ are automatically logged for each measurement.

Parameter Management: Relevant parameters can be defined and manipulated via the respective instances of a graphical user interface (GUI). Alternatively, parameter sets can be loaded from a previously conducted measurement or from a predefined set of default parameter files.

Sets of relevant parameters may vary depending on the configuration settings of the program, which are characterized by four (currently active) essential configuration properties:

- Mode of operation: measurement & analysis (online), or analysis only (offline)
- Type of analysis: self-consistent or *d2a*
- Spatially resolved calibration of photoluminescence images: yes or no
- Gain configuration settings of current amplifiers: manual or automatic

Graphical User Interface: The GUI of *QSSL Lifetime* consists of one main frame for the core tasks of the program and miscellaneous subframes for specific applications. The main control panel essentially features directory management, naming, loading,

[3] Signal amplitudes are recorded as *net* photocurrents, i.e. an individually measured dark photocurrent is subtracted from net amplitudes $\Phi(t_i)$ and $G(t_i)$, respectively.

and execution capabilities. Further, vital control parameters such as the modulation time $\mathcal{T}$, the number of modulation cycles to be averaged, the number of samples per modulation cycle, the irradiance, and (optionally) the gain settings of the current amplifiers can be controlled from within the main control panel. All other relevant parameters are arranged in groups of specimen parameters, laser control parameters, and technical parameters for the numerical analyses.

The graphical output of the GUI comprises gain settings of current amplifiers, maximal and dark signal levels (voltages) of both channels $\Phi(t)$ and $G(t)$, measured relative signals, a plot to monitor the proper execution of the numerical data analysis, and (optionally) a spatially resolved effective lifetime map (cf. Sect. 8.3).

Tools & Applications: The most frequently used add-on applications of *QSSL Lifetime* are

- an effective lifetime calibration of luminescence images via harmonically modulated luminescence measurements (detailed in Sect. 8.3),
- broad range injection-dependent lifetime measurements via steady-state luminescence (detailed in Sect. 8.2),
- a class of harmonically modulated luminescence based methods to determine net dopant concentration n_0 (detailed, in Sect. 9.1),
- and a tool for an appropriate synchronization of the two channels, i.e. a calibration of their response time (cf. Sect. 7.2.1).

7.1.2 Light Sources

Two single-mode and continuous wave high power semiconductor laser light sources operating at different wavelengths (791 and 880 nm) were used in this work. They were both connected to a current source controlled by the software-modulated laser control voltage. Photon current densities up to an equivalent of 50 suns (cf. Appendix A.6) on a specimen area of 4 cm^2 are achieved. Modulation up to frequencies of $\mathcal{T}^{-1} < 300\,s^{-1}$ is possible. For quasi-steady-state operation (as favored throughout this work), the latter range is sufficient regardless of the specimen effective lifetime. However, the analysis of effective lifetimes in the range of some 10^{-6} s in the intermediate to quasi-transient dynamic regime would require a faster modulation capability.

7.1.3 Filters—Optics—Apertures

The major experimental challenge of a photoluminescence setup is to ensure a negligible non-luminescent fraction of light in the luminescence detector. The art of engineering optimal optical filter conditions for silicon luminescence investigations [1] exceeds the scope of this work. Rather than attempting to capture as much luminescence as possible at a sufficient attenuation of non-luminescent light, this work solely focused on securing sufficient attenuation. Therefore, the maximization of the capture of luminescence both spectrally and in terms of the solid angle of detection remains to be accomplished.

Before impinging upon the silicon substrate under investigation, laser light is shaped via homogenization optics and diverging lenses in order to ensure lateral homogeneity of irradiance on the substrate area of interest. As the laser light was found to feature long wavelength contributions indistinguishable from the luminescence of the substrate under investigation (cf. Sect. 7.2.2), optical short pass filters ensuring an attenuation of such radiation by a minimum of five orders of magnitude in the wavelength range between 930 − 1,200 nm were mounted.

The optical path between the substrate under investigation (which itself serves as an optical long pass filter of laser light) and the luminescence detector comprises a dielectric long pass filter (with a specified optical density >6 for wavelengths less than 980 nm and beyond the spectrum of significant photodetector sensitivity), another dielectric short pass filter intended to block sub band gap optical transitions of the substrate at an optical density >4 in the wavelength range of 1,185 − 1,700 nm, an unpassivated 200 μm thick polished monocrystalline silicon wafer, an array of lenses focusing the luminescent radiation on the detector active area, and another dielectric long pass filter of identical optical properties as above. One might argue on the basis of ab initio estimations of luminescence intensity [2], that an optical density >12 should suffice in order to ensure negligible non-luminescent light for 200 μm thick 1 Ω cm substrates featuring an effective lifetime of $\tau_{eff} = 10^{-6}$ s. However, the breakdown of dielectric light attenuation at non-perpendicular incidence rendered an additional *absorption* filter with isotropic light attenuation necessary.

Three apertures for defined illuminated or radiating substrate areas are available in the current setup. A circular aperture featuring a diameter of 40 mm as well as two square apertures with edge lengths of 20 and 10 mm, respecctively. Apertures can be mounted between the light source and the substrate in order to define an illuminated area, or between the substrate and the luminescence detector in order to define a radiating area. The relative variation of local irradiance (impinging photon current density) within the largest aperture (circular 40 mm) was found to amount to less than ±2 %. For measurements without any aperture above or beneath the substrate, the photoluminescence detector captures radiative recombination from a circular substrate area with a diameter of 45 mm.

Apertures are of vital importance for the effective lifetime calibration of photoluminescence images via modulated luminescence, because this application requires a defined illuminated area. Also, for accurate lifetime calibrations, knowledge of the relative lateral detector sensitivity $\mathcal{S}_{\Phi}$ with respect to a particular substrate location is material. As both, irradiation and luminescence detection feature a cylindrical symmetry with respect to the optical axis of the setup, it appears reasonable to express local *radial* sensitivity $\mathcal{S}_{\Phi}(r)$ as a function of the perpendicular distance r from the optical axis. The function $\mathcal{S}_{\Phi}(r)$ was determined via measurements of the luminescence intensity of a float zone wafer covered by a pinhole with a diameter of 2 mm at defined radii r. Two cross sections (from west to east and from south to north) were performed, respectively. Between the two measurement runs, the

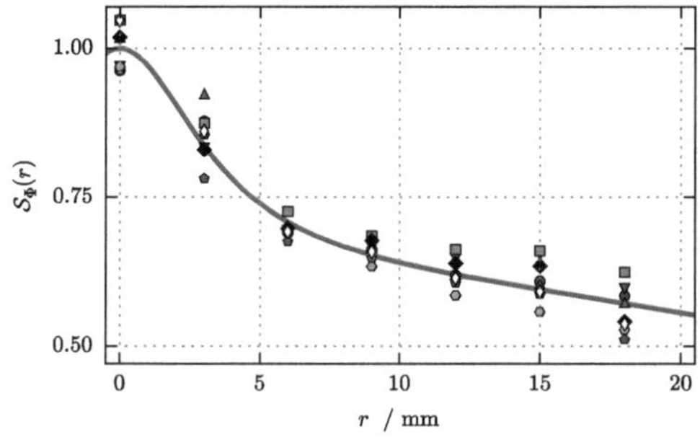

Fig. 7.2 Depiction of both, measurements and a fit of the radial sensitivity function $\mathcal{S}_\Phi(r)$ of the modulated luminescence experimental setup in its default measurement configuration. The measured data represent branches of two successive cross sections (from west to east and from south to north). The positioning uncertainty amounts to $\delta r = \pm 1$ mm. The measured data were normalized to an average maximum of unity. Details on the fit are to be found in the text

detection optics were unmounted in order to obtain data on the precision of the measured curve $\mathcal{S}_\Phi(r)$ in the presence of a respective scattering impact. The resulting eight curves $\mathcal{S}_\Phi(r)$ and a fit are shown in Fig. 7.2.

For an idealized detection geometry with a point-like luminescence detector on the optical axis of the setup and with the silicon substrate active area concentrically positioned on the optical axis at a vertical distance z from the detector, isotropic radiative recombination would suggest a radial sensitivity of the form

$$\mathcal{S}_\Phi(r) = \frac{z^2}{z^2 + r^2}. \tag{7.1}$$

While the real detector (circular InGaAs substrate with a diameter of 5 mm) does not substantially depart from the above idealized assumption, the involved optics bring about substantial deviations from this idealized relation, as can be seen from the experimental data plotted in Fig. 7.2. Therefore, the fit shown in Fig. 7.2 represents a weighted superposition of two functions of the above type

$$\mathcal{S}_\Phi(r) = \sum_{i=1,2} c_i \frac{z_i^2}{z_i^2 + r^2}, \tag{7.2}$$

with weights c_i meeting the condition $\sum c_i = 1$ and *virtual* distances z_i.[4] The fit shown in Fig. 7.2 features the parameter set $c_1 = 0.62$, $z_1 = 54$ mm, and $z_2 = 3.5$ mm, as implemented in *QSSL Lifetime*.

---

[4] The physical distance between substrate and luminescence detector in the default measurement configuration amounts to $z = 95$ mm.

7.1.4 Detectors and Amplifiers

The default light sensors of the modulated luminescence experimental setup used at Fraunhofer ISE are InGaAs photodiodes featuring a non-negligible spectral response in the wavelength range between 770−1,700 nm. They may be Peltier-cooled, but commonly they are operated at room temperature.[5] The luminescence detector features a circular active area with a diameter of 5 mm. It is concentrically positioned on the optical axis of the experimental setup (defined by the direction of the irradiating Poynting flux). The irradiance detector features a circular active area with a diameter of 1 mm. Other than indicated in the schematic of Fig. 7.1, it currently captures light directly deflected from the laser output. At present, the conversion of measured irradiance (in relative units) into a total excess carrier generation rate is accomplished via a measurement of the impinging photon current density at the location of the substrate under investigation.[6]

A previous configuration of the harmonically modulated luminescence experimental setup used silicon photodiodes, which feature a significant inherent limitation due to the different characteristic minority carrier diffusion times of irradiation light and silicon luminescent light in the detector bulk, to be detailed in Sect. 7.2.1. However, either if appropriate corrective measures are taken into account, or for steady-state applications (as e.g. detailed in Sect. 8.2), these diodes are available in the current experimental setup as well.

Measured photocurrents are amplified via variable gain low noise current amplifiers featuring gain settings in the range of $10^3 - 10^9\,\mathrm{V\,A^{-1}}$. Gain settings are software-controlled via a Universal Serial Bus (USB) control interface, thereby allowing for an automatic determination of optimal gain settings depending on the specific requirements of each measurement. Signal transfer is accomplished exclusively via bayonet coaxial cables.

7.2 Microsecond Lifetimes

The smaller carrier lifetimes of silicon solar cells are, the greater is the relative error of predictions of solar cell voltages or currents from erroneous lifetime measurements [3–5]. In spite of the very advantageous properties of harmonically modulated luminescence lifetime techniques, a broad deployment in photovoltaic research had previously been hampered. This is likely to be due to a relatively poor sensitivity limit in terms of carrier lifetime, if inappropriate detector hardware is used and if no signal

[5] The room temperature (298 K) thermal noise of the photodiodes was found not to be the dominant source of noise for modulated luminescence experiments.

[6] The irradiance detector is planned to capture light directly branching off the fiber to connect the light source to its shaping optics in a final setup configuration. That way, the captured light intensity is independent of backscattering of light from the substrate, leading to a substantial simplification of the calibration of the substrate's excess carrier generation rate $G(t)$.

synchronization measures are taken. A systematic analysis of the sensitivity limiting mechanisms led to a substantial upgrade of sensitivity with regard to accessible lifetimes in the course of this work—tackling the range of and beyond microsecond effective lifetimes. This section addresses the key cautions to be taken into account for harmonically modulated lifetime measurements via current amplifiers.[7]

7.2.1 Signal Synchronization

The amplification of a dynamic (e.g. harmonically modulated) photocurrent with an operational amplifier is known to induce a phase lag. Such phase lags would not affect harmonically modulated lifetime analyses, if their difference between both channels $G(t)$ and $\Phi(t)$ was negligible compared to harmonically modulated effective lifetime $\tau_{eff,h}$, and if they amounted to time shifts very much less than the modulation time $\mathcal{T}$ (in order not to substantially deform the course of signals). However, the first of these two conditions cannot be regarded as a priori satisfied under typical measurement conditions.

Previous modulated quasi-steady-state photoluminescence lifetime analyses appeared to feature an approximate lower bound of effective lifetime in the range of some 10^{-5} s, even for substrates with lifetimes known to be substantially shorter than 10^{-6} s. Also, this lower bound substantially varied depending on the type and arrangement of photodiodes and current amplifiers, as well as on amplification settings. This observation gave rise to a systematic experimental investigation, to be detailed in the following.

7.2.1.1 An Experiment on Signal Synchronization

The following experiment aimed at the identification of causes to a non-negligible difference between phase artifacts imposed on measured signals $G(t)$ and $\Phi(t)$ upon detection and amplification. A current amplifier comprises a resistor and an operational amplifier. A variable amplification factor is typically realized through variation of the resistance. The amplification-induced phase lag of a photocurrent depends on the product of photodiode capacitance and the current amplifier's resistance, but also on the response time of the operational amplifier. In order to ensure negligible deviations between phase artifacts imposed on measured signals $G(t)$ and $\Phi(t)$ upon detection and amplification, both the product of photodiode capacitances and resistances as well as the response times of operational amplifiers may only differ much less than harmonically modulated effective lifetime of an investigated substrate.

---

[7] The account given in this section is adapted from the corresponding original publication [5].

Table 7.1 Comparison of time shifts Δt between measured signals $G(t)$ and $\Phi(t)$, as encountered in different configurations of two (identical) silicon photodiodes (I & II) and two current amplifiers (A and B) operated at identical gain settings, which are exchanged, respectively, in the four possible detection configurations of interest

Configuration	Φ-Configuration	G-Configuration	$\Delta t/10^{-6}$ s
(i)	$\Phi \rightarrow I \rightarrow A$	$G \rightarrow II \rightarrow B$	46.0 ± 0.9
(ii)	$\Phi \rightarrow I \rightarrow B$	$G \rightarrow II \rightarrow A$	10.5 ± 1.0
(iii)	$\Phi \rightarrow II \rightarrow A$	$G \rightarrow I \rightarrow B$	47.5 ± 1.5
(iv)	$\Phi \rightarrow II \rightarrow B$	$G \rightarrow I \rightarrow A$	10.4 ± 1.1

In the experiment presented here, two current amplifiers and two (identical) silicon photodiodes[8] were exchanged, respectively, in order to compare resulting phase shifts between $G(t)$ and $\Phi(t)$ between all 2^2 possible configurations. Signal levels were adjusted via apertures in front of detectors in order to enable identical amplification settings (of $10^8\,\mathrm{V\,A^{-1}}$) for both channels. A $d = 365\,\mu\mathrm{m}$ thick p-type $1.3\,\Omega\,\mathrm{cm}$ multicrystalline test wafer was used, featuring as-sawn interfaces and a minority carrier bulk diffusion length $L \ll d$. An ab initio cautious estimate of its steady-state effective carrier lifetime from a photoluminescence image [2] yielded $\tau_{eff,s} < 1.3 \times 10^{-7}$ s at an irradiance equivalent to one sun (cf. Appendix A.6). The resulting time shifts Δt between maxima of generation rate $G(t)$ and luminescence intensity $\Phi(t)$ are shown in Table 7.1.

Notably, the time shifts encountered are all much greater than to be expected from the cautiously estimated upper bound of wafer lifetime. It is deemed extremely unlikely that this mismatch could be attributed to either a differential lifetime artifact (cf. Sect. 6.1) or a dynamic artifact (cf. Sect. 6.2). Further, time shifts feature a substantial and significant variation upon exchange of current amplifiers (comparing configurations $(i) \leftrightarrow (ii)$ and $(iii) \leftrightarrow (iv)$, respectively), whereas they prove to only vary insignificantly upon exchange of (identical) photodiodes (comparing configurations $(i) \leftrightarrow (iii)$ and $(ii) \leftrightarrow (iv)$, respectively). The average time shift over all four configurations amounts to $\Delta t = 28.6 \pm 0.5 \times 10^{-6}$ s.

7.2.1.2 Implications of the Synchronization Experiment

The first essential implication of the above experiment is that amplification-imposed phase artifacts matter on timescales relevant for lifetime investigations of silicon wafers. The (possibly variable) impact of current amplifier response times must be taken into account in an appropriate signal synchronization procedure— as detailded later in this section.

The second essential implication of the above experiment relates to the use of *silicon* photodiodes for dynamic luminescence-based lifetime investigations of *sili-*

[8] Note that the previous experimental setup—as used for the experiment described herein—featured silicon photodiodes, which were relatively thick compared to typical wafer thicknesses ($d_{PD} \gtrsim 400\,\mu\mathrm{m}$).

con wafers. In the following, this will prove to represent the cause to the net (average) time shift of $\Delta t = 28.6 \pm 0.5 \times 10^{-6}$ s (contrasting the cautious upper bound of the test wafer's effective lifetime of $\tau_{eff,s} < 1.3 \times 10^{-7}$ s) as found in the synchronization experiment above (cf. Table 7.1). Deviations between Δt and steady-state effective lifetime may be caused by both differential lifetime artifacts and dynamic lifetime artifacts, which were detailed in the previous chapter. Yet, none such artifacts could amount to the observed mismatch. Hence, another mechanism has to be considered: It is the dependence of photodiode response time on the spectrum (and consequently of the generation depth profile) of the detected light.

Detection of silicon luminescence via silicon photodetectors naturally brings about relatively large light penetration depths of the order and greater than typical detector thickness [6] (cf. Sect. 4.1). Consequently, depth profiles of excess carrier generation in a photoluminescence detector are rather homogeneous. In the case of charge carrier collection at the illuminated interface of a detector device, this implies that minority carriers undergo longer diffusion times for detection of silicon interband luminescence (900−1,250 nm spectrum, cf. Fig. 4.1) than for detection of laser light at a wavelength of (here) 880 nm corresponding to a room temperature light penetration depth $\alpha_\gamma^{-1} \leq 2.5 \times 10^{-3}$ cm [6]. The impact of the effect described above is modeled here via numerical solution of the continuity equation for the silicon photodiode operated under short circuit conditions. According to Eq. 3.19, and with excess carrier generation rates $G_\Phi(z, t)$ and $G_G(z, t)$ denoting detection of silicon interband luminescence (Φ) and 880 nm laser radiation (G), respectively, the continuity equation reads

$$\frac{\partial \Delta n_k (z, t)}{\partial t} - D_a \frac{\partial^2 \Delta n_k (z, t)}{\partial z^2} + \frac{\Delta n_k (z, t)}{\tau_{PD}} = G_k (z, t), \tag{7.3}$$

with an index $k = \Phi, G$ representing the spectrum of detected light, and a photodiode effective lifetime τ_{PD}.[9] Emitter and front surface recombination are expected to be without any impact due to a negligibly thin emitter and due to short circuit operation. Boundary conditions are $\Delta n_k(z, t = 0) = 0$, and $\Delta n_k(z = 0, t) = 0$. Harmonically modulated excess carrier generation rate within the photodiode can be described according to Eq. 3.55 as

$$G_k (z, t) = g_k (z) \sin^2 \left(\frac{\pi}{T} (t - t_0)\right), \tag{7.4}$$

with a depth-dependent part

$$g_k (z) = \xi_k \int_0^\infty d\varepsilon_\gamma \frac{dj_\gamma (\varepsilon_\gamma)}{d\varepsilon_\gamma} \alpha_\gamma (\varepsilon_\gamma) e^{-\alpha_\gamma(\varepsilon_\gamma)z}. \tag{7.5}$$

[9] Further assuming negligible back surface recombination velocity of the photodiode ($S_{PD,d} = 0$).

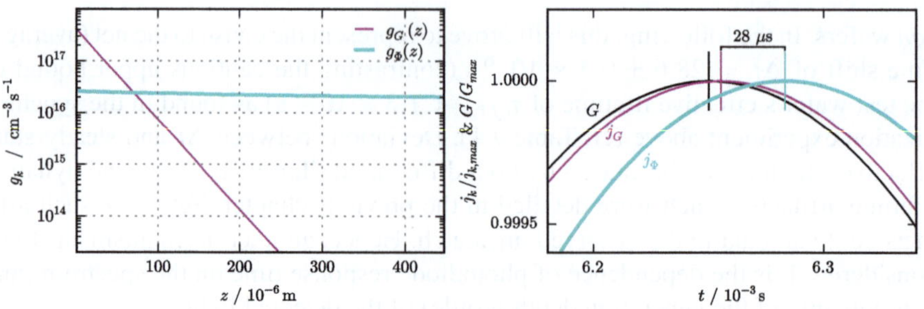

Fig. 7.3 Illustration of the difference between response times of identical silicon photodiodes at different spectra of detected light intensity. Monochromatic laser irradiation at a wavelength of 880 nm yields a very shallow room temperature depth profile of excess carrier generation $g_G(z)$, whereas interband luminescence yields the very homogeneous depth profile $g_\Phi(z)$, as depicted *on the left*. *On the right*, maxima of simulated photocurrents $j_k(t)$ according to the model described in the text (with $k = \Phi, G$) are depicted, suggesting that the average time shift $\Delta t = 28.6 \pm 0.5 \times 10^{-6}$ s encountered in the above described synchronization experiment is most likely to be caused by differences in photodiode response time due to different depth profiles $g_k(z)$. Relevant model parameters: photodiode bulk thickness $d = 450\,\mu$m, photodiode effective lifetime $\tau_{PD} = 130\,\mu$s, diffusion coefficient in photodiode bulk $D_a = 11.8\,\text{cm}^2\,\text{s}^{-1}$ (1.5 Ω cm n-type), internal back surface reflectivity $R_{PD,d} = 0.8$, and modulation time $\mathcal{T} = 1.25 \times 10^{-2}$ s. Adapted after [5]

Here, the dimensionless factor ξ_k ensures identical cumulative excess carrier generation rates for $k = \Phi, G$. The expression $\frac{dj_\gamma(\varepsilon_\gamma)}{d\varepsilon_\gamma}$ corresponds to the silicon interband luminescence spectrum for $k = \Phi$ (cf. Eq. 4.15),[10] whereas it is represented by a delta function peaking at the photon energy of the laser in the case $k = G$.

The time-dependent short circuit photocurrents

$$j_k(t) \propto \left. \frac{d\Delta n_k(z,t)}{dz} \right|_{z=0} \qquad (7.6)$$

were then calculated for both cases $k = \Phi, G$ on the basis of a numerical finite element solution of the continuity equation (Eq. 7.3, cf. Appendix A.3.2, the boundary condition $\Delta n_k(z = 0, t) = 0$ was implemented via $S_0 \to \infty$). As can be seen from the plots of $j_k(t)$ in Fig. 7.3, it is possible to explain the encountered net time shift of $\Delta t = 28.6 \pm 0.5 \times 10^{-6}$ s via the mechanism discussed herein with realistic photodiode parameters.

Advantageously, the wavelength dependence of photodiode response time as identified here is virtually independent of the effective lifetime of the silicon wafer under investigation. This fact allows a lifetime-independent correction by the experimentally determined mean time shift Δt (cf. transformation given in Eq. 7.7). Minor corrections due to differing substrate reflectivities and due to reabsorption of lumi-

---

[10] Also considering the reabsorption of luminescence and the photon energy dependence of the test wafer interface reflectivity. Other than indicated in Eq. 7.5, one reflection of luminescent radiation at the photodiode's back surface was also considered in the calculation of $g_\Phi(z)$.

nescence [7–10] could be taken into account as well. However, the complexity of the sum of the addressed corrective measures calls for alternative—faster—detector hardware. With InGaAs detectors, such hardware was implemented.

7.2.1.3 Signal Synchronization in the Present Setup

With a statistical uncertainty estimated to amount to $\pm 5 \times 10^{-7}$ s and possible further minor systematic errors of corrective measures related to luminescence reabsorption and varying substrate reflectivity, the harmonically modulated lifetime analysis via silicon photodiodes proved to be substantially less precise than a harmonically modulated lifetime analysis via InGaAs photodiodes. As direct band gap semiconductors, such devices feature much faster response times due to minority carrier diffusion than thick silicon diodes. Another advantage of InGaAs diodes relates to the greater overlap between the spectral range of significant sensitivity of these detectors and the silicon interband luminescence spectrum, thereby reducing the measurement time required in order to achieve a given signal-to-noise ratio. Also, InGaAs-based luminescence detection is more flexible with regard to optical filtering, as the spectral ranges of detection and irradiation are much better separable than for silicon detectors.

As can be concluded from the synchronization experiment above, phase lags induced by signal detection and amplification are to be expected even for InGaAs detectors (featuring recombination lifetimes in the range of some tens of nanoseconds). Yet, the essential difference is that the relatively fast response time of InGaAs detectors due to minority carrier diffusion in the detector only negligibly depends on the detected light spectrum. Therefore, the synchronization or response time calibration of InGaAs photodiodes is relatively straightforward: A harmonically modulated calibration measurement featuring a direct irradiation of the photoluminescence detector with laser light (no silicon substrate, no filters) yields two time-dependent photocurrents, which are possibly shifted with respect to each other because of differing signal delays during amplification. Ideally, the synchronization of a harmonically modulated lifetime measurement is performed through a calibration measurement at the exact same gain settings and signal voltages of current amplifiers. This calibration yields a time shift $\Delta t'$, by which the measured photoluminescence intensity Φ_m of the lifetime measurement has to be shifted such that the synchronized photoluminescence intensity Φ is[11]

$$\Phi(t) = \Phi_m \left(t + \Delta t'\right). \tag{7.7}$$

[11] The essential condition for this transformation to accurately account for phase artifacts induced by signal detection and amplification is again a quasi-steady-state condition: it is $\Delta t' \ll \mathcal{T}$, and it ensures that the total finite response times (here assumed to range within the same order of magnitude as the difference of response times $\Delta t'$) cannot substantially deform measured signals.

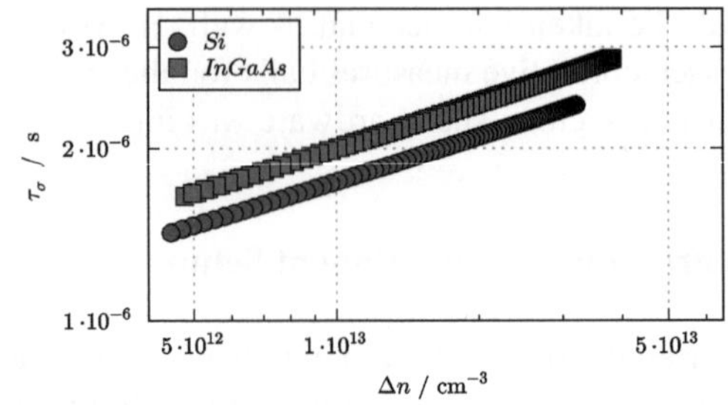

Fig. 7.4 Self-consistent analyses of lifetime $\tau_\sigma(\Delta n)$ of harmonically modulated quasi-steady-state photoluminescence measurements conducted on a $2 \times 2\,\mathrm{cm}^2$ active area of a $d = 165\,\mu\mathrm{m}$ thick upgraded metallurgical grade multicrystalline silicon wafer featuring a silicon nitride surface passivation. Different sets of detectors, involving InGaAs and silicon, were used. While the InGaAs signal synchronization is straightforward the comprehensive signal synchronization procedure for silicon detectors had to also account for finite minority carrier diffusion times in the detectors. The plotted results agree within the specified statistical measurement uncertainty of 2×10^{-7} s for the InGaAs detectors and 5×10^{-7} s for the silicon detectors. Adapted after [5]

For InGaAs detectors, the uncertainty of the time shift $\Delta t'$ was estimated to amount to $\delta\Delta t'_{\mathrm{InGaAs}} = \pm 2 \times 10^{-7}$ s,[12] which is a substantial improvement compared to the $\delta\Delta t'_{\mathrm{Si}} = \pm 5 \times 10^{-7}$ s encountered for silicon detectors. Figure 7.4 shows self-consistent lifetime interpretations $\tau_\sigma(\Delta n)$ of harmonically modulated luminescence measurements conducted both, via InGaAs photodiodes and via silicon photodiodes (with the latter featuring a correction of the photodiode response time due to minority carrier diffusion). As can be seen, lifetimes agree reasonably within the specified uncertainty range, confirming the hypothesis and the above numerical quantification of the spectral dependence of photodiode response times.

The synchronization of measured signals is implemented in the software *QSSL Lifetime* via default time shifts $\Delta t'_{ij}$ for any combination of gain settings of both current amplifiers i and j. However, there is experimental evidence that $\Delta t'_{ij}$ may not only depend on the distinct configuration of the current amplifiers' gain settings, but also on the combination of measured signal amplitudes (particularly for relatively high signals within the dynamic range of each gain setting). Therefore, the synchronization of signals was individually performed on the basis of (nearly) identical signal amplitudes throughout this work.

7.2.2 Attenuation of Nonluminescent Light

Apart from phase lags associated with finite response times in the course of signal detection and amplification, another source of systematic artifacts of pronounced

[12] Shot noise of luminescent radiation, fluctuation noise induced through detection and processing of signals, and interferences with the surrounding electronics may give rise to this uncertainty.

impact (particularly at low lifetimes and luminescence intensities) is the detection of non-luminescent light in the luminescence photodetector.

7.2.2.1 Irradiation in the Luminescence Spectral Range

Optical transitions along the light path prior to irradiation of the silicon wafer under test would be detrimental to harmonically modulated luminescence measurements, because if their energy was sufficiently low, such photons could penetrate optical long pass filters between the substrate under investigation and the luminescence detector and contribute to a photocurrent within the latter. The impact of such optical transitions would be particularly harmful, if the photocurrent induced by them was not much less than the photocurrent induced by substrate luminescence. In the worst case, the harmonically modulated lifetime analysis would yield the characteristic relaxation time of these optical transitions rather than the carrier lifetime of the wafer under test.

In order to investigate whether relevant optical transitions occured along the laser light path, two experiments were conducted. With the same test wafer as used for the above synchronization experiment, in a first step, the occurrence of optical transitions with relatively long relaxation times in the range between $10^{-4} - 10^{-2}$ s was tested via square waveforms of excess carrier generation at modulation times in the range between $10^{-2}\,\mathrm{s} \leq \mathcal{T} \leq 1\,\mathrm{s}$. With a simultaneous abrupt switch-off of both signals $G(t)$ and $\Phi(t)$, no such transitions could be identified. In a second step, the occurrence of faster optical transitions along the laser light path (with relaxation times in the range between $10^{-6} - 10^{-4}$ s) was tested by placing an optical short pass filter with an edge wavelength of 900 nm between the laser output and the wafer under test and measuring the time shift Δt between the two signals upon harmonic modulation ($\mathcal{T} = 1.25 \times 10^{-2}$ s) with and without this short pass filter. Similarly, Δt remained invariant, suggesting optical transitions along the laser light path to be negligible down to the microsecond range of transition times.

However, mounting the 900 nm short pass filter between laser and silicon substrate as described above enabled the identification of another artifact, consisting of a steady-state long wavelength background radiation of the light source above 900 nm, which almost completely dominated the photoluminescence detector's photocurrent, outrunning the test wafer's photoluminescence intensity by almost an order of magnitude, as can be seen in Fig. 7.5. This background radiation, which is unlikely to stem from optical activity along the light path, has a detrimental influence on any harmonically modulated lifetime analysis (i.e. a substantial overestimation of true lifetime). The cause to this is a higher order differential lifetime artifact, as e.g. discussed in Sect. 5.2.3. Therefore, optical short pass filters eliminating virtually any radiation from the light source above the laser wavelength (within the sensitivity range of the photoluminescence detector) were deemed inevitable, and were thus permanently installed.

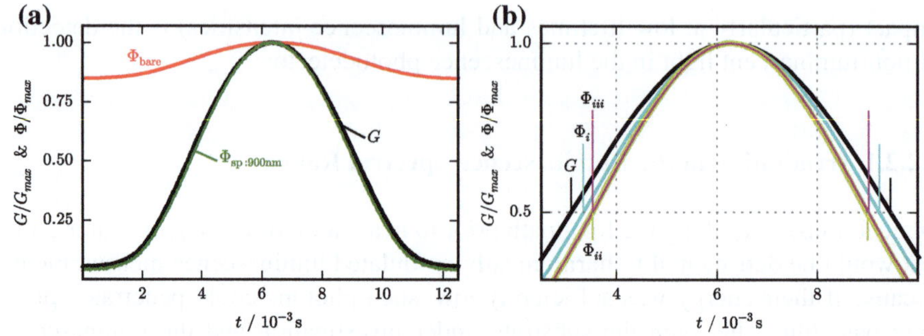

Fig. 7.5 On the attenuation of non-luminescent light: *On the left* (**a**), relative luminescence signals of a test wafer featuring a steady-state effective lifetime $<1.3 \times 10^{-7}$ s with ($\Phi_{sp:900nm}$) and without (Φ_{bare}) an additional 900 nm short pass filter mounted between laser and the test wafer are shown. The luminescence signal without the short pass filter is much dominated by steady-state non-luminescent light. *On the right* (**b**), the optical density of different filter configurations with respect to laser irradiation is investigated in an analysis of the FWHM ratio between excess carrier generation rate $G(t)$ and luminescence intensity $\Phi(t)$. The positive slope of the lifetime curves of the test wafer used here (lifetime shown in Fig. 7.4) imply a FWHM ratio less than unity. The different luminescence curves feature differnt filter combinations: Φ_i: $2 \times$ dielectric long pass 980 nm only; Φ_{ii}: additional 144 µm silicon bulk; Φ_{iii}: additional 365 µm silicon bulk. The *curves* shown here suggest that for wafers featuring an effective lifetime in the range of a microsecond, filter configuration (ii) sufficiently attenuates laser light. Adapted after [5]

7.2.2.2 Attenuation of Laser Light

Last, the possibility of insufficient filtering of irradiation light between the substrate and the luminescence detector was investigated. The occurrence of significant contributions of irradiation light in the measured photoluminescence signal could be unambiguously excluded for photoluminescence intensities corresponding to a substrate effective lifetime greater than 10^{-6} s, a wafer thickness greater or equal 160 µm, and a dopant concentration of 10^{16} cm^{-3}. This is to be concluded from an analysis of the full width at half maximum (FWHM) ratio between luminescence intensity $\Phi(t)$ and excess carrier generation rate $G(t)$ upon variation of the configuration of long pass filters. Using the substrate shown in Fig. 7.4, a positive slope of $\tau_\sigma(\Delta n)$ allows the conclusion of a FWHM ratio less than unity. Sufficient filter radiation would be evident, if the FWHM ratio corresponding to a long pass filter configuration did not differ from the FWHM ratio corresponding to a filter combination with a much higher optical density at the laser wavelength. Solely using two dielectric long pass filters (edge wavelength: 980 nm) proved to insufficiently attenuate laser irradiation with a FWHM ratio of 0.926, because with an additional 144 µm thick monocrystalline silicon wafer filter[13] (bringing about an additional attenuation of 880 nm laser light by a factor of 150) the FWHM ratio decreased to 0.862. Adding absorption filter bulk equivalent to another 4 orders of magnitude of

[13] Wafer filters feature no surface passivation and negligible luminescence—as they are only very weakly illuminated compared to the substrate under investigation.

attenuation of laser light only led to a further reduction of the FWHM ratio to 0.860. This suggests that the combination of the dielectric long pass filters with the 144 μm thick monocrystalline silicon wafer filter yields less than 2 % non-luminescent light in the photodetector for the given test wafer with an effective lifetime of little more than 2×10^{-6} s. The FWHM ratio analysis is also illustrated in Fig. 7.5. With a 200 μm thick monocrystalline silicon absorption filter in the default measurement configuration, lifetime analyses well below the microseccond range on $d \gtrsim 150\,\mu m$ thin substrates with dopant concentrations substantially less than $10^{16}\,cm^{-3}$ should be virtually unaffected by artifacts due to insufficient attenuation of laser light in the luminescence detector.

7.3 Carrier Lifetimes of Metalized Substrates

While for unmetalized silicon wafers, there are numerous lifetime measurement techniques avaliable, the number of lifetime measurement techniques allowing for metalized wafers (e.g. solar cell precursors or solar cells) is disproportionately small.[14] This is to be attributed mainly to two causes: First, the relatively high free carrier concentration of metallic layers (as opposed to the silicon bulk) largely limits the application of photoconductance-based techniques in terms of both, the dynamic range of instruments and a limited capacity to resolve the relatively faint light-induced excess conductivity. Second, the lack of optical transparency in the near infrared spectral range substantially constrains the scope of many optical techniques.

Access to recombination properties of pn-junction devices, and thereby implicitly to effective carrier lifetime, is typically provided by techniques sensitive to the device current [12–14] or open circuit voltage [15, 16] upon a defined steady-state or quasi-steady-state illumination. With the emergence of camera-based optical characterization techniques, several approaches to resolve recombination properties (effective minority carrier diffusion length or dark saturation current) of solar cells have been put forth [9, 17–19].

This section elaborates the required experimental setup modifications of effective lifetime measurements of solar cells and metalized cell precursors (even without a pn-junction) via harmonically modulated luminescence. Further, experimental evidence of the equivalent lifetime accuracy of an accordingly modified *reflective* measurement configuration[15] down to the microsecond timescale is provided. The key motivation for effective lifetime measurements on metalized substrates via harmonically modulated luminescence is to establish effective lifetime as *one* consistent and comparable recombination quality control parameter in any stage of a solar cell production process—including solar cells and metalized cell precursors (with or even without a pn-junction).

---

[14] The account given in this section is adapted from the corresponding original publication [11].

[15] As compared to the *transmissive* measurement configuration sketched e.g. in Fig. 7.1

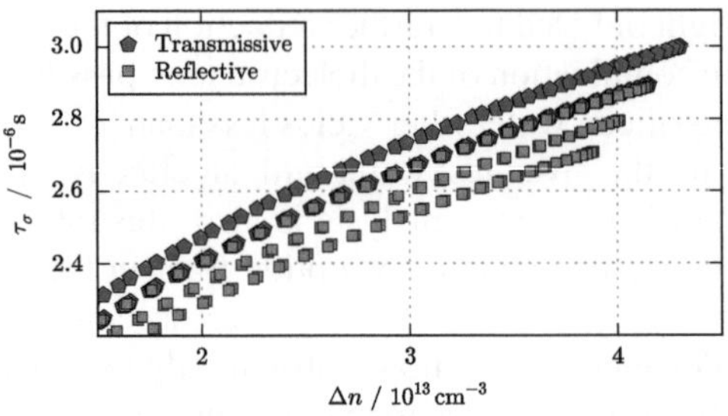

Fig. 7.6 Self-consistent lifetime analyses of the same wafer active area as shown in Fig. 7.4, conducted in both, a *transmissive* and a *reflective* measurement configuration of the luminescence detector with respect to the direction of irradiation. Lifetime results from three measurements in each configuration are shown, respectively (2×10^3 modulation cycles, $\mathcal{T} = 1.25 \times 10^{-2}$ s). Measured signals of both configurations were individually synchronized (cf. Sect. 7.2.1). The resulting lifetimes agree well within the statistical measurement uncertainty of 2×10^{-7} s as reported for InGaAs photodiodes in Sect. 7.2.1

7.3.1 Modifications of Setup and Analysis

For capture of luminescence of silicon wafers featuring opaque metallic layers, several modifications of the default experimental setup as described in Fig. 7.1 are necessary:

7.3.1.1 *Reflective* Positioning of the Luminescence Detector

The luminescence detector has to be positioned on the optically irradiated side of the device under investigation, to be referred to as a *reflective* measurement configuration herein (as opposed to the default *transmissive* measurement configuration). On the one hand, this brings about more rigorous requirements with regard to the capacity of the optical filter setup to attenuate laser irradiation, because filtering via the investigated substrate drops out. On the other hand, one might argue that the luminescence detector is then positioned such that only scattered laser light reaches the detector, whereas in the *transmissive* configuration the luminescence detector is positioned directly along the optical axis of irradiation (cf. Fig. 7.1). The light attenuation capacity of the *transmissive* configuration is quantified in Sect. 7.2.2. For the *reflective* setup, experimental evidence of equivalent accuracy is provided in Fig. 7.6 on the basis of lifetime measurements on the same (unmetalized) wafer as used for the quantification given in Sect.7.2.2. Note that the exact same arrangement of optical filters is used in both, the *transmissive* and the *reflective* measurement configuration. In the *reflective* measurement configuration, captured luminescence amplitudes Φ are typically reduced by a factor of two compared to the *transmissive* measurement configuration.

7.3.1.2 Luminescence Capture via an InGaAs Detector

InGaAs detectors are to be favored over silicon detectors, as outlined in Sect. 7.2.1. This particularly applies to the *reflective* measurement configuration, provided that silicon wafers are used as absorption filters. The reason for this is that capture of silicon luminescence via InGaAs detectors proceeds at significantly higher wavelengths compared to silicon detectors. With the latter, silicon absorption filters would disproportionately reduce captured luminescence amplitude.[16]

7.3.1.3 Definition of the Active Measurement Area via Apertures

In the *transmissive* measurement configuration, luminescence detection is confined to the active measurement area defined by the aperture of focusing optics and optical filters (if no further apertures are used). In contrast, in the *reflective* measurement configuration, the definition of an active (irradiated) measurement area via apertures plays a more important role, because the natural cofinement of an active measurement area is then accomplished only via the local sensitivity S of the luminescence detector with respect to the substrate,[17] and via the lateral distribution of laser irradiance on the substrate.[18] Particularly the lifetime calibration of luminescence images to be detailed in Sect. 8.3 requires a precise definition of the irradiated area, which should coincide with the measurement area.

7.3.1.4 Note on a Previous Data Correction

In the original publication about harmonically modulated luminescence lifetime measurements on metalized silicon wafers and solar cells [11], a systematic error likely to be induced by optical activity of a setup device other than the investigated substrate was encountered. In [11], this effect was presumed to be caused by optical transitions in the aperture coating. More recent evidence suggests that this conjecture might not have been robust, because the corresponding artifact completely vanished upon exchange of the silicon absorption filter. The initially used filter featured rough interfaces, which are most unvavorable for optical applications due to light scattering. The exact cause of the above described artifact has not been resolved: It

---

[16] Note that alternative filtering approaches on the basis of direct compound semiconductor filter hardware have been suggested elsewhere [1]. The use of such filters has not yet been evaluated or implemented in the current harmonically modulated luminescence experimental setup at Fraunhofer ISE.

[17] Note that the radial symmetry of the local sensitivity function S with respect to the optical axis of irradiation (cf. Sect. 7.1.3) is violated in the *reflective* measurement configuration due to apparent topological constraints.

[18] In the present experimental setup, for substrate active areas greater than the greatest aperture available, the above specified lateral homogeneity of irradiance (with a variance $<2\,\%$) is not valid anymore. Rather, irradiance steadily decreases upon further departure from the optical axis.

could be that the scattering properties of non-perpendicularly incident light might have amplified the capture of a faint fluorescence of the aperture coating. However, it appears more likely that the replaced filter itself caused this artifact.

As a matter of fact, the *reflective* measurements shown in Fig. 7.6 (conducted with the presently used, polished silicon absorption filter) do not feature the corresponding data correction as originally proposed in [11]. Rather, the lifetime analysis is now independent of the measurement configuration. This brings about a substantial simplification and a unification of the lifetime analysis, leaving minor changes of the local sensitivity functions $\mathcal{S}$ in the case of spatially resolved lifetime calibrations (cf. Sect. 8.3) as the only remaining difference.

7.3.2 Experimental Evidence of Equivalent Accuracy

In Fig. 7.6, effective lifetime measurements conducted in both *transmissive* and *reflective* measurement configurations on an unmetalized wafer are compared. Lifetime results from the same wafer active area are also shown in Fig. 7.4. Three measurements were conducted in each configuration, respectively, thereby providing an indication of the measurement precision. Also, measured signals of both configurations were individually synchronized (cf. Sect. 7.2.1). The resulting lifetimes coincide within the estimated statistical uncertainty of the synchronization measurement of 2×10^{-7} s as reported for the present experimental setup in Sect. 7.2.1. The fact that the *reflective* lifetime results appear 10^{-7} s less than the *transmissive* results could be an indication of a systematic error, but it could likewise be a result of the statistical uncertainty of the synchronization measurement. Even though a systematic error of either configuration cannot be ascertained on the basis of the measurements shown in Fig. 7.6, such error could be due to

- differing local sensitivity profiles $\mathcal{S}$ of both configurations (no radial symmetry in the *reflective* geometry, cf. Sect. 7.1.3),
- insufficient attenuation of laser light in either of the compared measurement configurations, or
- a remanent contribution of optical transitions of the aperture coating or filters to luminescence amplitude Φ.

Concerning the attenuation of laser light in the luminescence detector, independent experimental evidence of a sufficient attenuation in order to resolve microsecond lifetimes on 1 Ω cm material was gained by comparing luminescence amplitudes Φ obtained from a standard silicon solar cell featuring a very homogeneous lateral distribuation of effective carrier lifetime in the range of $2{-}3 \times 10^{-6}$ s. Luminescence amplitudes with and without a silver bus bar in the active measurement area were found to differ exactly by the proportion of metal coverage within the active measurement area, indicating a negligible contribution of laser light reflected from the bus bar to luminescence amplitude Φ.

Together with the results reported in Sect. 7.2, the agreement of lifetimes obtained in the different measurement configurations justifies to state a sufficient accuracy of both measurement configurations for a resolution of microsecond lifetimes (for $\sim 1\ \Omega\,\text{cm}$ material).[19]

7.3.3 Discussion of Limitations

Silicon wafers featuring highly doped emitters or metal coatings that are electrically connected to the wafer bulk may encounter effects of substantial lateral redistribution of excess carrier density. Related phenomena have been addressed and investigated for lifetime measurements on diffused wafers in the past [20–22], and they may to some extent even affect harmonically modulated lifetime measurements, particularly if a locally confined irradiation coincides with injection-dependent lifetime. Highly conductive lateral structures may bring about a drift away from the illuminated area such that excess carrier density within the illuminated area is substantially less than to be expected in the case of a vanishing lateral carrier drift. Although this does not a priori falsify the determination of harmonically modulated effective lifetime, significant errors may result if lifetime is injection-dependent due to the erroneous (too low) excess carrier density level assigned to a measured lifetime. Highly recombinative wafer spots electrically connected to a highly conductive lateral structure (emitter, metalization) may aggravate this effect. The same applies to local shunts on solar cells, shunted substrate edges lacking edge isolation, or highly recombinative substrate edges lacking surface passivation.

7.4 Dynamic Electroluminescence

For devices with contacting electrodes and a junction, injection of excess carriers is not restricted to optical generation. It may likewise take place electrically via application of a forward bias to the junction.[20] Electroluminescence is a well-established solar cell characterization technique [17]. The combination of the methodology of harmonically modulated luminescence lifetime techniques with electrical carrier injection can be advantageous in several respects:

1. The cost of instrumentation is substantially lowered when using electrical current sources compared to high-power lasers.
2. Filtering of irradiation light in the luminescence detector is not required in the case of electrical injection of excess carriers. This brings about the a priori exclusion of artifacts related to an insufficient attenuation of optical irradiation, and a higher flexibility with respect to the positioning of the luminescence detector

---

[19] Solar cell lifetime results obtained in a *reflective* measurement configuration are shown and validated in Sect. 8.3.

[20] The account given in this section is adapted from the corresponding original publication [23].

(enhanced luminescence capture due to an enhanced solid angle of detection, shorter measurement times, and an enhanced sensitivity).

3. Photoluminescence may feature some geometrical constraints and limitations that could be overcome via electrical injection (cf. Sect. 7.3.3). For instance, locally confined luminescence detection with full area carrier injection in a solar cell is easily achieved electrically, while it is virtually impossible via optical generation due to an opaque (metal-coated) interface.

In the following, key aspects of combining the methodology of harmonically modulated lifetime techniques with electrical carrier injection are outlined, focusing on differences in terms of both theory and instrumentation.

Harmonically modulated lifetime analyses[21] do not a priori distinguish between different ways of excess carrier injection. Therefore, such analyses should be applicable to both electrical and optical injection. Restricting ourselves to the simplified case of a depth-averaged treatment of the continuity equation,[22] optical excess carrier generation rate as of Eq. 3.58 corresponds to the absorbed photon current density $j_\gamma(t)$ divided by the substrate thickness d. In the case of laterally homogeneous electrical injection of excess carriers via application of a forward bias to a solar cell, the depth-averaged carrier injection rate per volume can be equivalently expressed as

$$G_{EL}(t) = \frac{j_Q(t)}{qd}, \tag{7.8}$$

with the electrical charge current density $j_Q(t)$ and the elementary charge q.[23]

7.4.1 Experimental Setup Modifications

The major distinctions between the harmonically modulated electro- and photoluminescence experimental setup relate to:

1. **Carrier Injection:** Optical irradiation has to be replaced by the injection of a harmonically modulated electrical charge current of density $j_Q(t)$ through a solar cell.

2. **Detection of Injection Rate:** The detection of relative excess carrier generation rate via a photodetector has to be replaced by a precise time-resolved detection of the solar cell electrical current.

[21] i.e. self-consistently or according to any of the phase-sensitive approaches outlined in Sect. 8.1

[22] Conditions for equivalent dynamic lifetime analyses between optical and electrical carrier injection are sketched in Sect. 3.2.1.

[23] Note that the lateral homogeneity of this injection rate depends on the series resistance (particularly, the emitter sheet resistance) of the solar cell. The higher the injected current at a given sheet resistance, the closer is the current confined to the immediate surroundings of metal contacts. Lateral uniformity of carrier injection can be validated via electroluminescence imaging.

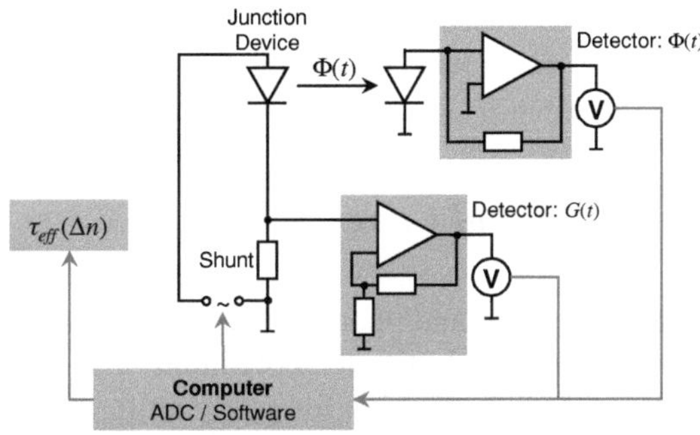

Fig. 7.7 Scheme of the modulated electroluminescence setup developed in the course of this work. The main circuit (*depicted on the left*) consists of a harmonically modulated current source (forward-biased with respect to the investigated pn-junction device), a shunt to precisely quantify the time-dependent current injected into a pn-junction device, and the radiating pn-junction device. The current in the main circuit is software-controlled. A voltage amplifier enables the precise measurement of the current in the main circuit from the voltage drop across the shunt, thereby providing access to excess carrier generation rate $G(t)$. The electroluminescent emission of the device under investigation is captured with a photodetector connected to a current amplifier. Both signals (voltages) are simultaneously digitized via an ADC unit (analog–digital conversion), before they are synchronized and analyzed in terms of carrier lifetime τ_{eff} (Δn). Adapted after [23]

The latter setup modification is realized via a resistor (shunt) that is series connected to the solar cell under investigation (cf. Fig. 7.7). An accurate indication of resistance $(10 \pm 0.1\,\text{m}\Omega)$ together with accurately given amplification factors of a connected voltage amplifier (cf. Fig. 7.7) featuring an experimentally validated uncertainty of less than $\pm 0.5\,\%$ yield current measurements at an accuracy of $\pm 1.5\,\%$.

7.4.1.1 Modified Synchronization Procedure

Because of the finite response time of the signal processing electronics as detailed in Sect. 7.2.1, a signal synchronization is mandatory. In the case of electrical excess carrier injection, the synchronization procedure described in Sect. 7.2.1 is complicated by the fact that the electrical excess carrier injection rate $G_{EL}(t)$ cannot be measured optically.

Therefore, a fundamentally different synchronization procedure was implemented: The junction device under investigation (solar cell) was replaced by a 3 mW AlGaAs light-emitting diode (LED) with a maximum emission at a wavelength of 940 nm. Further, the shunt of Fig. 7.7 was replaced by a resistor of sufficiently high resistance in order to compensate for the lower LED current compared with the solar cell (thereby yielding a nearly identical voltage drop across the shunt compared with the solar cell). As a direct band gap semiconductor, this LED is known to feature negligible electronic relaxation times compared with the microsecond timescale of interest. Therefore, the time shift $\Delta t'_{EL}$ between the LED light signal Φ_{LED} measured by the luminescence detector and the electrical current through the LED measured via the voltage amplifier can be used for a synchronization of luminescence intensity according to Eq. 7.7, if the quasi-steady-state condition $\Delta t'_{EL} \ll T$ is satisfied.

7.4.2 Experimental Proof of Concept

Harmonically modulated electro- and photoluminescence measurements on a solar cell were conducted under identical injection conditions. The investigated solar cell featured an area of $50 \times 50\,\text{mm}^2$, allowing for nearly uniform optical excess carrier generation throughout the entire solar cell area with the dynamic luminescence setup. Further, the cell featured an n-type base with a resistivity of $10\,\Omega\,\text{cm}$, an aluminum p^+ rear emitter and a phosphorous n^+ front surface field (with selective n^{++} regions under plated contacts). The solar cell's parallel resistance amounted to $8 \times 10^5\,\Omega\,\text{cm}^2$.

Injection-dependent effective carrier lifetime $\tau_\sigma(\Delta n)$ was determined self-consistently for both electro- and photoluminescence lifetime analyses. Individually determined synchronization time shifts were $\Delta t'_{EL} = 8.7 \pm 0.2 \times 10^{-6}$ s and $\Delta t'_{PL} = 5.5 \pm 0.3 \times 10^{-6}$ s at the amplification settings of interest, respectively. The resulting injection-dependent lifetime curves are shown in Fig. 7.8.

The curves shown in Fig. 7.8 represent the average of 25 successive measurements, respectively. At an excess carrier density of $2 \times 10^{14}\,\text{cm}^{-3}$, average effective lifetimes certainly coincide within the standard error, whereas toward higher excess carrier densities, a slightly lower lifetime as inferred from the electroluminescence measurement is observed. Given the standard error of mean lifetimes (rather than the standard deviation plotted in the insert of Fig. 7.8), this deviation is to be regarded as systematic. The observed systematic mismatch can be attributed to an electroluminescence artifact caused by the emitter sheet resistance: Toward higher injected currents, current paths (and excess carrier density) concentrate around metal fingers. With a higher excess carrier density in the vicinity of metal fingers than to be expected on the basis of an assumed laterally uniform electrical current injection, the excess carrier density level assigned to the measured lifetime by the self-consistent lifetime analysis is too low. The onset of such an effect of a concentration of current paths in the respective injection range could be qualitatively verified via electroluminescence images.

7.4.3 Discussion of Limitations

The sheet resistance effect of a concentration of electrical current paths toward an increasing electrical injection of excess carriers represents a limitation with regard to the scope of dynamic electroluminescence lifetime investigations. Yet, electro- and photoluminescence lifetime analyses are notably found to coincide at a sufficiently low injection level in spite of diverging (and erroneous) lifetime interpretations at higher injection (cf. Fig. 7.8).

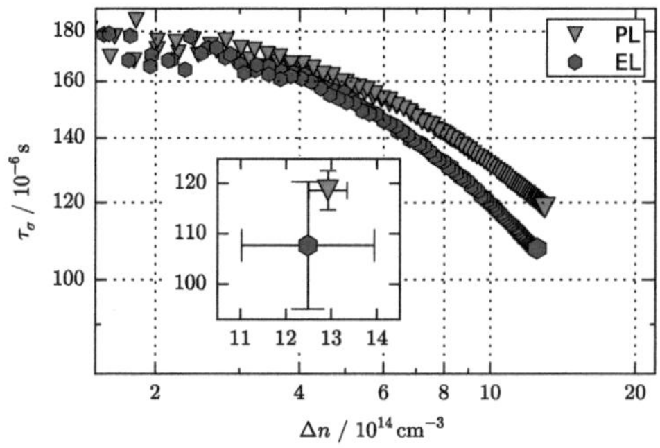

Fig. 7.8 Self-consistent analysis of effective lifetime $\tau_\sigma(\Delta n)$ from harmonically modulated electro- and photoluminescence measurements on a silicon solar cell, carried out at a modulation time $\mathcal{T} = 1.25 \times 10^{-2}$ s (the cell concept and relevant properties are detailed in Sect. 7.4.2). The measurement uncertainty specified in the inserted plot relates to the standard deviation of 25 successive measurements, respectively. Adapted after [23]

It should be noted here that local shunts of solar cells could possibly cause artifacts in dynamic electroluminescence lifetime measurements. By lowering the cell voltage at a given injected charge current density j_Q, shunts may lead to an overestimation of excess carrier generation rate (and thereby of excess carrier density). Also, the lateral variation of effective carrier lifetime brings about a lateral inhomogeneity of excess carrier injection rate. This is because in a parallel network of diodes with different saturation currents, those (worst quality) diodes which feature the highest saturation currents also allow for the highest current injection at a given (constant) forward bias applied to the device. However, both of the addressed artifacts could likewise occur in photoluminescence measurements on pn-junction devices due to lateral redistribution of excess carriers through emitter and metalization.

Another limitation related to a modulation of junction capacitance had previously been raised by Cuevas and Recart [24]. Therein, a modulation of the space charge of a junction device is argued to affect the dynamic lifetime interpretation—particularly for low base dopant concentrations and fast modulation (e.g. short modulation times $\mathcal{T}$). However, with typical dopant concentrations $n_0 \gtrsim 10^{15}\,\text{cm}^{-3}$ and modulation times $\mathcal{T} > 10^{-2}$ s, the models shown in [24] suggest a negligible impact of such capacitive dynamic artifacts for excess carrier densities $\Delta n > 10^{12}\,\text{cm}^{-3}$. Throughout this work, lower excess carrier densities are only accessed via a steady-state application (cf. Sect. 8.2) which is apparently unaffected by such dynamic artifacts.

References

1. B. Steuer, Aufbau eines Photolumineszenz-Messplatzes zur Charakterisierung von Wafern und Solarzellen aus kristallinem Silizium, Master thesis, Universität Konstanz (2011)
2. J.A. Giesecke, M. Kasemann, W. Warta, Determination of local minority carrier diffusion lengths in crystalline silicon from luminescence images. J. Appl. Phys. **106**, 014907 (2009)
3. J. Isenberg, J. Dicker, W. Warta, Averaging of laterally inhomogeneous lifetimes for one-dimensional modeling of solar cells. J. Appl. Phys. **94**, 4122–4130 (2003)
4. B. Michl, M. Rüdiger, J.A. Giesecke, M. Hermle, W. Warta, M.C. Schubert, Efficiency limiting bulk recombination in multicrystalline silicon solar cells. Solar Energy Mater. Solar Cells **98**, 441–447 (2012)
5. J.A. Giesecke, W. Warta, Microsecond carrier lifetime measurements in silicon via quasi-steady-state photoluminescence. Prog. Photovoltaics: Res. Appl. **20**, 238–245 (2012)
6. M.A. Green, M.J. Keevers, Optical properties of intrinsic silicon at 300 K. Prog. Photovoltaics: Res. Appl. **3**, 189–192 (1995)
7. T. Trupke, Influence of photon reabsorption on quasi-steady-state photoluminescence measurements on crystalline silicon. J. Appl. Phys. **100**, 063531 (2006)
8. M. Rüdiger, T. Trupke, P. Würfel, T. Roth, S.W. Glunz, Influence of photon reabsorption on temperature dependent quasi-steady-state photoluminescence lifetime measurements on crystalline silicon. Appl. Phys. Lett. **92**, 222112 (2008)
9. P. Würfel, T. Trupke, T. Puzzer, E. Schäffer, W. Warta, S.W. Glunz, Diffusion lengths of silicon solar cells from luminescence images. J. Appl. Phys. **101**, 123110 (2007)
10. J.A. Giesecke, M. Kasemann, M.C. Schubert, P. Würfel, W. Warta, Separation of local bulk and surface recombination in crystalline silicon from luminescence reabsorption. Prog. Photovoltaics: Res. Appl. **18**, 10–19 (2010)
11. J.A. Giesecke, B. Michl, F. Schindler, M.C. Schubert, W. Warta, Minority carrier lifetime of silicon solar cells from quasi-steady-state photoluminescence. Solar Energy Mater. Solar Cells **95**, 1979–1982 (2011)
12. P. A. Basore, Extended spectral analysis of internal quantum efficiency, in *Proceedings of the 23rd IEEE PVSC* (Louisville, 1993), pp. 147–152
13. W. Warta, J. Sutter, B.F. Wagner, R. Schindler, Impact of diffusion length distributions on the performance of mc-silicon solar cells, in *Proceedings of the 2nd WC PVEC* (Vienna, 1998), pp. 1650–1653
14. J. Carstensen, G. Popkirov, J. Bahr, H. Föll, CELLO: An advanced LBIC measurement technique for solar cell local characterization. Solar Energy Mater. Solar Cells **76**, 599–611 (2003)
15. M. Wolf, H. Rauschenbach, Series resistance effects on solar cell measurements. Adv. Energy Convers. **3**, 455–479 (1963)
16. R.A. Sinton, A. Cuevas, A quasi-steady-state open-circuit voltage method for solar cell characterization, in *Proceedings of the 16th EU PVSEC* (Glasgow, 2000), pp. 1152–1155
17. T. Fuyuki, H. Kondo, T. Yamazaki, Y. Takahashi, Y. Uraoka, Photographic surveying of minority carrier diffusion length in polycrystalline silicon solar cells by electroluminescence. Appl. Phys. Lett. **86**, 262108 (2005)
18. D. Hinken, K. Bothe, K. Ramspeck, S. Herlufsen, R. Brendel, Determination of the effective diffusion length of silicon solar cells from photoluminescence. J. Appl. Phys. **105**, 104516 (2009)
19. M. Glatthaar, J. Haunschild, M. Kasemann, J. Giesecke, W. Warta, S. Rein, Spatially resolved determination of dark saturation current and series resistance of silicon solar cells. Phys. Stat. Sol. (RRL) **4**, 13–15 (2010)
20. M.D. Abbott, J.E. Cotter, T. Trupke, R.A. Bardos, Investigation of edge recombination effects in silicon solar cell structures using photoluminescence. Appl. Phys. Lett. **88**, 114105 (2006)
21. F.W. Chen, J.E. Cotter, M.D. Abbott, T.T.A. Li, K.C. Fisher, The influence of parasitic effects on injection-level-dependent lifetime data. IEEE Trans. Eletrron Dev. **54**, 2960–2968 (2007)
22. M. Kessler, T. Ohrdes, P.P. Altermatt, R. Brendel, The effect of sample edge recombination on the averaged injection-dependent carrier lifetime in silicon. J. Appl. Phys. **111**, 054508 (2012)

23. J.A. Giesecke, M.C. Schubert, W. Warta, Carrier lifetime from dynamic electroluminescence. IEEE J. Photovoltaics **3**, 1012–1015 (2013)
24. A. Cuevas, F. Recart, Capacitive effects in quasi-steady-state voltage and lifetime measurements of silicon devices. J. Appl. Phys. **98**, 074507 (2005)

Chapter 8
Conceptual Advances: Recombination Properties

This chapter details the novel techniques and applications to resolve and quantify recombination properties via modulated luminescence arising from this work. A class of three closely related phase-sensitive basic approaches to injection-dependent effective carrier lifetime via modulated luminescence is presented—complementing Trupke's self-consistent approach [1]. Applications involve a technique to measure injection-dependent effective carrier lifetime over a very broad injection range, a spatially resolved carrier lifetime calibration of steady-state luminescence images via harmonically modulated luminescence measurements, and a combined approach to access bulk lifetime and surface recombination velocity of silicon ingots, which could likewise be used in order to access bulk lifetime and the minority carrier diffusion coefficient.

8.1 Phase-Sensitive Carrier Lifetime Approaches

A class of phase-sensitive carrier lifetime techniques based on harmonically modulated experiments is described in this section. This class complements the self-consistent carrier lifetime calibration as proposed by Trupke et al. [1] and as detailed in Sect. 5.2, while allowing for a broader range of application—including accurate lifetime measurements at intermediate or high-level injection conditions without a priori information about net dopant concentration.

8.1.1 Self-Consistency and Net Dopant Concentration

It has been reasoned in Sect. 5.2.3 that under low-level injection conditions the self-consistent lifetime analysis is to some extent independent of net dopant concentration n_0. However, for intermediate to high-level injection conditions, self-consistent

J. Giesecke, *Quantitative Recombination and Transport Properties in Silicon from Dynamic Luminescence*, Springer Theses, DOI: 10.1007/978-3-319-06157-3_8,
© Springer International Publishing Switzerland 2014

carrier lifetime critically depends on the choice of n_0.[1] From a rigorous theoretical standpoint, it could be argued that the self-consistency condition (cf. Eq. 5.12) could only be satisfied if both a scaling factor c_Φ and net dopant concentration n_0 were adequately chosen. Therefore, an accurate self-consistent lifetime interpretation would not be complicated by the uncertainty of n_0. Rather, at intermediate to high-level injection conditions, the self-consistent lifetime analysis should accurately yield both n_0 and τ_σ.

In practice though, measured photocurrents suffer from both statistical noise and minor inaccuracies.[2] In fact, the self-consistent hysteresis (cf. Eq. 5.10) features a very faint dependence on n_0. Consequently—given the above-mentioned sources of uncertainty—it is only possible to roughly determine the order of magnitude of n_0 via the self-consistent approach. As quantitatively substantiated in Sect. 9.1, self-consistent hysteresis contrast is not even under intermediate injection conditions sufficient for a precise determination of both n_0 and τ_σ. Thus, if n_0 is unknown or uncertain and low injection conditions are not satisfied, it is deemed mandatory to obtain the injection-dependent carrier lifetime without making use of the relation between luminescence intensity Φ and excess carrier density Δn as given in Eq. 4.39.[3]

8.1.1.1 Accessing Actual Lifetime from Differential Phase Information

Let us recall some previously derived fundamentals of the class of phase-sensitive lifetime techniques to be described in the following: In Sect. 3.2.3, the quasi-steady-state harmonically modulated time shift between excess carrier density and generation rate was shown to correspond to effective carrier lifetime (cf. Eq. 3.87) in the special case of injection-independent lifetime. However, the commonly encountered injection-dependence of lifetime brings about systematic deviations from the above identity, which are quantitatively explained in Sect. 6.2—revealing harmonically modulated time shift measurements as *differential* rather than *actual* measurements of efffective lifetime. The class of lifetime techniques described in the following provides access to *actual* injection-dependent effective lifetime $\tau_{eff}(\Delta n)$ from phase-sensitive measurements of *differential* lifetime $\tau_m(G)$[4]—notably independent of net dopant concentration n_0.

---

[1] The reasoning outlined here may to some extent accord with [2].

[2] E.g. due to a finite response time of the signal processing electronics.

[3] Secondary benefits of lifetime techniques independent of Eq. 4.36 might be the determination of n_0 (cf. Sect. 9.1), a measurement of the injection dependence of the integral coefficient of radiative recombination B [3], or a validation of Eq. 4.36 (as postulated by Roosbroeck and Shockley [4]), which had been subject to challenge [5].

[4] For quasi-steady-state harmonically modulated experiments, the differential lifetime τ_m corresponds to harmonically modulated effective lifetime $\tau_{eff,h}$.

8.1.1.2 The Iterative Procedure

Having measured *differential* lifetimes $\tau_m(G_i)$ at excess carrier generation rates G_i, the determination of *actual* lifetime $\tau_{eff}(G_i)$ requires

1. an algorithm capable of an injective assignment of an expected measurement result $\tau_m^j(G_i)$ to a virtual set of test lifetimes $\tau_{eff}^j(G_i)$,[5] and
2. another algorithm capable of an iterative optimization of $\tau_{eff}^j(G_i)$ based on the comparison of the virtually expected measurement result $\tau_m^j(G_i)$ with the actual measurement result $\tau_m(G_i)$.

The latter iterative procedure represents the common ground of the class of phase-sensitive lifetime analyses elaborated here, whereas they differ with regard to the first algorithm ($\tau_{eff}^j(G_i) \mapsto \tau_m^j(G_i)$). With an approximation of *actual* effective lifetime $\tau_{eff}^j(G_i)$ of order j and a corresponding virtual *differential* lifetime $\tau_m^j(G_i)$, the said iterative procedure would take the form

$$\tau_{eff}^{j+1}(G_i) = \tau_{eff}^j(G_i) + \epsilon_1 \left(\tau_m(G_i) - \tau_m^j(G_i)\right), \tag{8.1}$$

with a sufficiently small number $\epsilon_1 > 0$ ensuring convergence of the iteration. With this iterative method, the approximation $\tau_{eff}^j(G_i)$ is expected to approach true actual lifetime $\tau_{eff}(G_i)$ for sufficiently high j. The quality of the approximation $\tau_{eff}^j(G_i)$ may be numerically evaluated by a measure of the difference between $\tau_m(G_i)$ and $\tau_m^j(G_i)$, defining a stop criterion of the iteration, e.g. via the sum of squared residuals

$$\sum_i \left(\tau_m(G_i) - \tau_m^j(G_i)\right)^2 \overset{!}{<} \epsilon_2, \tag{8.2}$$

with another small number $\epsilon_2 > 0$.

As for the algorithm performing the assignment $\tau_{eff}^j(G_i) \mapsto \tau_m^j(G_i)$, originally a numerical approach based on the finite-element solution of the continuity equation (cf. Appendix A.3.1) was pursued, to be detailed in Sects. 8.1.2 and 8.1.3. This approach was later replaced by a much faster analytical approach based on the theory of light-biased decay time (cf. Sect. 6.2), to be detailed in Sect. 8.1.4.

8.1.2 The Peak Shift Approach

In the so-called peak shift (or maximum shift) approach [6], *differential* lifetime $\tau_m(G_i)$ is determined from the time shifts between maxima of luminescence intensity $\Phi(t)$ and excess carrier generation rate $G(t)$ of a minimum of two harmonically

---

[5] Here, j is to be perceived as an iteration index rather than as an exponent.

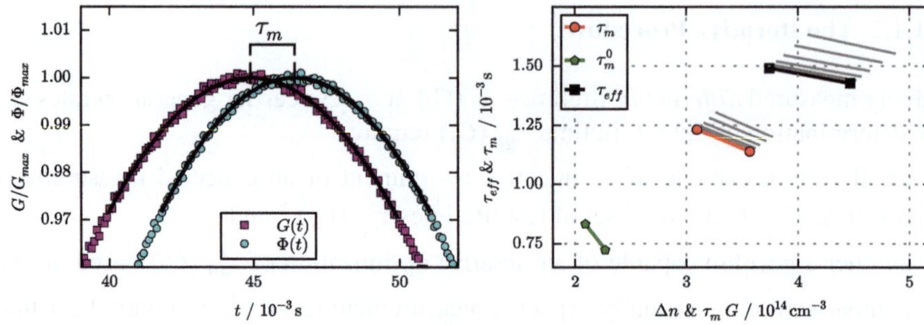

Fig. 8.1 On the left, the determination of *differential* lifetime $\tau_m(G_i)$ as the time shift between parabolic fits to maxima of photoluminescence intensity $\Phi(t)$ and excess carrier generation rate $G(t)$ is illustrated. On the right, the iteration method of Eq. 8.1 is visualized for the peak shift approach. Peak shifts $\tau_m(G_i)$ (*differential* lifetimes) are measured at two adjacent excess carrier generation rates G_i, respectively. Using $\tau_m(G_i) = \tau_{eff}^0(G_i)$ as an initial approximation of *actual* effective lifetime, one obtains the corresponding virtual initial approximation of *differential* lifetime $\tau_m^0(G_i)$ via a numerical solution of the continuity equation. Higher order iterations are shaded, finally converging toward true *actual* effective carrier lifetime $\tau_{eff}(G_i)$. Adapted after [6]

modulated measurements, which are to be conducted at a series of adjacent maximum excess carrier generation rates G_i, respectively. The measured *differential* lifetimes $\tau_m(G_i)$ are used as an initial approximation of *actual* effective lifetimes

$$\tau_{eff}^0(G_i) = \tau_m(G_i), \tag{8.3}$$

to be used for a calculation of corresponding virtual *differential* lifetimes $\tau_m^0(G_i)$ according to the numerical solution of the continuity equation (cf. Appendix A.3.1).

For two measured peak shifts $\tau_m(G_i) = \tau_{eff}^0(G_i)$, a solution of the continuity equation may be approximated with a linear lifetime curve $\tau_{eff}^0\left(\tau_{eff}^0 G\right)$ intersecting the points $\left(\tau_{eff}^0(G_i), \tau_{eff}^0(G_i)\, G_i\right)$. For more than two measured peak shifts, a smooth curve $\tau_{eff}^0\left(\tau_{eff}^0 G\right)$ can be fitted to the set of points $\left(\tau_{eff}^0(G_i), \tau_{eff}^0(G_i)\, G_i\right)$ in order to be subsequently used for the numerical solution of the continuity equation.

$\tau_{eff}^0(G_i)$, $\tau_m(G_i)$, and the set of $\tau_m^0(G_i)$ (resulting from the solution of the continuity equation) may then be inserted into Eq. 8.1, thereby obtaining a first order approximation of *actual* effective lifetime $\tau_{eff}^1(G_i)$. The above set of operations may be iterated until the stop criterion (Eq. 8.2) is fulfilled. The working principle of the peak shift approach is visualized in Fig. 8.1 on the basis of two photoluminescence measurements conducted at adjacent excess carrier generation rates on a float zone wafer (wafer details are reported in the caption of Fig. 8.2).

In Fig. 8.2 the injection-dependent lifetime curves obtained via the peak shift lifetime analysis shown in Fig. 8.1 are compared with different measurement techniques, involving QSSPC [7, 8], a self-consistent photoluminescence lifetime analysis

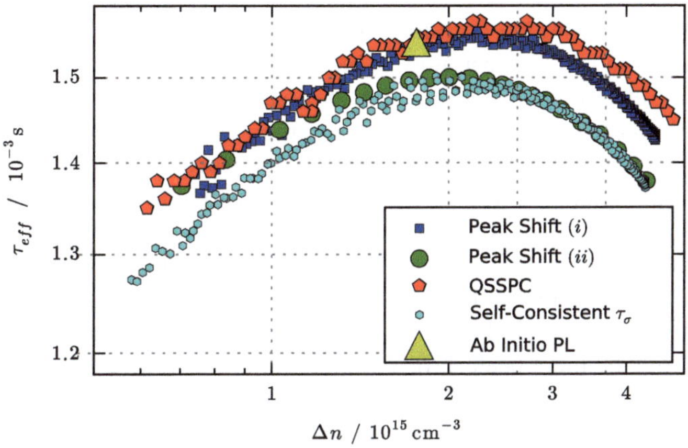

Fig. 8.2 Comparison of different lifetime analyses of a laterally homogeneous and planar d = 247 μ m thick p-type 1.1 Ω cm float zone wafer featuring a silicon nitride surface passivation. The different peak-shift analyses are distinguished in the text. Adapted after [6]

(cf. Sect. 5.2), and an ab initio calculation of effective lifetime from a steady-state photoluminescence image [9]. Figure 8.2 shows two different injection-dependent lifetime analyses based on the peak shift approach:

i. A peak shift lifetime analysis featuring solely two adjacent measurement points $\tau_m(G_i)$ allows for the calculation of a scaling factor c_Φ according to Eq. 4.36, which then allows the determination of injection-dependent lifetime from the measured amplitudes $\Phi(t)$ and $G(t)$. For an adequate determination of the injection-dependent course of lifetime under intermediate to high-level injection conditions, this analysis would require information about net dopant concentration n_0.
ii. A peak shift lifetime analysis featuring a set of more than two measurement points $\tau_m(G_i)$ may reveal the injection-dependent course of lifetime without having to rely upon signal amplitudes, thereby rendering the requirement of net dopant concentration n_0 obsolete.

Reasonable agreement of all lifetime techniques gathered in Fig. 8.2 can be stated. Notably, this would not be the case if peak shifts τ_m were interpreted as *actual* lifetimes, as can be seen from the corresponding peak shifts reported in Fig. 8.1.

8.1.3 The Self-Sufficient Approach

Good agreement between peak shift and self-consistent lifetime analyses was achieved for surface passivated 1 Ω cm monocrystalline wafers featuring very high effective carrier lifetimes in the range of 10^{-3} s. Yet, poor agreement was observed on substrates featuring lower carrier lifetimes and a light-induced instability of carrier lifetime [10–12]. This lack of accuracy of the peak shift approach is likely to be caused by the combination of a minimum of two successively conducted measurements. Both the statistical uncertainty of measurement results as well as a systematic variation of the electronic response time in the course of measurements (e.g., optically or thermally induced) may lead to crucial perturbations of this sensitive iterative procedure. With an iteration procedure that combines two peak shifts obtained at adjacent irradiances, but which is at the same time very sensitive to the difference between these peak shifts, a statistical artifact could be systematically amplified via the proposed iterative procedure (cf. Eq. 8.1). This systematic amplification would not occur if injection-dependent time shifts between generation rate and luminescence intensity were extracted from one single harmonically modulated luminescence measurement rather than from the mere peak shifts of a minimum of two measurements. This will be substantiated both theoretically and experimentally in the following.[6]

8.1.3.1 Generalized Time Shifts

Here, instructions on how to obtain injection-dependent time shifts from one single harmonically modulated luminescence measurement—consisting of two time-dependent sequences of measured amplitudes $\Phi(t)$ and $G(t)$—are given. First, the *differential* lifetime $\tau_m\,(G_0)$ corresponding to the maximum G_0 of the sequence $G(t)$ is determined from the peak shift, as illustrated in Fig. 8.1. Due to the harmonic shape of $G(t)$, every generation rate $G_i < G_0$ occurs once in each of the two injective rising (r) and falling (f) branches of $G(t)$. Thus, the times $t_r\,(G_i)$ and $t_f\,(G_i)$ can be explicitly assigned to any available generation rate $G_i < G_0$. The corresponding *differential* lifetime (time shift) $\tau_m\,(G_i)$ is then determined by shifting the measured luminescence intensity $\Phi'\,(t) = \Phi\,(t + \tau_m\,(G_i))$ such that the condition

$$\Phi'\,(t_r\,(G_i)) \stackrel{!}{=} \Phi'\left(t_f\,(G_i)\right) \tag{8.4}$$

is satisfied, as illustrated in Fig. 8.3. Due to the fact that measured signals are superimposed by statistical noise, the determination of $\tau_m\,(G_i)$ is performed via linear fits within adequately chosen fitting ranges.

Analogous to the peak shift approach, a smooth fit of the thereby obtained injection-dependent set of differential lifetimes $\tau_m\,(G_i)$ may be used as an initial approximation of *actual* effective lifetime $\tau^0_{\mathit{eff}}\,(G_i)$ according to Eq. 8.3, in order to obtain a corresponding set of virtual *differential* lifetimes $\tau^0_m\,(G_i)$ from a numerical solution of the continuity equation. Subsequent insertion into Eq. 8.1 yields a first order approximation of *actual* effective lifetime $\tau^1_{\mathit{eff}}\,(G_i)$—converging toward the true value of *actual* effective lifetime upon iteration. Figure 8.4 shows the result of this iteration procedure for a quasi-steady-state harmonically modulated photoluminescence measurement on a silicon nitride passivated multicrystalline silicon wafer. As can be seen, using a smooth fit of the final lifetime curve for a finite-element solution

6 The account given here is adapted from the corresponding original publication [2].

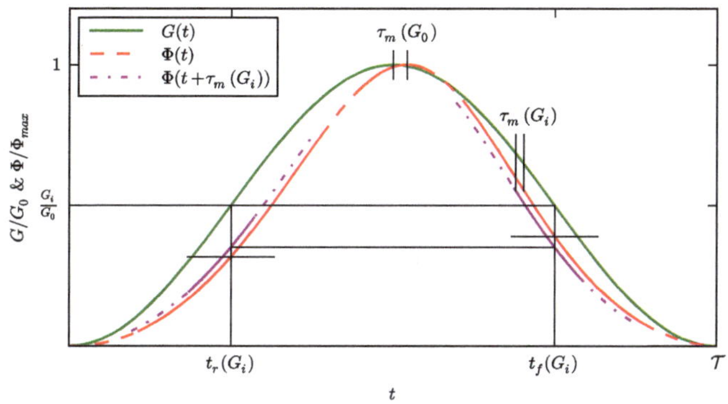

Fig. 8.3 Illustration of the injection-dependent determination of the time shift between excess carrier generation rate $G(t)$ and luminescence intensity $\Phi(t)$. The time shifts $\tau_m\,(G_i)$ corresponding to generation rates $G_i < G_0$ are determined by shifting the measured luminescence intensity $\Phi'\,(t) = \Phi\,(t + \tau_m\,(G_i))$ such that the condition $\Phi'\,(t_r\,(G_i)) \overset{!}{=} \Phi'\,(t_f\,(G_i))$ is satisfied. Here, measurement conditions departing from quasi-steady-state were deliberately chosen for illustrative purposes. Adapted after [2]

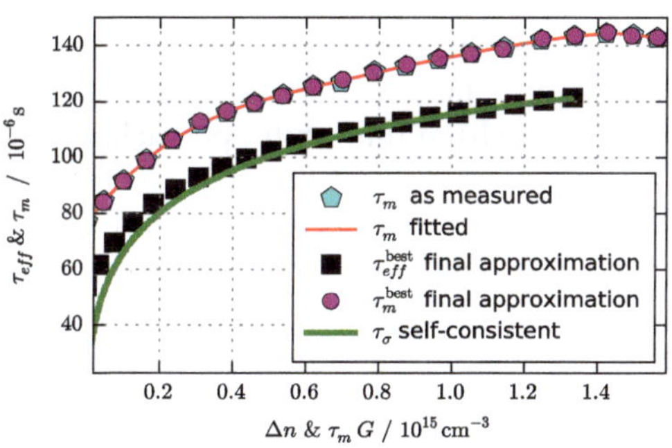

Fig. 8.4 *Self-sufficiet* lifetime analysis of a $d = 200\,\mu m$ thick silicon nitride passivated multicrystalline silicon wafer. The best approximation of effective lifetime τ_{eff}^{best} largely agrees with the self-consistent lifetime analysis τ_σ. This best approximation is determined via alignment of measured injection-dependent differential lifetimes τ_m with a best approximation thereof τ_m^{best} according to a finite-element solution of the time-dependent continuity equation. Adapted after [2]

of the continuity equation, the measured *differential* lifetimes $\tau_m\,(G_i)$ are apparently reproduced. The approach elaborated herein is referred to as *self-sufficient*, because it requires no a priori information about material parameters other than reflectivity and thickness (which are straightforwardly attainable). A self-consistent lifetime analysis of the same measurement is in good accordance with the *self-sufficient* lifetime analysis, as also shown in Fig. 8.4.

Table 8.1 Comparison of the **precision** of the *self-sufficient* versus the peak shift approach. 10 short (50 modulation cycles at $\mathcal{T} \approx 5 \times 10^{-2}$ s) quasi-steady-state harmonically modulated photoluminescence measurements were conducted on a silicon nitride passivated float zone wafer at a maximal irradiance equivalent to one sun. For the peak shift analysis, secondary measurements were conducted at 93 % of that maximal irradiance, respectively. The short measurement time was chosen in order to demonstrate the superiority of the *self-sufficient* approach in coping with statistical noise

Approach	τ_{eff}^{best} (μs)
peak shift	620 ± 170
self-sufficient	519 ± 5

8.1.3.2 Considerations of Accuracy and Precision

Besides the substantial experimental simplifcation of requiring only one (rather than a minimum of two harmonically modulated measurements), the *self-sufficient* approach is distinguished by a significant improvement of both measurement precision and accuracy, to be motivated in the following.

The major operational distinction between the generalized time shift approach and the peak shift approach relates to the manner in which they cope with perturbations. As stated above, the default peak shift approach (denoted (*i*) in Fig. 8.2) features an inherent tendency to systematically amplify relative perturbations of measured peak shifts, regardless whether these perturbations are of statistical or systematic origin. Likewise, in the generalized time shift approach, injection-dependent time shifts may be subject to scatter. However, here a sufficiently large number of time shifts is smoothly fitted prior to feeding this fit into the lifetime evaluation algorithm. This effectively reduces error amplification through Eq. 8.1. A tremendous improvement of measurement precision is also confirmed experimentally by the scattering of representative lifetime results shown in Table 8.1.

As for systematic variations of carrier lifetime during measurements (e.g., optically or thermally induced), at typical irradiation conditions significant lifetime changes may take place on a timescale much greater than modulation time (typically $10^{-2}\,\text{s} < \mathcal{T} < 10^{-1}\,\text{s}$), but not necessarily greater than the total duration of a harmonically modulated measurement (amounting to 5–100 s due to averaging of modulation cycles). The iteration method given by Eq. 8.1 is very sensitive to the slope of measured time shifts with respect to excess carrier generation rate. The bias imposed on that slope by a systematic variation of carrier lifetime during measurements is very different for the *self-sufficient* approach than for the peak shift approach. For the latter, the slope is systematically biased by the lifetime change on the timescale of the total duration of a measurement, while for the first it is only biased by the lifetime change on a timescale significantly shorter than modulation time $\mathcal{T}$. Through careful approximation of the latter timescale with $\frac{\mathcal{T}}{5}$, the resulting impact yet varies by a factor >250 (>10^3 in most practical cases) between these methods. This estimate theoretically corroborates, that systematic lifetime changes during measurements are of negligible impact for the *self-sufficient* approach.[7]

This superiority of the *self-sufficient* approach in terms of measurement **accuracy** is also confirmed by simulations of light-induced changes of carrier lifetime. For this purpose, two realistic (bilinear) scenarios of lifetime degradation

$$\tau_{eff}(\Delta n, t) = \tau_{eff}(\Delta n = 0, t = 0) + \frac{d\tau_{eff}}{d\Delta n}\Delta n + \frac{d\tau_{eff}}{dt}t \qquad (8.5)$$

were modeled: (a) fast lifetime degradation related to B–O complex dissociation, modeled on the basis of a time sequence of lifetime measurements shown in [11], and assuming an approximately linear lifetime decay (lifetime change per time interval) of $\frac{d\tau_{eff}}{dt} = -5 \times 10^{-7}$; (b) relatively slow Fe–B re-pairing after light-induced dissociation, modeled according to data given in [13], assuming an approximately linear lifetime decay of $\frac{d\tau_{eff}}{dt} = -5 \times 10^{-8}$. In a simplified model, the injection dependence of carrier lifetime was set to a constant $\frac{d\tau_{eff}}{d\Delta n} = 1.3 \times 10^{-20}\,\text{cm}^3\,\text{s}$. This is a reasonable assumption, as such values are often encountered in the respective injection range [14]. The time-dependent photoluminescence intensity was then modeled for both scenarios with a modulation time $\mathcal{T} = 1.25 \times 10^{-2}$ s, peaking at an excess carrier generation rate of $G_0 = 10^{19}\,\text{cm}^{-3}\,\text{s}^{-1}$.[8] Assuming a linear decay of carrier lifetime during measurements at the above stated characteristic lifetime degradations, two successive single-cycle measurements with an intermediate time interval of 12.5 s (in order to account for typically 10^3 modulation cycles) were simulated. Net dopant concentration was set to $n_0 = 1.5 \times 10^{16}\,\text{cm}^{-3}$ for the simulation of photoluminescence intensity. The thereby modeled photoluminescence intensities were subsequently analyzed and interpreted in terms of injection-dependent carrier lifetime via both, the peak shift approach and the *self-sufficient* approach.

The resulting injection-dependent carrier-lifetime curves are depicted in Fig. 8.5 and compared to the true carrier lifetime at the time of the first of the two measurements, respectively. As the peak shift analysis combines two consecutive measurements featuring an intermediate time interval of 12.5 s, it should yield the carrier lifetime 6.25 s after the first measurement. Therefore, in case (*a*) one would expect a degradation of $\Delta\tau_{eff} = 3\,\mu\text{s}$ with respect to the initial lifetime curve, whereas in case (*b*) one would expect a degradation of $\Delta\tau_{eff} = 300$ ns after 6.25 s. However, in case (*a*) the peak shift lifetime is $\Delta\tau_{eff} = 50\,\mu\text{s}$ lower than true initial lifetime as opposed to the expected $\Delta\tau_{eff} = 3\,\mu\text{s}$, and in case (*b*) the peak shift lifetime is $\Delta\tau_{eff} = 5\,\mu\text{s}$ lower than the true initial lifetime as opposed to the expected $\Delta\tau_{eff} = 300$ ns. In contrast, the *self-sufficient* lifetime exactly reproduces true initial lifetime in both cases, demonstrating the superiority of the *self-sufficient* approach over the peak shift approach in terms of measurement accuracy.

[7] Note that the same applies to the self-consistent lifetime analysis.

[8] A second simulated measurement, which is required for the peak shift lifetime analysis, was modeled at 95 % of that excess carrier generation rate.

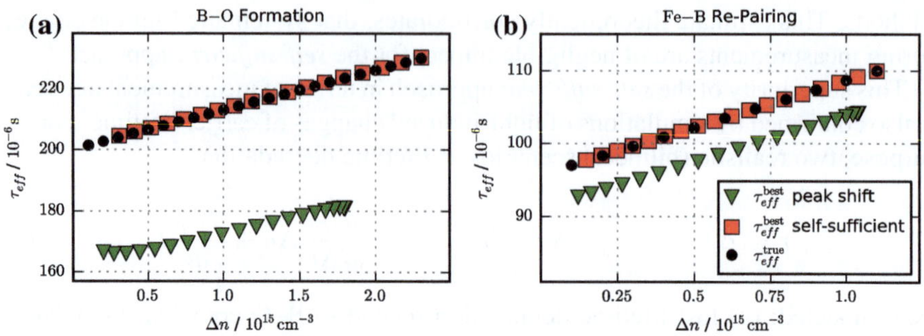

Fig. 8.5 Illustration of the effect of carrier lifetime degradation in the course of harmonically modulated photoluminescence measurements. Typical lifetime degradation rates were taken from published data **a** on a fast B–O related degradation [11] and **b** on slow Fe–B re-pairing [13]. Harmonically modulated photoluminescence intensity was based on the lifetime curves τ_{eff}^{true} as plotted, also accounting for a linear lifetime degradation during measurements. Modeled data were then analyzed in terms of effective lifetime via both, the peak shift approach as well as the *self-sufficient* approach—substantiating the superiority of the *self-sufficient* approach in terms of measurement accuracy. Adapted after [2]

8.1.4 *The* d2a *Approach*

Both peak shift and *self-sufficient* lifetime analyses perform the assignment $\tau_{eff}^{j}(G_i) \mapsto \tau_m^j(G_i)$ via the numerical finite element solution of the time-dependent continuity equation as described in Appendix A.3.1. With the theory of light-biased decay time (cf. Eq. 6.41), such extensive numerics can be replaced by a simple analytical relation between *differential* and *actual* lifetime. In fact, the combination of the generalized time shift determination with Eq. 6.41 in the limit of an infinitesimal bias light parameter $\xi \to 1$

$$\tau_m(G_i) = \frac{\tau_{eff}(G_i)}{1 - G_i \frac{d\tau_{eff}}{d\Delta n}\Big|_{\Delta n = \tau_{eff}(G_i)G_i}} \tag{8.6}$$

substantially reduces numerical effort and thereby accelerates the phase-sensitive determination of effective lifetime. This approach is herein denoted as the *differential-to-actual* lifetime technique (*d2a*).

8.1.4.1 Linear Lifetime: Algebraic Treatment

With measured *differential* lifetimes $\tau_m(G_i)$ at given excess carrier generation rates G_i, the sought quantity is *actual* effective lifetime $\tau_{eff}(G_i)$. If carrier lifetime is a (locally) linear function of excess carrier density, two measurements of *differential* lifetime allow a straightforward algebraic determination of *actual* lifetime. As

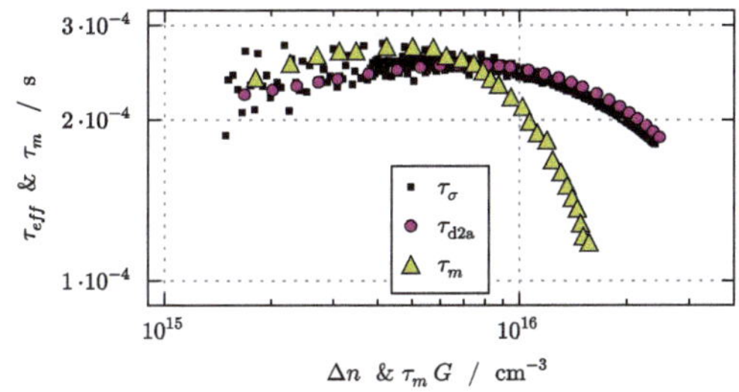

Fig. 8.6 Experimental validation of the *d2a* technique via a harmonically modulated photoluminescence measurement on a $d = 160\,\mu m$ thick surface-passivated silicon wafer featuring a net dopant concentration of $n_0 = 1.5 \times 10^{16}\,cm^{-3}$. *Differential* lifetime τ_m (as determined via the generalized time shift approach) considerably differs from the best approximation of actual lifetime τ_{d2a}, which virtually coincides with a self-consistent lifetime analysis τ_σ. Adapted after [15]

$\tau_{eff}(G_1)$ and $\tau_{eff}(G_2)$ can then be expressed as functions of each other and the *constant* derivative of lifetime with respect to excess carrier density $\frac{d\tau_{eff}}{d\Delta n}$, Eq. 8.6 represents a system of **two** equations featuring effectively **two** unknowns $\tau_{eff}(G_1)$ and the constant $\frac{d\tau_{eff}}{d\Delta n}$.

8.1.4.2 Generalized Analysis: Iterative Procedure

The fact that the assumption of linearity of $\tau_{eff}(\Delta n)$ is commonly violated complicates the straightforward algebraic determination of *actual* effective lifetime, because the number of unknowns always exceeds the number of available equations. However, applying the iterative procedure of Eq. 8.1 to a **smooth** fit of a minimum of three measured *differential* lifetimes yields a very accurate approximation of *actual* effective lifetime. This is experimentally validated in Fig. 8.6 via comparison with a self-consistent lifetime analysis of a wafer of accurately known net dopant concentration n_0.

8.1.4.3 Nonuniform Excess Carrier Density and Generation Rate

Systematic errors in the determination of *actual* lifetime from *differential* lifetime measurements in the case of nonuniform excess carrier density and generation rate have previously been addressed [16, 17]. The following considerations suggest that the *d2a* lifetime analysis is unaffected by systematic errors due to nonuniform excess carrier density and generation rate.[9] Nonuniform depth distributions occur if wafer thickness is significantly greater than both light penetration depth α_γ^{-1} and minority carrier diffusion length L. As lifetime analyses based on the time-dependent continuity equation (cf. Eq. 3.75) use depth-averaged values of excess carrier density and

[9] Proceeding along the lines of [15].

generation rate, both quantities are effectively underestimated in the nonuniform case due to a reduced electronically active effective substrate thickness $d_{eff} < d$ (bulk thickness featuring significant excess carrier density). This could be taken into account in a depth-averaged treatment by introducing a factor $\varkappa = \frac{d}{d_{eff}} \geq 1$ relating excess carrier generation rate to the electronically active bulk thickness d_{eff}. The time-dependent continuity equation would then read

$$\frac{d\Delta n\,(t)}{dt} + \frac{\Delta n\,(t)}{\tau_{eff}} = \varkappa G\,(t)\,. \tag{8.7}$$

With such enhanced depth-averaged excess carrier generation rate $\varkappa G(t)$, the resulting excess carrier density according to Eq. 8.7 is enhanced by the same factor $\varkappa$. The factor $\varkappa$ then cancels out in the expression of *differential* lifetime according to Eq. 6.38. Consequently, *actual* lifetime according to Eq. 8.6 is unaffected by $\varkappa > 1$.

The determination of a depth-averaged excess carrier density to be assigned to *actual* lifetime τ_{eff} is erroneous for a factor $\varkappa$ significantly in excess of unity, albeit corrigible through an intensity-weighted averaging of excess carrier density, as previously proposed, e.g. by Bowden and Sinton (cf. Eq. 8.14) [18].

8.1.4.4 Imbalance or Asymmetry of Generation Rate

The shape of excess carrier generation rate may deviate from an exact sinusoidal shape for technical reasons. This could bring about a departure from the symmetry of $G(t)$ with respect to its maximum. In order to ensure that the generalized time shift approach as elaborated in Sect. 8.1.3 is unaffected by such an imbalance, its potential impact was investigated via a numerically modeled measurement.

For this purpose, in the second half of a sinusoidal modulation cycle, harmonically modulated excess carrier generation rate $G(t)$ (cf. Eq. 3.83) was decorated with an imbalance $\delta G(t)$. With a symmetrical excess carrier generation rate

$$G_{\rm sym}\,(t) = G_0 \sin^2 \frac{\pi t}{T}, \tag{8.8}$$

an imbalance

$$\delta G\,(t) = \begin{cases} 0 & \forall \;\; t \leq \frac{T}{2} \\ -\frac{G_0}{20} \sin^2 \frac{2\pi t}{T} & \forall \;\; t > \frac{T}{2} \end{cases} \tag{8.9}$$

gives rise to an asymmetrical excess carrier generation rate $G_{\rm asym}(t) = G_{\rm sym}(t) + \delta G(t)$, as plotted in Fig. 8.7. Generalized time shifts were determined from numerically modeled photoluminescence measurements (cf. Appendix A.3.1) of symmetrical and asymmetrical configurations, respectively. Subsequently, a *d2a* lifetime analysis was performed. As can be seen from the resulting curves plotted in

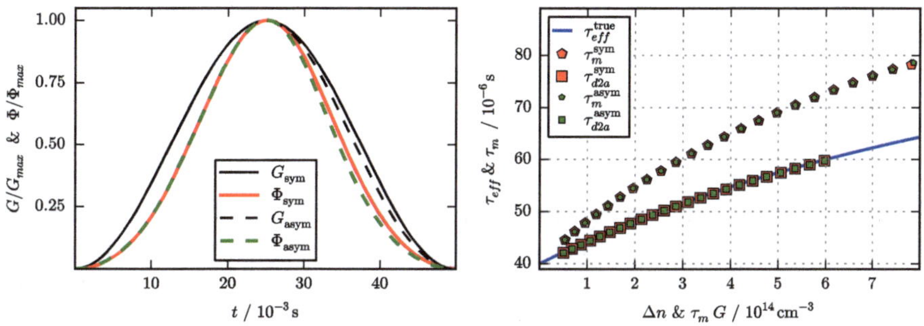

Fig. 8.7 Investigation of the impact of an imbalance of excess carrier generation rate on the generalized time shift analysis as proposed in Sect. 8.1.3: On the left, modeled relative excess carrier generation rate $G(t)$ (with $G_0 = 10^{19}\,\text{cm}^{-3}\,\text{s}^{-1}$) and photoluminescence intensity $\Phi(t)$ with and without an imbalance are plotted. $\Phi(t)$ was modeled on the basis of a net dopant concentration $n_0 = 1.5 \times 10^{15}\,\text{cm}^{-3}$ and of the lifetime curve τ_{eff}^{true} shown on the right. The generalized time shifts τ_m depicted on the right reveal no significant mismatch between symmetrical (G_{sym}) and asymmetrical (G_{asym}) shapes of excess carrier generation rate. Therefore, the phase-sensitive (*d2a*) lifetime analyses coincide as expected—even for the disproportionately large imbalance shown on the left

Fig. 8.7, the imbalance of $G(t)$ is of negligible impact for the generalized time shift approach—even in the case of a disproportionately large imbalance, as assumed here.

8.2 Broad Range Injection-Dependent Lifetime

In this section, the determination of effective lifetime via quasi-steady-state harmonically modulated luminescence is combined with a steady-state luminescence configuration capable of scanning a very broad range of excess carrier densities. Thereby, the advantages of harmonically modulated lifetime techniques are combined with the benefits of steady-state luminescence under extremely low injection conditions—particularly in terms of carrier trapping and statistical noise. Fluctuations of photocurrent are identified as the essential source of uncertainty under such low injection, and a measure to cope with such fluctuations is proposed and implemented. Lifetime measurements at excess carrier densities of $\Delta n \approx 10^8\,\text{cm}^{-3}$ are shown in detail, also revealing novel material-related insights. This approach is also exploited for broad range injection-dependent lifetime measurements on solar cells via harmonically modulated electroluminescence.[10]

[10] The account given here is adapted from the corresponding original publications [19, 20].

8.2.1 Combined Modulated and Steady-State Approach

Injection-dependent carrier lifetime measurements at very low injection conditions feature numerous applications in both fundamental research and quality control in an industrial environment. Such applications involve the determination of defect parameters such as defect energy levels and capture cross sections, the experimental validation of recombination theory [21, 22] (cf. Eq. 3.12), or the investigation of recombination mechanisms of novel solar cell technologies (e.g. of surface recombination concepts).

As previously stated, harmonically modulated luminescence lifetime methods feature numerous advantages due to a most immediate time domain access. This virtually eliminates signal amplitude artifacts due to the self-absorption of luminescence, the substrate interface morphology, or possibly erroneous assumptions about carrier mobilities.

Toward very low injection, the harmonically modulated time domain approach may feature several limitations: As elaborated in Sect. 3.1.6, trapping could cause artifacts if excess carrier density Δn becomes comparable or less than the trap density N_t. Further, if pn-junction devices are investigated, the modulation of junction capacitance may cause lifetime artifacts at very low injection [23]. Most importantly however, the above advantage due to the circumvention of signal amplitude artifacts turns into a disadvantage when it comes to measuring relatively weak signals at low signal-to-noise ratios. Amplitude information of such weak signals may remain precisely distinguishable, whereas phase information is already lost in noise. Steady-state luminescence is free of any of the addressed potential artifacts. This suggests the combination of a harmonically modulated luminescence lifetime measurement at sufficiently high injection with a steady-state broad range injection-dependent scan of luminescence intensities $\Phi(G_i)$.[11] The key equation of this approach is again the relation between photoluminescence intensity and excess carrier density (cf. Eq. 4.36)

$$\Phi(G_i) = c_\Phi \Delta n(G_i)(n_0 + \Delta n(G_i)). \tag{8.10}$$

The scaling factor c_Φ is determined self-consistently or via the phase-sensitive *d2a* approach. Subject to an iterative correction of c_Φ due to the injection dependence of the coefficient of radiative recombination (detailed in Sect. 8.2.3), this scaling factor can subsequently be used for a steady-state luminescence lifetime determination according to Eq. 8.10 at any injection level. In steady state, the relation

$$\Delta n(G_i) = \tau_{eff}(\Delta n) G_i \tag{8.11}$$

[11] Previous publications of lifetime measurements at very low injection conditions were based on quasi-steady-state modulated luminescence measurements [24, 25] calibrated via photoconductance [7, 8].

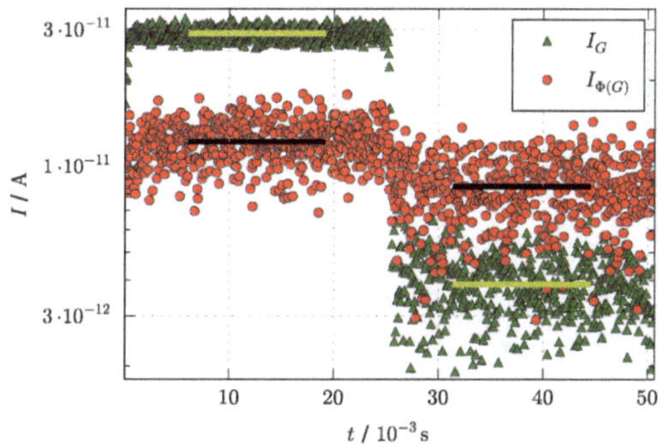

Fig. 8.8 Representative measurement of low injection luminescence-induced ($\propto\ \Phi(G_i)$) and irradiance-induced ($\propto\ G_i$) photocurrents I in the steady-state measurement configuration. Photocurrents vary between a state of finite steady-state illumination (*left half*) and a state without illumination (*right half*). The displayed data represent an average over 10^2 modulations at a modulation time $\mathcal{T} = 5.07 \times 10^{-2}$ s and a readout rate of 10^3 samples per modulation. Also, average photocurrent amplitudes are plotted within their respective averaging time interval. Adapted after [19]

is accurate, thereby allowing the straightforward determination of injection-dependent effective lifetime $\tau_{eff}(\Delta n(G_i))$ as the positive solution of the quadratic Eq. 8.10 upon measurement of a set of luminescence intensities $\Phi(G_i)$.

The essential modification of the measurement procedure needed for this combined approach was the implementation of a modulation of excess carrier generation rate via a rectangular or square waveform (rather than a harmonical modulation). As for the harmonically modulated lifetime measurement, both luminescence intensity $\Phi(G_i)$ and a reference signal (irradiance detector) proportional to the corresponding excess carrier generation rate G_i are recorded. During the first half of a modulation cycle, a finite and constant excess carrier generation causes constant light induced photocurrents in both detectors, whereas during the second half of a modulation cycle, dark photocurrents are measured. The net photocurrents (i.e. the difference between light-induced and dark photocurrents) are then to be evaluated according to Eq. 8.10. The total magnitude G_i of excess carrier density is obtained from the ratio of photocurrents of the irradiance detector between the harmonically modulated calibration measurement and the steady-state measurement of $\Phi(G_i)$. In order to obtain sufficiently precise net photocurrents, averaging over a sufficient amount of modulations of the square waveform is performed. An example photocurrent measurement is shown in Fig. 8.8

8.2.2 Fluctuation of Dark Photocurrent

The lowest injection lifetime measurement shown in the following (cf. Sect. 8.2.4) corresponds to a luminescence net photocurrent of $I_\Phi = 1.6 \times 10^{-14}$ A, which is significantly less than the statistical uncertainty of dark photocurrent due to fluctu-

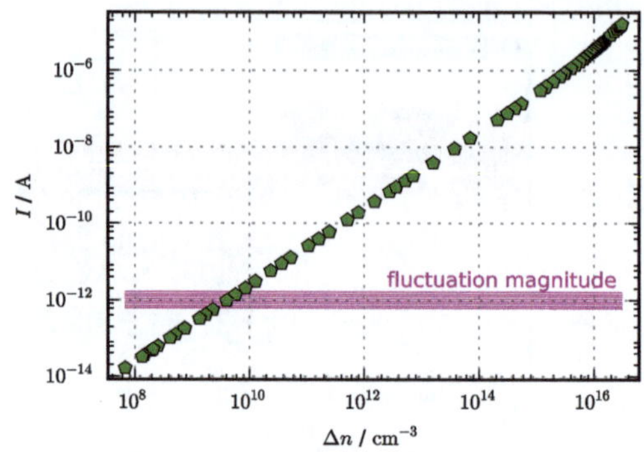

Fig. 8.9 Depiction of the luminescence detector's photocurrent I as a function of excess carrier density Δn. The data shown correspond to the lifetime measurement depicted in Fig. 8.11. The characteristic magnitude of fluctuations of dark photocurrent indicates that the lowermost two orders of magnitude in both I and Δn could not be resolved, if fluctuations of photocurrent were not taken into account. Adapted after [19]

ations. Elimination of this source of uncertainty requires a method, which accounts for the possibility of fluctuations of dark photocurrent on any relevant timescale.

Therefore, the characteristic magnitude of such fluctuations was quantified on all relevant accessible timescales via analysis of the scattering of measured dark photocurrents as a function of the measurement duration. From this, fluctuations of dark photocurrent were found to range between 0.6 and 1.6×10^{-12} A, featuring a slightly negative derivative with respect to the investigated timescale. Accordingly, the targeted resolution in terms of precisely measurable photocurrents is exceeded by these fluctuations by two orders of magnitude, as visualized in Fig. 8.9. Therefore, it is essential to find a measurement design unaffected by fluctuations to a degree such that the remaining measurement uncertainty would be significantly less than the minimum of measured net photocurrents. This was achieved by composing one measurement (represented by one point in the lifetime curve depicted in Fig. 8.11) out of a sequence of periodic cycles, each featuring one time interval of constant and finite excess carrier injection, and another time interval with zero carrier injection—equivalent to a periodic square waveform. If the duration of each time interval is much greater than both, effective carrier lifetime and any response time of the measurement system, steady-state measurement conditions are fulfilled (cf. Fig. 8.8).

A measurement design of the latter kind is suited to properly account for fluctuations of dark photocurrent on any relevant timescale:

- Fluctuations on a very short timescale compared to the duration of a modulation cycle are eliminated through averaging over a sufficient amount of samples within each time interval (typically 250–500 samples are averaged).
- Fluctuations on an intermediate and long timescale compared to the duration of a modulation cycle are eliminated through averaging over a sufficient number of modulation cycles.

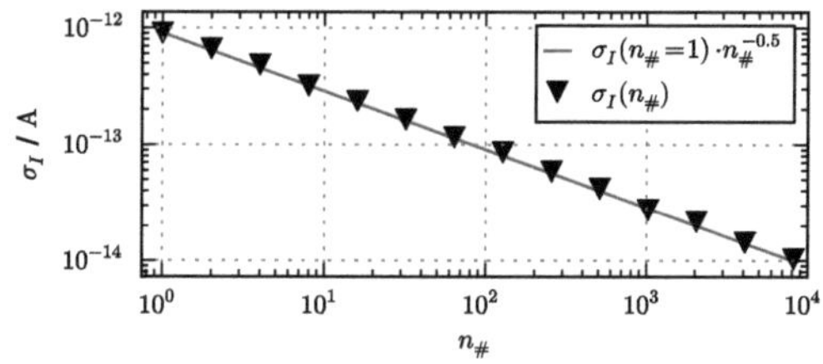

Fig. 8.10 Analysis of statistical noise: the standard deviation of the mean value of measured photocurrents σ_I is plotted here as a function of the number of averaged modulations $n_\#$ of a square waveform with a modulation time $\mathcal{T} = 1.25 \times 10^{-2}$ s. The nearly exact relation $\sigma_I \propto n_\#^{-\frac{1}{2}}$ confirms the purely statistical nature of measurement noise throughout the entire range of photocurrents investigated here. Adapted after [19]

Although the specific case of a large drift of photocurrent on the timescale of and greater than measurement time could still theoretically affect the measurement, this is not relevant in practice. This is because it is known from the above quantification of the magnitude of fluctuations of dark photocurrent that such a slow drift does not significantly exceed 10^{-12} A. By taking the average of typically $10^3 \lesssim n_\# \lesssim 2 \times 10^4$ modulations, the resulting error caused by such slow drift can be estimated to range significantly below 10^{-15} A, which is much less than the minimum of photocurrents depicted in Fig. 8.9.

Although weak systematic oscillatory artifacts have occasionally occurred in harmonically modulated luminescence measurements in the past,[12] it was possible to prove that such systematic perturbations are usually negligible when compared to statistical noise (i.e. fluctuation of photocurrent). The predominantly statistical nature of noise on any relevant scale of photocurrents is demonstrated in Fig. 8.10.

8.2.3 Data Corrections

Both effective lifetime and the excess carrier density assigned to it may be subject to data corrections to be addressed here:

8.2.3.1 Corrections of Carrier Lifetime

Several effects may cause systematic errors of a simplistic and straightforward steady-state lifetime determination based on Eq. 8.10. The most relevant effect is due to the injection dependence of the integral coefficient of radiative recombination B (cf. Sect. 4.2.2 and Appendix A.5), which is an essential constituent of the scaling factor c_Φ. Further, at shallow (i.e. highly nonuniform) excess carrier generation, the

[12] Such artifacts are likely to be induced by strong electromagnetic fields of other instruments.

factor c_Φ may slightly depend on the depth profile $\Delta n(z)$ due to self-absorption of luminescence in the specimen. If depth profiles between the calibration measurement and one particular steady-state lifetime measurement vary greatly due to differing carrier lifetimes, this could also affect the fraction ξ_α of detectable luminescence intensity, which is reabsorbed within the specimen [26, 27]. With the scaling factor obeying

$$c_\Phi \propto \left(1 - \xi_\alpha\left(\tau_{eff}\right)\right) B(\Delta n), \tag{8.12}$$

with a harmonically modulated lifetime calibration $\tau_\sigma(\Delta n_\sigma)$, and attempting to determine the effective carrier lifetime $\tau_{eff}(G_i)$ from the measured luminescence intensity $\Phi(G_i)$, the corresponding corrected luminescence intensity is

$$\Phi'(G_i) = \Phi(G_i)\,\frac{\left(1 - \xi_\alpha(\tau_\sigma)\right) B(\Delta n_\sigma)}{\left(1 - \xi_\alpha\left(\tau_{eff}(G_i)\right)\right) B(\Delta n(G_i))}. \tag{8.13}$$

As neither $\tau_{eff}(G_i)$ nor $\Delta n(G_i)$ are a priori given, this correction needs to be performed iteratively.[13] As for the correction of self-absorption, depth profiles $\Delta n(z)$ were calculated on the basis of the steady-state solution of the continuity equation (cf. Sect. 3.2.2).[14] The factor ξ_α was then determined via a simplified optical model, neglecting multiple reflections in the specimen and assuming planar interfaces. This correction is uncertain with regard to both, assumptions about surface recombination velocity, and a possibly rough surface morphology. Comparing this correction to the worst case scenarios as discussed by Trupke [26], it is likely to slightly overcompensate the true mismatch. In practice however, due to the fact that InGaAs detectors and silicon absorption filters are used in the setup, the typical impact of the reabsorption correction is virtually negligible compared to the typical impact of the correction due to $B(\Delta n)$.

8.2.3.2 Corrections of the Assigned Excess Carrier Density

Substantial deviations between depth-averaged excess carrier density (as used for the dynamic lifetime analysis) and the very excess carrier density giving rise to radiative recombination occur if the depth distribution $\Delta n(z)$ is nonuniform. With an analogous line of reasoning as provided in the discussion of the impact of nonuniform excess carrier density on the *d2a* lifetime analysis in Sect. 8.1.4, it can be shown that the determination of effective lifetime from Eq. 8.10 is unaffected by such errors of assigned excess carrier density. This is because lifetime is effectively determined as the ratio between excess carrier density and generation rate $\tau_{eff} = \frac{\Delta n}{G}$, while both

[13] Here, a possible correction of the assigned effective excess carrier density as detailed in the following paragraph should also be considered.

[14] The solution $\Delta n(z)$ was obtained with $\tau_{eff} = \tau$ and $S_0 = S_d = 0$, and provided minority carrier mobilities according to the Klaassen model [28, 29].

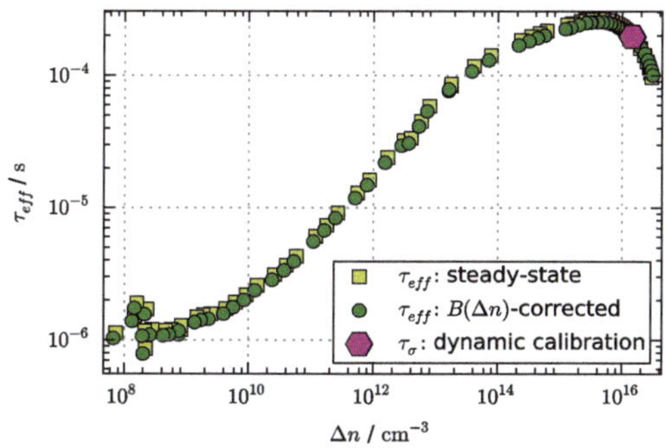

Fig. 8.11 Broad range injection-dependent lifetime measurement on a $2 \times 2\,\text{cm}^2$ large spot of a $d = 160\,\mu\text{m}$ thick multicrystalline p-type silicon wafer featuring a silicon nitride surface passivation. The lifetime analysis results from a steady-state photoluminescence scan according to Eq. 8.10, which was calibrated via harmonically modulated photoluminescence (here: self-consistent determination of lifetime τ_{σ}). The data correction due to the injection dependence of the coefficient of radiative recombination $B(\Delta n)$ is detailed in Sect. 8.2.3. The number of modulations per measured point was adapted to the noise level (with $20 \leq n_{\#} \leq 10^4$ modulations at a modulation time of $\mathcal{T} = 1.25 \times 10^{-2}$ s). Total measurement time was less than 21 min, while the lowermost 10 measured points (lowermost order of magnitude of Δn) took two thirds of that time. Adapted after [19]

quantities suffer from the exact same mismatch $\varkappa$ between depth-averaged and most appropriate effective values (cf. Eq. 8.7). Yet, the *effective* excess carrier density to be assigned to a correctly determined lifetime must be separately determined, if Δn suffers from substantial depth nonuniformity. For this purpose, the weighted average [18]

$$\Delta n_{eff} = \frac{\int dz\, \Delta n^2\,(z)}{\int dz\, \Delta n\,(z)} \tag{8.14}$$

is calculated on the basis of the depth profile $\Delta n\,(z)$ obtained from the steady-state solution of the continuity equation (cf. Sect. 3.2.2). The aim of this weighted average is to find a best approximation of the excess carrier density at the location of luminescence emission. Obviously, this correction is only useful, if effective lifetime approximately conincides with bulk lifetime. If this condition is not fulfilled (e.g. due to dominant surface recombination), the approximation of Δn_{eff} may be inadequate.[15]

[15] Note however, that the nonuniformity of $\Delta n\,(z)$ caused by dominant surface recombination may only lead to deviations less than $\frac{4}{3}$ between Δn_{eff} according to Eq. 8.14 and depth-averaged excess carrier density (cf. Sect. 9.2). In contrast, for very small bulk lifetimes of thick substrates, the corresponding effect may well amount to factors >10.

8.2.4 Experimental Results

Figure 8.11 depicts broad range injection-dependent lifetime measurement as elaborated above, spanning a range of excess carrier densities of $8 \times 10^7\,\text{cm}^{-3} < \Delta n < 3 \times 10^{16}\,\text{cm}^{-3}$, corresponding to an irradiance range equivalent to $5 \times 10^{-6} - 19$ suns.

Most prominently, the measurement shown in Fig. 8.11 contains information about the injection dependence of the surface recombination velocity of the investigated p-type silicon wafer ($n_0 \approx 2 \times 10^{16}\,\text{cm}^{-3}$) featuring a silicon nitride passivation. Whereas the injection dependence of lifetime in the range of $\Delta n \sim 10^{15}\,\text{cm}^{-3}$ is likely to be attributed to the dominant bulk recombination, one would expect bulk lifetime to stabilize at a constant low-level injection value at excess carrier densities several orders of magnitude below net dopant concentration [21, 22] (cf. Eq. 3.12). Yet, this stabilization does not occur, and the measured effective carrier lifetime keeps decreasing down to a constant value of $\tau_{eff} \approx 10^{-6}$ s at an injection range of $\Delta n < 10^9\,\text{cm}^{-3}$, suggesting a dramatic increase of surface recombination velocity as the cause.[16]

Assuming a homogeneous depth profile of excess carrier generation rate $G(z)$, this would imply the complete breakdown of surface passivation (i.e. invariance of excess carrier density upon further increase of surface recombination velocity) at the given wafer thickness of $d = 160\,\mu\text{m}$ [30].[17] However, with a room temperature penetration depth of $\alpha_\gamma^{-1} \approx 10\,\mu\text{m}$ for 791 nm irradiation in crystalline silicon [31], the depth-dependent solution of the steady-state continuity equation (cf. Sect. 3.2.2) yields an effective carrier lifetime of $\tau_{eff} \approx 3 \times 10^{-7}$ s for infinitely large surface recombination velocities. Thus, a complete breakdown of surface passivation can neither be unambiguously concluded nor excluded from this experiment. However, surface recombiation velocity can be estimated to feature a lower bound of $S > 10^4\,\text{cm}\,\text{s}^{-1}$ according to the steady-state solution of the continuity equation. The exclusion of the complete breakdown of surface passivation is not possible, because there is no unambiguous experimental counterevidence to the existence of a systematic artifact leading to a stabilization of a spurious measured effective lifetime τ_{eff} at a higher level than true τ_{eff}.[18]

Surface recombination velocity at c-Si/SiN_x interfaces has long been known to increase with decreasing excess carrier density [32–35]. Steingrube et al. recently suggested an explanation of this effect via surface damage [36].

---

[16] The lateral conductivity due to an inversion layer caused by surface charges is unlikely to give rise to such dramatic effects. It is estimated to amount to less than 1 % compared to typical emitters.

[17] Corresponding data on the minority carrier mobility were taken from the Klaassen model [28, 29].

[18] A stabilization of measured carrier lifetime at a higher level than true carrier lifetime could e.g. be caused by a significant fraction of irradiation light in the luminescence detector. Although experimental evidence suggests that this does not occur in the present modulated luminescence setup down to effective carrier lifetimes substantially less than 10^{-6} s (cf. Sect. 7.2.2), this test did not consider the operation of the used light sources below the laser threshold (LED operation with spectrally broadened irradiation).

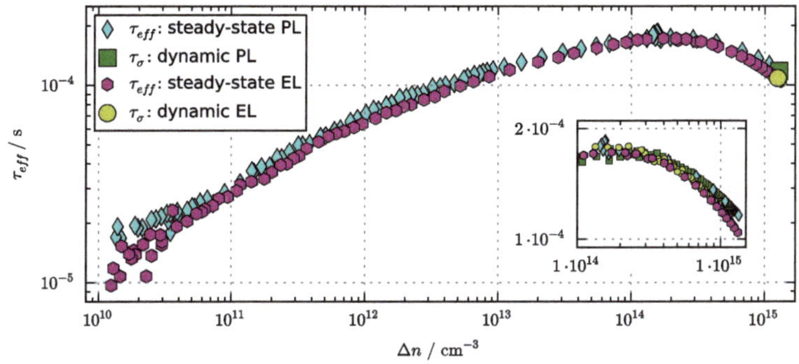

Fig. 8.12 Electroluminescence-based measurement results of broad range injection-dependent effective carrier lifetime τ_{eff} of a solar cell, whose harmonically modulated lifetime is also depicted in Fig. 7.4 (cf. inserted plot here). The dynamic (self-consistent) lifetime measurement τ_σ was used for a calibration of a set of steady-state electroluminescence measurements $\Phi(G_i)$. Adapted after [20]

Concerning the lowermost (leftmost) part of the injection-dependent lifetime curve shown in Fig. 8.11, it must be noted that there is some uncertainty due to statistical noise of very low measured photocurrents. In spite of the scattered measurement points in the vicinity of $\Delta n \sim 10^8$ cm^{-3}, the stabilization of carrier lifetime between excess carrier densities of 10^8–10^9 cm^{-3} can be deemed robust on the basis of the other measurement points available in this range, because these points feature photocurrents well above the statistical noise level as specified in Fig. 8.10 (with photocurrents as of Fig. 8.9). Concerning the statistical noise level, there is still room for improvement via an enhancement of the solid angle of luminescence capture, or via more efficient filter configurations [37]. Such improvements could bring about a higher measurement precision or significantly shorter measurement times.

Figure 8.12 compares broad range injection-dependent effective lifetime reults obtained via both electro- and photoluminescence measurements on a silicon solar cell (specifics of the electroluminescence experimental setup and of the investigated solar cell are to be found in Sect. 7.4), showing good agreement of both techniques throughout the investigated five orders of magnitude in Δn. This reference measurement demonstrates the applicability of the broad range injection-dependent lifetime approach to modulated electro- and photoluminescence measurement configurations alike.

8.3 Lifetimes from Photoluminescence Images

This section elaborates the effective lifetime calibration of steady-state luminescence images via quasi-steady-state harmonically modulated luminescence lifetime measurements. An averaging procedure of local luminescence intensities of wafers or solar cells featuring a laterally inhomogeneous distribution of lifetime is motivated, thereby providing the hitherto missing link between harmonically modulated lifetime and its spatially resolved constituents.[19]

8.3.1 Requirements and Constraints for Accurate Lifetimes

Common objectives of carrier lifetime measurements are process control as well as the prediction of cell performance in any stage of solar cell production. While process control often requires only relative lifetime information, the prediction of solar cell voltage or current calls for accurate quantitative carrier lifetime information. For laterally inhomogeneous distributions of carrier lifetime, this implies the need of both an accurate spatially resolved measurement technique (to be aspired here) and an appropriate averaging procedure accounting for the interaction of well and badly performing areas within a solar cell [39, 40].

The endeavor to reduce solar cell production cost fuels the demand for accurate lifetime measurements on low cost material such as multicrystalline upgraded metallurgical grade silicon, which often exhibits relatively low lifetimes and high trap densities. Also, such material features carrier mobilities that are not straightforwardly predictable due to the possibility of crystal defects acting as mobility barriers [41–50], and due to the mobility lowering effect of partially compensating opposing dopant species [28, 51]. The fact that such material does not allow a straightforward prediction of carrier mobilities also implies that conventional measurements of net dopant concentration via eddy current or resistive techniques must be questioned, as they rely on the validity of carrier mobility models. Therefore, lifetime techniques which substantially depend on assumptions about net dopant concentration or carrier mobility appear unfavorable in the light of the above mentioned restrictions. A lifetime technique which is unaffected by these restrictions is deemed highly relevant.

Camera-based luminescence images [52, 53] yield steady-state effective lifetime information at a high spatial resolution (typically $10^{-2} \times 10^{-2}\,\mathrm{cm}^2$), while being unaffected by trapping and not requiring a priori information about carrier mobilities.[20] Yet, steady-state luminescence amplitude is very sensitive to net dopant concentration n_0, and it may be substantially altered through nonplanar interface morphologies. These limitations complicate an ab initio lifetime determination and call for alternative access to spatially resolved lifetime. A photoconductance-based calibration of luminescence images [54] could resolve some of the addressed problems. However, it is very prone to trapping [55, 56], it requires very accurate information about both minority and majority carrier mobility, and it may suffer from the attenuation of eddy currents through crystal defects—particularly in multicrystalline material [50]. Dynamic lifetime calibration techniques of steady-state luminescence images [14, 57] or stand-alone spatially resolved dynamic luminescence lifetime

[19] The account given here is adapted from the corresponding original publications [14, 38].

[20] Note that the determination of very high surface recombination velocities from steady-state luminescence requires knowledge of minority carrier mobility.

techniques [58] appear to be promising alternatives—albeit not free of limitations (cf. Sect. 8.3.4)—less prone to severe artifacts.

The effective lifetime calibration of luminescence images via quasi-steady-state harmonically modulated luminescence to be elaborated in the following is independent of carrier mobilities and virtually unaffected by net dopant concentration n_0. As for trapping, a systematic error cannot be excluded from a theoretical standpoint (cf. Sect. 3.1.6). Empirically however, harmonically modulated luminescence appears to be substantially less affected by trapping than photoconductance (cf. Appendix C).

8.3.2 Derivation of Averaging Procedure

With a lateral resolution of harmonically modulated lifetime measurements on the order or greater than typical areas of homogeneous electronic material quality in multicrystalline silicon substrates, the question how such different substrate areas contribute to measured harmonically modulated lifetime τ_σ is to be addressed.[21] Therefore, the analytical core of the method derived here is an averaging algorithm, which relates an ensemble of possibly differing local effective lifetimes $\tau_{eff,i}$ to one aggregated measurable harmonically modulated lifetime τ_σ.

8.3.2.1 Steady-State Luminescence Imaging

Local steady-state luminescence intensity φ_i—as e.g. measurable in one pixel i of a charge coupled device (CCD) camera—can be expressed in terms of local effective carrier lifetime[22] $\tau_{eff,i}$ (cf. Eq. 4.36 or 8.10) as

$$\varphi_i = c_\varphi G_0 \tau_{eff,i} \left(G_0 \tau_{eff,i} + n_0 \right) . \tag{8.15}$$

Here, G_0 denotes a laterally homogeneous excess carrier generation rate, n_0 denotes net dopant concentration, and c_φ denotes a characteristic scaling factor, which is assumed to be laterally homogeneous throughout the investigated substrate. If both the specific scaling factor c_φ of a luminescence image and net dopant concentration n_0 were known, local effective lifetime $\tau_{eff,i}$ could be straightforwardly determined from Eq. 8.15. However, if c_φ and n_0 are uncertain or cannot be estimated at all for any reason, an appropriate calibration method is to be applied. An injection-dependent harmonically modulated luminescence lifetime analysis τ_σ (G)—conducted on a well defined substrate area, and involving the generation rate G_0 in its measurement

[21] Herein, the subscript σ (denoting a self-consistent lifetime analysis) is used for harmonically modulated lifetime. Yet, other phase-sensitive analyses of harmonically modulated *actual* effective lifetime (e.g. τ_{d2a}) could be used alike.

[22] Local effective carrier lifetime may be affected by both interface recombination as well as any recombination center within lateral reach of an effective minority carrier diffusion length.

Fig. 8.13 Illustration of a typical active measurement area of a harmonically modulated lifetime measurement (Ø: 40 mm) on a multicrystalline wafer, to be used for an effective lifetime calibration of a luminescence image. The averaging procedure derived here enables a lifetime calibration by adequately relating the locally measured luminescence intensities φ_i within the highlighted active area to a harmonically modulated lifetime τ_σ measured at the same excess carrier generation rate as the luminescence image. The inactive luminescence image area is shaded. The radial sensitivity profile $\mathcal{S}_\Phi$ was superimposed upon the luminescence intensities shown here. Adapted after [14]

domain—provides the sought information. Figure 8.13 highlights a typical active area of a harmonically modulated lifetime measurement in a luminescence image.

8.3.2.2 Decomposition of Harmonically Modulated Luminescence Intensity

In order to derive an appropriate averaging algorighm for a lifetime calibration of steady-state luminescence images φ_i via harmonically modulated lifetime τ_m, harmonically modulated luminescence intensity $\Phi(t)$ is to be decomposed in its constituents according to the pixels i of a corresponding luminescence image φ_i

$$\Phi(t) = \sum_i \Phi_i(t) . \tag{8.16}$$

With a local relative sensitivity $\mathcal{S}_{\Phi,i}$ (cf. Fig. 7.2), the local constituent of harmonically modulated luminescence intensity is

$$\Phi_i(t) = c_\Phi \mathcal{S}_{\Phi,i} \Delta n_i(t) \left(\Delta n_i(t) + n_0\right) . \tag{8.17}$$

With the phase-sensitive *differential* lifetime $\tau_{m,i} = \tau_{m,i}(G_0)$ and the corresponding *actual* effective lifetime $\tau_{eff,i} = \tau_{eff,i}(G_0)$ at pixel i, harmonically modulated excess carrier density Δn_i may be well approximated in the vicinity of its maximum as[23]

[23] Here, excess carrier generation rate of the steady-state luminescence image G_0 is only chosen identical to maximum generation rate of the modulated luminescence measurement for reasons of

$$\Delta n_i(t) = G_0 \tau_{eff,i} \sin^2 \left(\frac{\pi}{T} \left(t - \tau_{m,i} \right) \right). \tag{8.18}$$

8.3.2.3 Averaging Algorithm

A harmonically modulated lifetime measurement summing over a defined ensemble of substrate locations corresponding to pixels i of a luminescence image yields the aggregated phase-sensitive *differential* lifetime $\tau_m = \tau_m(G_0)$, while a lifetime analysis yields the *actual* lifetime $\tau_\sigma = \tau_\sigma(G_0)$. By definition, the *differential* lifetime τ_m must fulfill the condition (cf. Fig. 6.8)

$$\left. \frac{d\Phi(t)}{dt} \right|_{t=\frac{T}{2}+\tau_m} \stackrel{!}{=} 0. \tag{8.19}$$

Insertion of $\Delta n_i(t)$ as of Eq. 8.18 into Eq. 8.17, evaluation of its derivative with respect to time at $t = \frac{T}{2} + \tau_m$, and insertion into Eq. 8.16 then yields

$$0 \stackrel{!}{=} \sum_i S_{\Phi,i} \tau_{eff,i} \left(2G_0 \tau_{eff,i} \cos^3 \alpha \sin \alpha + n_0 \cos \alpha \sin \alpha \right), \tag{8.20}$$

with $\alpha = \frac{\pi}{T}\left(\tau_m - \tau_{m,i}\right)$. Now, the quasi-steady-state condition (cf. Sect. 5.1)

$$\pi |\tau_m - \tau_{m,i}| \ll T \tag{8.21}$$

comes into play once again—allowing accurate first order Taylor approximations of all trigonometric functions such that

$$0 = \sum_i S_{\Phi,i} \tau_{eff,i} \left(2G_0 \tau_{eff,i} + n_0 \right) \left(\tau_m - \tau_{m,i} \right). \tag{8.22}$$

This finally allows the expression of aggregated phase-sensitive *differential* lifetime as

$$\tau_m = \frac{\sum_i S_{\Phi,i} \tau_{m,i} \tau_{eff,i} \left(2G_0 \tau_{eff,i} + n_0 \right)}{\sum_i S_{\Phi,i} \tau_{eff,i} \left(2G_0 \tau_{eff,i} + n_0 \right)}. \tag{8.23}$$

Equation 8.23 describes the manner in which local harmonically modulated time shifts $\tau_{m,i}$ aggregate to one measurable global time shift τ_m. Yet, we are interested in *actual* local effective lifetimes $\tau_{eff,i}(G_0)$ to be deduced from *actual* lifetime $\tau_\sigma(G_0)$ resulting from a harmonically modulated lifetime analysis. If all contributing local *differential* lifetimes $\tau_{m,i}$ differed from local *actual* lifetimes $\tau_{eff,i}$ by one invariable

(Footnote 23 continued)
simplicity. Note that this is not required for the method. Steady-state generation rate must solely lie within the domain of an injection-dependent dynamic analysis of *actual* lifetime $\tau_\sigma(G)$.

factor, the conclusion

$$\tau_\sigma = \frac{\sum_i S_{\Phi,i}\tau_{eff,i}^2\left(2G_0\tau_{eff,i}+n_0\right)}{\sum_i S_{\Phi,i}\tau_{eff,i}\left(2G_0\tau_{eff,i}+n_0\right)}. \tag{8.24}$$

would be exact. From a rigorous standpoint, the invariance of the mismatch between $\tau_{m,i}$ and $\tau_{eff,i}$ is seldom fulfilled throughout all contributing locations i of a laterally inhomogeneous substrate, so Eq. 8.24 is not exact. However, the uncertainty brought about by this slightly inaccurate conclusion by analogy can be argued to be faint. This is because—as can be concluded from Eq. 8.23—τ_m is much dominated by a narrow range of relatively large lifetimes, for which the approximation of a constant mismatch between $\tau_{m,i}$ and $\tau_{eff,i}$ is in most cases a good one.

Equation 8.24 defines the averaging procedure, which effectively enables a lifetime calibration of steady-state luminescence images via quasi-steaty-state harmonically modulated luminescence. Given a measured luminescence image φ_i, Eq. 8.15 is to be solved for positive local effective lifetimes $\tau_{eff,i}$ and inserted into Eq. 8.24 for all pixels i of the luminescence image, which contribute to the modulated luminescence measurement. Thus, with an analysis of harmonically modulated *actual* effective lifetime $\tau_\sigma(G_0)$, the scaling factor c_φ (or in low-level injection, the product $c_\varphi n_0$) can be determined. This is equivalent to a lifetime calibration of the luminescence image.[24]

8.3.2.4 On the Interpretation of the Averaging Algorithm

The averaging procedure of Eq. 8.24 can be roughly interpreted as a weighting of local lifetimes $\tau_{eff,i}$ by local luminescence intensity, i.e. an intensity weighted average. Accordingly, the relatively high lifetimes within the active area of a harmonically modulated luminescence measurement contribute more strongly to the measured lifetime τ_σ than the relatively low ones. Under rigorous low-level injection conditions, Eq. 8.24 exactly represents an intensity-weighted average, while under intermediate- to high-level injection conditions, the quadratic relation $\Phi(\Delta n)$ leads to a departure from the intensity-weighted average such that the globally measured time shift τ_m is even greater than to be expected if local excess carrier densities $\Delta n_i(t)$ rather than local luminescence intensities $\Phi_i(t)$ were superposed [14].

Low-Level Injection: Low-level injection conditions bring about a substantial simplification of Eq. 8.24, which is worth to be discussed here, as it provides access to an intuitive understanding of the intensity-weighted average. Assuming rigorous low-level injection ($G_0\tau_{eff,i} \ll n_0$) and a uniform local sensitivity $S_{\Phi,i} = 1$, Eq. 8.24 becomes

[24] The software *QSSL Lifetime* applies Newton's iteration method for a most accurate determination of the scaling factor c_φ.

$$\tau_\sigma = \frac{\sum_i \tau_{eff,i}^2}{\sum_i \tau_{eff,i}}. \tag{8.25}$$

Exploiting the low-level injection representation of Eq. 8.15

$$\varphi_i = c_\varphi G_0 n_0 \tau_{eff,i}, \tag{8.26}$$

Equation 8.25 may be expressed as

$$\tau_\sigma = \frac{\sum_i \tau_{eff,i}\varphi_i}{\sum_i \varphi_i}. \tag{8.27}$$

This is the exact intensity-weighted average of local steady-state effective lifetimes $\tau_{eff,i}$. Multiplication of the latter equation with the factor $c_\varphi G_0 n_0$ yields the virtual local luminescence intensity

$$\varphi_\sigma = c_\varphi G_0 n_0 \tau_\sigma = \frac{\sum_i \varphi_i^2}{\sum_i \varphi_i} \tag{8.28}$$

to be expected in the case of a virtual local lifetime $\tau_{eff,i} = \tau_\sigma$. The resulting low-level injection map of steady-state effective lifetime is then

$$\tau_{eff,i} = \frac{\tau_\sigma}{\varphi_\sigma}\varphi_i. \tag{8.29}$$

While in most cases it is deemed safest to go with the generally valid expression of Eq. 8.24, the latter special case of low-level injection provides a straightforward understanding of the averaging procedure motivated here.

The Impact of Dopant Concentration: Note that while under general injection conditions, the assumption of net dopant concentration n_0 only has a weak impact on the calibration of a lifetime map $\tau_{eff,i}$, this impact vanishes completely under low-level injection conditions. The impact of a typical uncertainty of net dopant concentration (e.g. in upgraded metallurgical grade silicon) may easily be quantified via variation of n_0 as used for the spatially resolved effective lifetime analysis. For the wafers shown in Fig. 8.16, the variation of the assumed value of n_0 by $\pm 30\,\%$ (typical mismatch encountered in umg material) led to lifetime variations of less than $0.3\,\%$.

Some Illustrative Examples: In order to illustrate the difference between the intensity-weighted averaging of local luminescence intensities φ_i motivated here (as opposed to an arithmetical averaging of local luminescence intensities φ_i), three different exemplary and realistic lifetime configurations are discussed here.

Assuming low-level injection, injection-independent lifetime, and a uniform sensitivity $\mathcal{S}_{\Phi,i}$ throughout, 16 pixels of a virtual luminescence image are separated in two groups of pixels with possibly different sizes (areas $A_{1/2}$) and lifetimes $\tau_{eff,1/2}$, respectively. For each configuration, the intensity weighted average lifetime τ_σ, the

Table 8.2 Calculation of the intensity weighted average lifetime τ_σ (according to Eq. 8.27), the arithmetically averaged lifetime $\langle\tau\rangle$, and the mismatch between the assumed local lifetimes $\tau_{eff,i}^{\mathrm{true}}$ and an erroneous lifetime interpretation based on the arighmetically averaged lifetime $\tau_{eff,i}^{\langle\tau\rangle}$ for the three exemplary lateral lifetime configurations sketched in Fig. 8.14

Configuration	$\tau_{eff,1}/\tau_{eff,2}$	A_2/A_1	τ_σ	$\langle\tau\rangle$	$\tau_{eff,i}^{\mathrm{true}}/\tau_{eff,i}^{\langle\tau\rangle}$
I	1	1	$\tau_{eff,1}$	$\tau_{eff,1}$	1
II	10	1	$\frac{101}{110}\tau_{eff,1}$	$\frac{11}{20}\tau_{eff,1}$	$\frac{202}{121} \approx 1.67$
III	10	16	$\frac{115}{250}\tau_{eff,1}$	$\frac{5}{32}\tau_{eff,1}$	$\frac{368}{125} \approx 2.94$

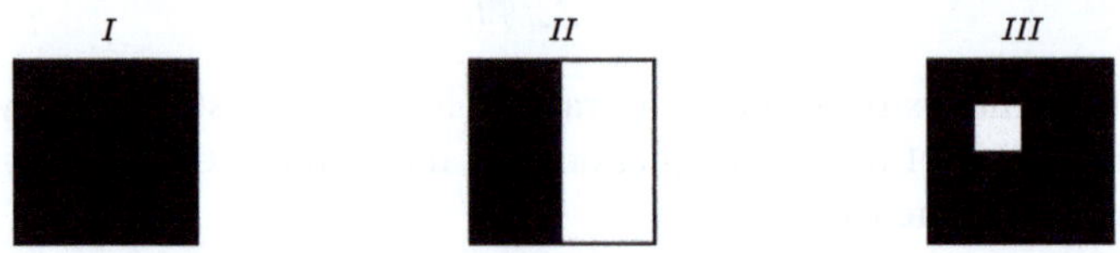

Fig. 8.14 Depiction of the three exemplary lateral lifetime configurations discussed in Table 8.2. Different local effective lifetimes are indicated in black and white, respectively. Configuration *I* is homogeneous, configuration *II* features two equal areas of different lifetime, and configuration *III* features both different areas and different lifetimes

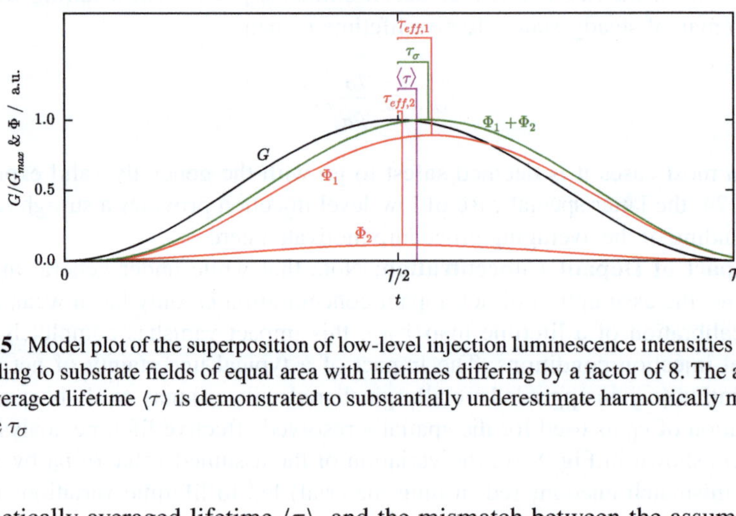

Fig. 8.15 Model plot of the superposition of low-level injection luminescence intensities $\Phi(t)$ corresponding to substrate fields of equal area with lifetimes differing by a factor of 8. The arithmetically averaged lifetime $\langle\tau\rangle$ is demonstrated to substantially underestimate harmonically modulated lifetime τ_σ

arithmetically averaged lifetime $\langle\tau\rangle$, and the mismatch between the assumed local lifetimes $\tau_{eff,i}^{\mathrm{true}}$ and an erroneous lifetime interpretation based on the arithmetically averaged lifetime $\tau_{eff,i}^{\langle\tau\rangle}$ are calculated and displayed in Table 8.2.

Also, schemes of idealized lifetime distributions are depicted in Fig. 8.14. From the simple but realistic examples of laterally inhomogeneous lifetime distributions, it must be concluded that the intensity-weighted lifetime average is inevitable for a harmonically modulated lifetime calibration of luminescence images.

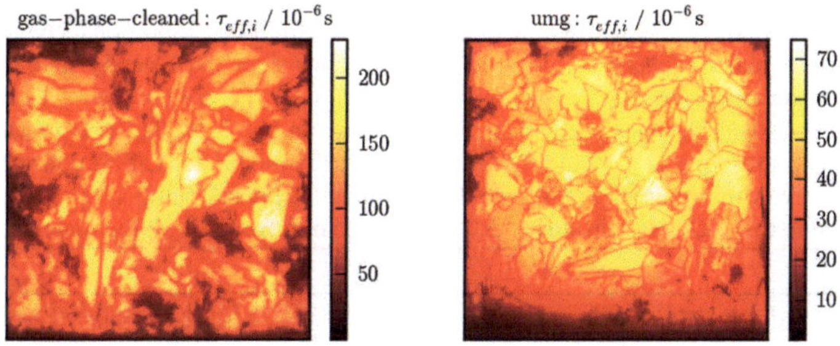

Fig. 8.16 Effective lifetime calibrations (performed via quasi-steady-state harmonically modulated photoluminescence) of photoluminescence images of $d = 210\,\mu$m thick $125 \times 125\,\text{mm}^2$ p-type multicrystalline silicon wafers featuring phosphorous gettering and a silicon nitride surface passivation (no emitter). On the left, lifetimes $\tau_{eff,i}$ of a conventionally gas-phase-cleaned wafer with a dopant concentration $n_0 \approx 1.6 \times 10^{16}\,\text{cm}^{-3}$ are shown. On the right, lifetimes of an upgraded metallurgical grade wafer with a dopant concentration $n_0 \approx 3.6 \times 10^{16}\,\text{cm}^{-3}$ are shown. Measurements were conducted at an irradiance equivalent to one sun (cf. Appendix A.6). 880 nm laser light sources were used for both, steady-state and harmonically modulated measurements. Adapted after [14]

Another illustrative example is given in Fig. 8.15, depicting the mismatch between arithmetical and intensity-weighted lifetime averages via the superposition of harmonically modulated luminescence intensities. Here, time-dependent low-level injection luminescence intensities $\Phi_{1/2}(t)$ of two virtual substrate fields of equal size, but with injection-independent lifetimes $\tau_{eff,1/2}$ differing by a factor of 8 are plotted. The time shift τ_σ[25] of the superposition $\Phi_1 + \Phi_2$ with respect to excess carrier generation is to be compared with the arithmetical average of lifetimes $\langle\tau\rangle$. As expected, $\langle\tau\rangle$ substantially underestimates τ_σ, thereby corroborating the relevance of the intensity-weighted lifetime averaging algorithm.[26]

8.3.3 Experimental Results

Here, the spatially resolved effective lifetime calibration of photoluminescence images via harmonically modulated photoluminescence $\{\varphi_i\,(G_0)\,,\,\tau_\sigma\,(G_0)\} \mapsto \tau_{eff,i}\,(G_0)$ is demonstrated and independently validated on both, wafers and solar cells.

25 Here, *differential* and *actual* lifetimes coincide due to the assumed injection independence.

26 Note that departure from quasi-steady-state was deliberately chosen here in order to better depict time shifts.

Table 8.3 Comparison between different spatially resolved lifetime calibration techniques on multicrystalline wafer material classes. Calibrations via self-consistent quasi-steady-state harmonically modulated photoluminescence $\langle\tau\rangle_\sigma$ and via QSSPC $\langle\tau\rangle_{QSSPC}$ [7] are compared via arithmetical average values of local lifetimes $\tau_{eff,i}$ of the wafers shown in Fig. 8.16. The given standard deviation indicates the scattering of five calibrations. Harmonically modulated calibration spots involve a circular active area with a diameter of 45 mm (Ø: 40 mm in the present setup, cf. Fig. 8.13), centered in the wafer center as well as at distances Δx, $\Delta y = \pm 25$ mm from the wafer center, respectively

Material class	$\langle\tau\rangle_\sigma$	$\langle\tau\rangle_{QSSPC}$
gas-phase-cleaned	117 ± 8	102 ± 6
umg	44 ± 3	27 ± 2

8.3.3.1 Wafers: Comparison to QSSPC

Harmonically modulated lifetime calibrations of steady-state photoluminescence images were conducted on representative[27] wafers of several silicon material classes featuring laterally inhomogeneous lifetimes—comprising gas-phase-cleaned and upgraded metallurgical grade cast silicon (cf. Fig. 8.16). The thereby obtained local lifetimes $\tau_{eff,i}$ were compared to local lifetimes obtained via a QSSPC calibration [7, 8].[28] Relevant considerations for a QSSPC calibration of photoluminescence images are detailed in Appendix B. Most importantly, the corresponding averaging algorithm and a measurement of the lateral sensitivity distribution of that technique are elaborated therein.

Steady-state photoluminescence images were acquired via a CCD camera at a lateral resolution of about $10^{-2} \times 10^{-2}\,\text{cm}^2$. Wafers were exposed to a $360\,\text{Wm}^{-2}$ Xe spectrum for 18 hours prior to lifetime measurements in order to make sure that no significant light induced degradation would take place in the course of measurements [10–12]. Calibrations via modulated luminescence were conducted according to Eq. 8.24. Net dopant concentrations were measured via eddy current, but are deemed to only have a very faint impact on the lifetime result. Both harmonically modulated and QSSPC calibrations were carried out on five different spots of each wafer (cf. Appendix C.1) in order to be able to assess the significance of the quantitative comparison shown in Table 8.3. The lifetime maps $\tau_{eff,i}$ shown in Fig. 8.16 represent the average of these five calibrations, respectively.

The results shown in Table 8.3 reveal significant deviations between harmonically modulated luminescence and QSSPC lifetime calibrations for both investigated wafers. In the case of the (compensated) upgraded metallurgical grade wafer, the deviation is very pronounced. A number of two representative wafers certainly does not provide sufficient experimental evidence in order to draw robust conclusions about systematic artifacts. Yet, in anticipation of the results of a comprehensive

27 A comprehensive statistical comparative analysis of QSSPC and harmonically modulated photoluminescence is given in Appendix C.

28 A similar calibration approach (exploiting steady-state photoconductance) had been proposed by Herlufsen et al. [54].

comparative statistical analysis of the two calibration methods (cf. Appendix C), the combination of three major effects provides a reasonable quantitative explanation to the observed deviations. These effects are associated with

1. the inherently incomplete mobility model of QSSPC,
2. the resistive impact of crystal defects on eddy currents, and
3. lateral optical blurring of effective lifetime information in luminescence images.

The Incomplete Mobility Model of QSSPC: The photoconductance amplitude is proportional to the sum of minority and majority carrier mobilities. These are determined through a bulit-in measurement of wafer resistivity via dark eddy current, which is converted into carrier mobilities through mobility models which are valid for monocrystalline uncompensated silicon wafers [59, 60]. Departure from the validity of such models—particularly due to compensation of dopants—cannot be adequately taken into account. At an acceptor dopant concentration N_A in the range of $1.0 - 1.6 \times 10^{17}$ cm^{-3} and a compensating donor concentration N_D in the range of $4 - 7 \times 10^{16}$ cm^{-3}, the resulting mismatch between Klaassen's mobility model [28, 29] (taking the coexistence of complementary dopant species into account) and conventional one-species mobility models valid for monocrystalline silicon [59–63] may amount to 15–25 %. Further, coherent recent findings of several researchers [64] indicate that even Klaassen's model would not sufficiently account for the mobility lowering effect of compensation [64–66], explaining another mismatch in the range of 5–15 % in the case of the umg wafer under investigation. Therefore, the above figures amount to a total overestimation of carrier mobility by the mobility model implemented in QSSPC of 30 ± 10 %, and a corresponding 30 ± 10 % underestimation of effective lifetime by QSSPC. This figure confirms the difference of deviations $\langle\tau\rangle_\sigma/\langle\tau\rangle_{\text{QSSPC}}$ between compensated umg material and uncompensated gas-phase-cleaned material.

The Resistive Impact of Crystal Defects on Eddy Currents: In spite of the fact that the scientific discourse on the role of crystal defects (e.g. grain boundaries) as electrical potential barriers has been going on for a long time [41–49], its implications appear to be controversial and unclear. Recent findings by Spitz and Rein [50] suggest a tremendous resistive impact of grain boundaries (particularly upon damage etching in combination with air drying). Grain boundaries acting as potential barriers could weaken eddy currents and thereby lead to a systematic underestimation of effective lifetime via QSSPC. This effect might explain the mismatch encountered for the gas-phase-cleaned wafer. Also, it might explain a residual fraction of the mismatch encountered for the umg wafer. This error source is further elaborated in Appendix C.

Optical Blurring of Lifetime Information in Luminescence Images: Lateral optical spreading of luminescence light in the investigated substrate and in the detector may lead to deviations between the *actual* distribution of true effective carrier lifetime $\tau_{eff,i}$ in a wafer and the *apparent* lateral distribution of excess carrier densities inferred from measured luminescence intensities φ_i [67, 68].[29]

[29] A similar effect has been known to cause significant artifats in lifetime measurements based on free carrier emission or absorption [69].

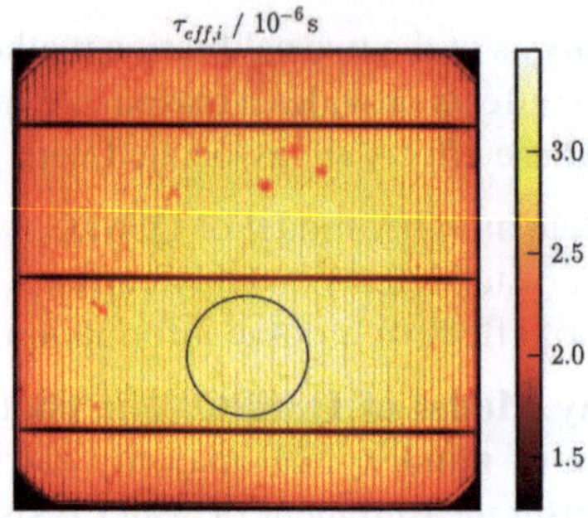

Fig. 8.17 Effective lifetime calibration of a photoluminescence image of a $d = 260\,\mu m$ thick $156 \times 156\,mm^2$ upgraded metallurgical grade silicon solar cell ($n_0 \approx 8 \times 10^{16}\,cm^{-3}$, featuring an Al-BSF), performed on the highlighted active area via quasi-steady-state harmonically modulated photoluminescence. Measurements were conducted at an irradiance equivalent to one sun (cf. Appendix A.6). 880 nm laser light sources were used for both, steady-state and harmonically modulated measurements. Adapted after [38]

This effect is most unlikely to affect the QSSPC lifetime calibration of luminescence images. The reason for this lies in the amplitude-sensitive nature of QSSPC, which approximately brings about an arithmetical averaging of local lifetimes. However, for the harmonically modulated lifetime calibration of luminescence images, blurring may matter. This is due to the averaging procedure of Eq. 8.24. Recalling Eq. 8.24, relatively high effective lifetimes within the active area of a modulated luminescence measurement feature a more pronounced contribution to measured lifetime τ_σ than relatively low effective lifetimes. The lateral distribution of effective lifetimes is inferred from a luminescence image. If the characteristic dimension of blurring is on the order of or greater than the characteristic dimension of electronically uniform spots within a wafer, the optical blurring effect may artificially reduce *apparent* local effective lifetime in regions of relatively high lifetime, while it may substantially enhance *apparent* local effective lifetime in regions of relatively low lifetime. Due to the averaging algorithm of Eq. 8.24, particularly the resulting reduction of relatively high luminescence intensities could cause a systematic error of a harmonically modulated lifetime calibration of photoluminescence images. With a too low local luminescence intensity φ_σ being assigned to the measured lifetime τ_σ, a systematic overestimation of local effective lifetime $\tau_{eff,i}$ results. Experimental evidence of this effect is to be found in Appendix C.3. Note that the impact of optical blurring does not represent an inherent limitation of the method elaborated here. Effective countermeasures for measurement [70] and (most importantly) for the image interpretation [67, 71] have been proposed. Their implementation is deemed to resolve the observed luminescence-inherent artifacts.

8.3.3.2 Solar Cells: Comparison to Open Circuit Voltage

Figure 8.17 shows the spatially resolved lifetime calibration of an example solar cell, to be investigated in the following. Lateral luminescence contrast due to shunts or series resistance could be erroneously interpreted as lifetime contrast, here. Yet, the harmonically modulated lifetime calibration is unlikely to suffer from such nonrecombinative local effects, as long as they do not substantially alter the injection level of the lifetime analysis (as major shunts would do). Here, the obtained lifetimes are compared to open circuit voltages, which were simultaneously measured at the cell electrodes (between Al back contact and Ag bus bars). Associating the separation of quasi-Fermi energies with the device voltage $qU_{oc} = \eta_{\sum}$, open circuit voltage may according to Eq. 2.29 be expressed in terms of the product of minority and majority carrier densities as

$$U_{oc} = \frac{k_B T}{q} \ln\left(\frac{1}{n_{i,eff}^2} \Delta n_{oc} \left(\Delta n_{oc} + n_0\right)\right), \tag{8.30}$$

while $\Delta n_{oc} = G_0 \tau_{oc}$ represents an appropriate average of open circuit excess carrier densities.

Thus, an appropriately averaged effective lifetime result τ_{oc} is to be inserted into Eq. 8.30 in order to compare the resulting open circuit voltage U_{oc}^{τ} to the measured open circuit voltage of

$$U_{oc}^{\text{meas}} = 613.7 \pm 0.4\,\text{mV}. \tag{8.31}$$

Measurement temperature was $T = 296.1 \pm 0.3\,\text{K}$, net dopant concentration as measured via FTIR measurements [72, 73] on sister wafers amounted to $n_0 = 7.8 \pm 0.4 \times 10^{16}\,\text{cm}^{-3}$, an intrinsic carrier concentration of $n_{i,eff} = 8.5 \pm 0.5 \times 10^9\,\text{cm}^{-3}$ was assumed,[30] and total excess carrier generation rate amounted to $G_0 = 9.7 \pm 0.5 \times 10^{18}\,\text{cm}^{-3}\,\text{s}^{-1}$. The remaining quantity was thus an appropriately averaged effective lifetime τ_{oc}. According to Isenberg et al. [39], this average value is expected to lie between the arithmetical and the harmonic average of local effective lifetimes $\tau_{eff,i}$. With a relatively homogeneous lateral distribution of effective lifetimes, there was only little difference between these average values ($\langle\tau\rangle_{\text{arithm.}} = 2.78\,\mu\text{s}$ vs. $\langle\tau\rangle_{\text{harm.}} = 2.74\,\mu\text{s}$). Insertion into Eq. 8.30 led to the open circuit voltage

$$U_{oc}^{\tau} = 614 \pm 2\,\text{mV}, \tag{8.32}$$

which is well in accordance with the measured open circuit voltage. It must be mentioned that such good agreement did not sustain throughout since this initial

[30] For the determination of $n_{i,eff} = n_i e^{\frac{\Delta\varepsilon_G}{k_B T}}$ according to Altermatt et al. [74], n_i was approximated by a parametrization given by Sproul and Green [75] with a band gap of $\varepsilon_G\,(T = 296.1\,\text{K}) = 1125 \pm 1\,\text{meV}$ [76]. Further, band gap narrowing was estimated to amount to $\Delta\varepsilon_G\,(n_0) = 9 \pm 2\,\text{meV}$ [74, 77].

evaluation. In fact, measured lifetimes occasionaly systematically underestimated measured open circuit voltages. This could be due to an inadequate depth-averaging of excess carrier density, or to a systematic error in the temperature measurement (largely affecting $n_{i,eff}$). However, the most likely cause is an overestimation of the actual injection level in combination with a pronounced injection-dependence of lifetime in the presence of shunts, remote recombination centers, or locally confined illumination.

8.3.4 *Discussion of Limitations*

Beyond the impact of optical blurring on steady-state luminescence images which was addressed in the discussion of results in Sect. 8.3.3, several other potential limitations and relevant cautions of a harmonically modulated effective lifetime calibration of photoluminescence images are mentioned in the following. Considerations include the dynamic (rather than steady-state) impact of blurring, lateral variations of net dopant concentration and interface morphology, higher order differential lifetime artifacts, dynamic lifetime artifacts, and reabsorption.

8.3.4.1 The Dynamic Impact of Blurring

Ramspeck et al. have previously pointed out that dynamic lifetime measurements are more likely to suffer from artifacts of lateral optical blurring (caused by rough substrate interfaces) than steady-state lifetime measurements [78]. Although initially encountered for free carrier absorption measurements, such reasoning also holds for dynamic luminescence lifetime techniques in some cases. Corresponding artifacts may arise if lifetime calibration proceeds in a device area of relatively low lifetime, while an adjacent area of relatively high lifetime is also illuminated. On the one hand, such effects could be avoided by exclusively illuminating the designated calibration area. On the other hand, such exclusive local illumination may be unfavorable in some situations (e.g. solar cells featuring substantial lateral redistribution of excess carriers through emitter and metalization).

8.3.4.2 Lateral Variation of Net Dopant Concentration and Interface Morphology

A harmonically modulated effective lifetime calibration of luminescence images is based on the assumption of a uniform net dopant concentration n_0 throughout the substrate under investigation. Even though the lifetime result $\tau_{eff,i}$ only weakly depends on the assumption about n_0, substantial errors may result if n_0 is laterally nonuniform across the investigated device. The same applies to lateral variations of interface morphology, which can likewise substantially impact on luminescence amplitude.

Note that alternative dynamic stand-alone imaging approaches exist, which do not rely on the assumption of lateral invariance of net dopant concentration and interface morphology [58, 79].

8.3.4.3 Higher Order Artifacts of Differential Lifetime

In Sect. 8.3.2, a possibly slightly inaccurate conclusion by analogy in the derivation of the averaging algorithm of Eq. 8.24 has been stressed. It relates to the fact that not all local time shifts $\tau_{m,i}$ (*differential* lifetimes as detailed in Sect. 6.2) contributing to a harmonically modulated lifetime measurement τ_σ differ from *actual* local lifetimes by the same factor. Yet, this inaccuracy is deemed minor, as τ_m is mostly dominated by a narrow band of relatively large lifetimes, for which the approximation of a constant mismatch between $\tau_{m,i}$ and $\tau_{eff,i}$ is good.

8.3.4.4 Dynamic Versus Steady-State Effective Lifetime

In the case of very high surface recombination velocities, the harmonically modulated lifetime calibration of **steady-state** luminescence images may be inaccurate due to the deviation between dynamic and steady-state effective lifetime as elaborated in Sect. 6.1. Notably, in the special case of a harmonically modulated lifetime calibration, this does not necessarily imply an overestimation of local lifetime.

8.3.4.5 Reabsorption

As opposed to harmonically modulated luminescence, steady-state luminescence imaging obtains relative effective lifetime information from signal amplitudes. This fact renders the lifetime calibration $\{\varphi_i(G_0), \tau_\sigma(G_0)\} \mapsto \tau_{eff,i}(G_0)$ prone to local reabsorption artifacts [26, 27].

8.4 Recombination Properties of Silicon Ingots

Building on the findings about harmonically modulated effective lifetime derived in Sect. 6.1, this section is devoted to a resolution of bulk lifetime of silicon ingots from (self-consistent) harmonically modulated lifetime measurements. The very pronounced contrast of harmonically modulated effective lifetime with respect to bulk lifetime (when compared to steady-state techniques) is addressed and exploited for an accurate combined analysis of bulk lifetime and surface recombination velocity. Likewise, bulk lifetime and the minority carrier diffusion coefficient could be simultaneously determined.[31]

[31] The account given here is adapted from the corresponding original publication [80].

8.4.1 Review of Relevant Facts and Literature

Information about minority carrier bulk lifetime of silicon ingots is essential for an assessment of the electronic material quality and the efficiency potential prior to wafering. Several bulk lifetime measurement techniques for silicon ingots have been published. They involve steady-state or quasi-steady-state photoconductance [7, 8, 18, 81, 82], transient photoconductance decay [82–84], as well as steady-state [68, 85] or transient [86] photoluminescence.

The aforementioned photoconductance-based techniques infer bulk lifetime from measurements of effective carrier lifetime, either as the ratio of cumulative excess carrier density and generation rate (in quasi-steady-state), or from the transient decay time of excess carrier density. Lacking surface passivation, silicon ingots usually feature relatively high surface recombination velocities, which renders effective lifetime very weakly dependent on surface recombination velocity. This may be beneficial for the determination of bulk lifetime from effective lifetime measurements, if minority carrier mobility [28, 29, 64, 87] is known, and if the depth profile of excess carrier generation rate is both known and not too confined to the surface [18, 81, 82].

A recently published photoluminescence-based approach by Mitchell et al. [68, 85] provides immediate access to bulk lifetime via the ratio of steady-state photoluminescence images capturing different ranges of the luminescence spectrum [88]. It is demonstrated that—provided that surface recombination is sufficiently high—this intensity ratio essentially depends on minority carrier bulk diffusion length, thereby enabling the determination of bulk lifetime, if minority carrier mobility is known.

A common feature of any of the above-mentioned techniques is the requirement of a priori information about both minority carrier mobility (or the minority carrier diffusion coefficient) and surface recombination velocity. As for the latter, it is reasoned that for silicon ingots it is usually sufficiently high in order to assume its quasi-infinity (greater than thermal carrier drift velocity). Nevertheless, with surface recombination velocities of polished silicon ingots estimated to range between $10^4\,\mathrm{cm\,s^{-1}} \lesssim S \lesssim 10^6\,\mathrm{cm\,s^{-1}}$, an accurate determination of surface recombination velocity could further enhance the accuracy of lifetime techniques. Beyond, a robust technique to quantify surface recombination velocity of silicon ingots is deemed of interest for miscellaneous R&D purposes and applications. Concerning minority carrier mobility, one might argue that established mobility models feature sufficient accuracy. Yet, for compensated silicon this statement is unlikely to hold [89–92], giving rise to the need for a simultaneous determination of bulk lifetime and minority carrier mobility from a practical standpoint.

8.4.2 Combined Access to Recombination and Diffusion

The approach to be detailed here aims at a dynamic bulk lifetime determination in silicon ingots via harmonically modulated photoluminescence. Either bulk lifetime $\tau|_{S,D}$ may be determined on the basis of assumptions about both surface recombination velocity S and the minority carrier diffusion coefficient D, or the presented approach may be combined with a two-wavelength steady-state photoluminescence analysis sensitive to both, surface recombination rate and the diffusion coefficient [93], thereby yielding $\{\tau, S\}|_{D_a}$ or $\{\tau, D_a\}|_S$, respectively.

8.4.2.1 On the High Bulk Lifetime Contrast of Harmonically Modulated Lifetime

Recalling the essential conclusion of Sect. 6.1, a substantial discrepancy is encountered between harmonically modulated *actual* effective lifetime τ_σ[32] and steady-state effective lifetime $\tau_{eff,s}$ (as the ratio of cumulative excess carrier density and generation rate) in the case of thick or quasi-semi-infinite wafers (e.g. ingots) and shallow excess carrier generation much confined to highly recombinative interfaces. This discrepancy robustly sustains in the quasi-steady-state limit $\mathcal{T} \to \infty$, and it can be quantitatively explained by a nonlinear weighting of the different modes of excess carrier decay through harmonically modulated lifetime techniques (cf. Sect. 6.1.2). The most essential implication for harmonically modulated lifetime measurements with shallow excess carrier generation on semi-infinite substrates featuring a very high surface recombination velocity is a tremendous mismatch $\tau_\sigma \gg \tau_{eff,s}$, as already demonstrated by modeled luminescence intensities shown in Fig. 6.1, which are based on the harmonically modulated Green's function solution of the continuity equation for ingots, as given in Sect. 3.2.4. Figure 8.18 illustrates the mismatch between harmonically modulated effective lifetime τ_σ and steady-state effective lifetime $\tau_{eff,s}$ as a function of both, bulk lifetime τ and surface recombination velocity for p-type $1\,\Omega\,$cm silicon material and an inverse light penetration depth of $\alpha_\gamma = 417\,\text{cm}^{-1}$.

A phenomenological explanation of the mismatch between τ_σ and $\tau_{eff,s}$ beyond the mathematical reasoning related to the weighting of different modes of excess carrier decay is linked to the dynamic nature of harmonically modulated lifetime: With most excess carriers recombining in the very vicinity of a highly recombinative interface, the few carriers reaching deeply into the bulk may virtually survive the duration of a bulk lifetime τ there. As these carriers are the only ones that do not quickly recombine at the interface, it is mainly their fundamental decay mode which contributes to harmonically modulated effective lifetime τ_σ.

The fact of $\tau_\sigma \gg \tau_{eff,s}$ brings about a substantial advantage of a dynamic—and particularly a harmonically modulated—lifetime approach over a steady-state approach due to a much higher contrast of effective lifetime with respect to bulk lifetime.[33] This high contrast enables a most accurate resolution of bulk lifetime in spite of an excess carrier density largely constrained by surface recombination. Notably,

[32] Note that in Sect. 6.1 harmonically modulated effective lifetime was denoted as $\tau_{eff,h}$. However, in order to keep the notation simple, and in order not to get confused with the distinction between *actual* and *differential* lifetime, here the notation τ_σ is chosen (as motivated for self-consistent lifetime in Sect. 5.2).

[33] Here, the term *contrast* denotes the relative change of effective lifetime upon a change of bulk lifetime.

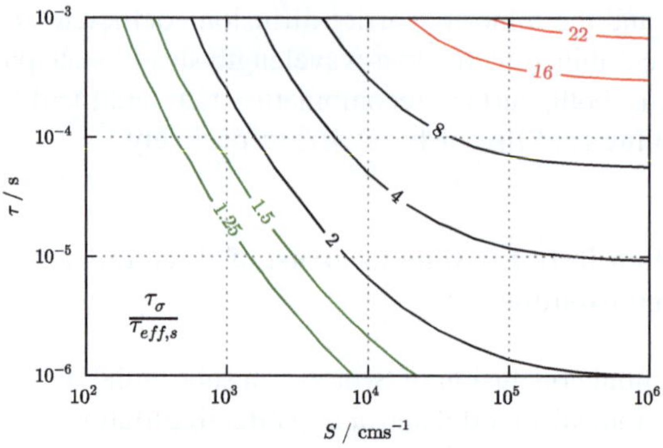

Fig. 8.18 Contour plot of the ratio between self-consistent harmonically modulated effective lifetime τ_σ and steady-state effective lifetime $\tau_{eff,s}$, plotted as a function of both, the surface recombination velocity S and the bulk lifetime τ of a silicon ingot. Harmonically modulated lifetime data were inferred from the self-consistent analysis of the Green's function solution of the continuity equation as of Eq. 3.93, while steady-state lifetime data were obtained from the ratio of cumulative excess carrier density and generation rate, with $\Delta n(z)$ according to the steady-state ingot solution of Eq. 3.72. Model assumptions: minority carrier diffusion coefficient $D_a = 27\,\text{cm}^2\text{s}^{-1}$, inverse light penetration depth $\alpha_\gamma = 417\,\text{cm}^{-1}$, and modulation time $\mathcal{T} = 0.1$ s. Adapted after [80]

this benefit also relates to highly recombinative interfaces within the bulk (e.g. grain boundaries). Even though such interfaces substantially reduce excess carrier density in their vicinity, their impact on dynamic (here: harmonically modulated) effective lifetime remains virtually negligible, provided that the characteristic dimension of adjacent substrate regions of homogeneous lifetime significantly exceeds minority carrier bulk diffusion length.

8.4.2.2 Access to Surface Recombination and Diffusion

A steady-state analysis of low-level injection photoluminescence intensity ratios upon monochromatic irradiation with different wavelengths provides accurate access to either surface recombination velocity or the minority carrier diffusion coefficient, respectively. This approach traces back to Baek et al. [93]. Different irradiation wavelengths imply a different effective penetration depth α_γ^{-1} of light in crystalline silicon. Once excited, excess carriers undergo a random walk through the crystal bulk. If recombination probability for excess carrier density at a surface site is enhanced compared to a site within the bulk, this implies a dependence of excess carrier density on α_γ. The higher α_γ, the closer is excess carrier generation confined to the interface, and the higher is the recombination probability at the interface. Therefore, the ratio R_B of photoluminescence intensities upon irradiation with different wavelengths is a function of both surface recombination velocity and the minority carrier diffusion coefficient.

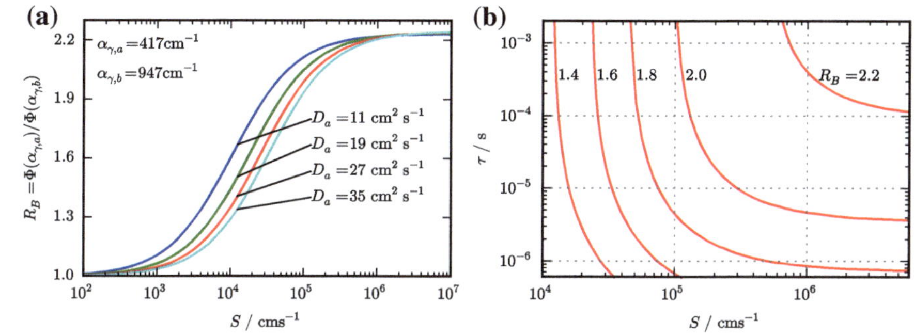

Fig. 8.19 On the left (**a**), the ratio R_B of photoluminescence intensities measured with different irradiation wavelengths α_γ of 417 and 947 cm^{-1} is shown as a function of surface recombination velocity S for different minority carrier diffusion coefficients D_a. Here, R_B is modeled according to the steady-state solution of the continuity equation $\Delta n(z)$ for ingots as of Eq. 3.72, and according to photoluminescence intensity as of Eq. 4.31. Bulk lifetime was set to $\tau = 10^{-3}$ s. On the right (**b**), injective curves $\tau(S)|_{R_B,D_a}$ defined by a measured luminescence intensity ratio R_B at a given minority carrier diffusion coefficient $D_a = 27\,\mathrm{cm^2\,s^{-1}}$ are plotted. Adapted after [80]

Particularly at surface recombination velocities ranging between $10^4\,\mathrm{cm\,s^{-}1} < S < 10^6\,\mathrm{cm\,s^{-}1}$, which is roughly the expected range for polished silicon ingots, a relatively high contrast of R_B with respect to S (and also with respect to the minority carrier diffusion coefficient D_a) is encountered—as can be seen in Fig. 8.19a. Here, intensity ratios R_B were modeled based on the steady-state solution of the continuity equation as given in Eq. 3.72, while also considering the setup-specific depth-dependent luminescence detection probability $p_\Phi(z)$ as introduced in Eq. 4.31 and as plotted in Fig. 4.2.

As already pointed out by Baek et al. [93], the ratio R_B also features a dependence on bulk lifetime, which is faint for high bulk lifetimes, but becomes pronounced toward lower ones. In the harmonically modulated photoluminescence setup, two lasers with individually determined inverse penetration depths α_γ of 417 and 947 cm^{-1} are used.[34] A measured intensity ratio R_B may at a known diffusion coefficient D_a be exploited to reduce the two-dimensional space of qualifying recombination properties to a curve $\tau(S)|_{D_a}$, as depicted in Fig. 8.19b. Likewise, R_B could define a curve $\tau(D_a)|_S$ at a known surface recombination velocity S.

8.4.2.3 Technical Procedure of Measurement and Analysis

Harmonically modulated self-consistent effective lifetime τ_σ in a silicon ingot can be calculated as a function of surface recombination velocity S, of the minority carrier diffusion coefficient D_a, and most distinctly of bulk lifetime τ via the Green's

[34] Effective penetration depths α_γ^{-1} were determined at perpendicular incidence and for a planar interface at a temperature of $T = 301$ K, cf. Appendix D.

function solution of the continuity equation as of Eq. 3.93 and time-dependent photoluminescence intensity as of Eq. 4.31. In the following, let us distinguish between a harmonically modulated lifetime measurement result $\tau_{\sigma m}$, modeled harmonically modulated lifetime τ_{σ} (according to the above Green's function solution), and bulk lifetime τ. With assumptions about S and D_a, a bulk lifetime determination from a measured lifetime $\tau_{\sigma m}$ via

$$\tau_{\sigma m} \stackrel{!}{=} \tau_{\sigma}\left(\tau\right)|_{S,D_a} \tag{8.33}$$

is straightforward. Yet, aiming at both, an enhanced accuracy of bulk lifetime τ as well as a new path to quantitatively access minority carrier mobility or surface recombination velocity, a simultaneous determination of S or D_a is deemed of interest. For this purpose, the harmonically modulated photoluminescence lifetime analysis is to be combined with a steady-state analysis of luminescence intensity ratios related to the two-wavelength approach of Baek et al. [93]. As the harmonically modulated photoluminescence setup is usually operated under a quasi-steady-state condition, it is justified to use the ratio of maximal modulated photoluminescence amplitudes measured at different irradiation wavelengths for a steady-state analysis according to Eqs. 3.72 and 4.31.

The combined approach requires two measurements: a self-consistent effective lifetime $\tau_{\sigma m}$ and an intensity ratio R_B. These may be used to either determine bulk lifetime and surface recombination velocity at a known diffusion coefficient $\{R_B, \tau_{\sigma m}\} \mapsto \{\tau, S\}|_{D_a}$, or to determine bulk lifetime and the diffusion coefficient at a known surface recombination velocity $\{R_B, \tau_{\sigma m}\} \mapsto \{\tau, D_a\}|_S$.

The technical procedure exploits Newton's iteration method:[35] A measured intensity ratio R_B defines an injective function $\tau\left(S\right)|_{D_a}$, as plotted in Fig. 8.19b. Starting with an initial approximation of bulk lifetime of $\tau^0 = \tau_{\sigma m}$,[36] a self-consistent effective lifetime $\tau_{\sigma}^0\left(\tau^0, S\left(\tau^0\right)|_{D_a}, D_a\right)$ to be expected according to Eqs. 3.93 and 4.31 is calculated. With a small real number $0 < \delta \ll 1$, the same calculation must be performed for a slightly varied bulk lifetime $\tau_{\delta}^0 = (1+\delta)\,\tau^0$, likewise yielding the self-consistent lifetime $\tau_{\sigma,\delta}^0\left(\tau_{\delta}^0, S\left(\tau_{\delta}^0\right)|_{D_a}, D_a\right)$ to be expected at a bulk lifetime τ_{δ}^0. Subsequently, starting from an initial approximation with an index $j = 0$, the iteration

$$\tau^{j+1} = \tau^j\left(1 + \delta\frac{\tau_{\sigma}^j - \tau_{\sigma m}}{\tau_{\sigma}^j - \tau_{\sigma,\delta}^j}\right) \tag{8.34}$$

[35] In the iteration method sketched in the following, S and D_a may be interchanged depending on which quantity is given and which quantity is sought.

[36] As in the iteration used in Sect. 8.1, the superscript index is not to be confused with a mathematical exponent here.

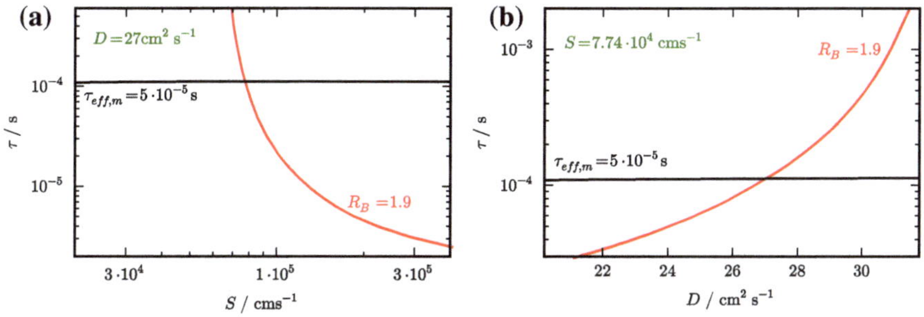

Fig. 8.20 Working principle of the combined approach to simultaneously determine ingot bulk lifetime τ and either surface recombination velocity S (case (**a**)) or the minority carrier diffusion coefficient D_a (case (**b**)). Measurements of a harmonically modulated (e.g. self-consistent) effective lifetime $\tau_{\sigma m}$ and of a luminescence intensity ratio R_B define curves $\tau(S)|_{D_a}$ or $\tau(D_a)|_S$, respectively. The sought recombination and transport properties are to be determined as the intercept of the curves defined by measurements of $\tau_{\sigma m}$ and R_B. Adapted after [80]

yields a rapidly converging bulk lifetime approximation. Empirically, two iteration steps suffice for a relative deviation of $\frac{|\tau_\sigma^j - \tau_{\sigma m}|}{\tau_{\sigma m}} < 10^{-3}$, thereby providing an accurate lifetime analysis within seconds.

The high contrast of this combined approach is illustrated in Fig. 8.20 through curves in the $\{\tau, S\}$-plane or in the $\{\tau, D_a\}$-plane. Each of the two measurement results defines a curve in either of the planes. This high contrast is caused by the very different magnitude of the slopes of the two curves at their intercept.

8.4.2.4 Spatially Resolved Bulk Lifetime Calibration

Bulk lifetime measurements inferred from a harmonically modulated lifetime analysis can be used for a bulk lifetime calibration of steady-state photoluminescence images, as depicted in Fig. 8.21. For this purpose, spatially resolved low-level injection ingot photoluminescence intensity φ_i is to be normalized with respect to relative variations of net dopant concentration by means of a linescan or a spatially resolved measurement of ingot resistivity ρ_i [94]

$$\varphi_i^{\rho_0} = \frac{n_0\left(\rho_0\right)}{n_0\left(\rho_i\right)}\varphi_i. \tag{8.35}$$

Subsequently, steady-state photoluminescence intensity is modeled as a function of bulk lifetime according to Eqs. 3.72 and 4.31 and normalized such that the intensity weighted average (cf. Eq. 8.27, but accounting for the local sensitivity $\mathcal{S}_\Phi(r)$ as well) of local photoluminescence intensities in the active harmonically modulated measurement area corresponds to the measured bulk lifetime $\tau_{\sigma m}$. The resulting ingot

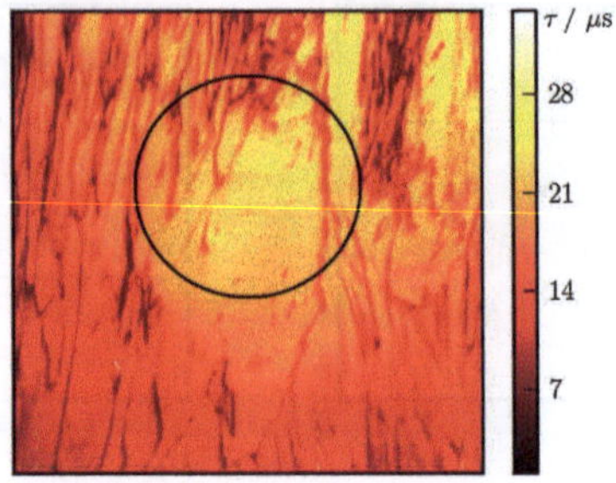

Fig. 8.21 Harmonically modulated bulk lifetime calibration of a (1 sun equivalent) steady-state photoluminescence image of a side face of a multicrystalline silicon ingot, assuming a surface recombination velocity of $S = 2.35 \times 10^5\ \mathrm{cm\,s^{-1}}$ (cf. Table 8.4). The circular active measurement area of the modulated photoluminescence lifetime measurement (Ø: 40 mm) is highlighted. Adapted from [80], ©2013 IEEE

bulk lifetime map allows a judgement on the bulk quality in areas of poor electronic quality.[37]

8.4.3 Experimental Results

With the measurement technique sketched in 8.4.2, a simultaneous determination of bulk lifetime τ and surface recombination velocity S on a polished side face of a directionally solidified multicrystalline silicon ingot was performed. The active measurement area of the measurement result presented here is highlighted in a photoluminescence image shown in Fig. 8.21. Small low lifetime areas or highly recombinative interfaces such as grain boundaries within the active measurement area are expected to be of negligible impact in the lifetime analysis. There are two reasons for this: First, as outlined in Sect. 8.4.2, harmonically modulated (dynamic) lifetime is much less affected by highly recombinative interfaces than steady-state lifetime, which is sensitive to the total magnitude of excess carrier density. Second, due to the relatively low excess carrier density of low bulk lifetime regions, their contribution to the measured harmonically modulated lifetime $\tau_{\sigma m}$ is accordingly weak (cf. Eq. 8.24). Therefore, both $\tau_{\sigma m}$ as well as the measured intensity ratio R_B are expected to represent recombination and transport properties within the laterally homogeneous grains featuring relatively high lifetimes.

Harmonically modulated photoluminescence measurements were carried out at a modulation time $\mathcal{T} = 1.25 \times 10^{-2}$ s, and at an impinging photon current density of $j_{\gamma,i} = 2.5 \times 10^{17}\ \mathrm{cm^{-3}}$. For measurements of harmonically modulated excess carrier generation rate $G(t)$ and photoluminescence intensity $\Phi(t)$, averaging over

[37] Local corrections of steady-state luminescence amplitude due to lateral optical blurring [67, 68] or due to free carrier absorption [95] have been addressed by Mitchell et al. and by Walter et al.

Table 8.4 Results and relevant data of the analysis of ingot bulk lifetime τ and surface recombination velocity S via harmonically modulated luminescence

Relevant data	
Inverse laser penetration depths α_γ	417 and 947 cm^{-1}
Ingot resistivity ρ	$1.13 \pm 0.04\,\Omega\,\mathrm{cm}$
Minority carrier diffusion coefficient D_a	$27.45(+0.16, -0.12)\,\mathrm{cm}^2\,\mathrm{s}^{-1}$
Measurement	
Self-consistent effective lifetime $\tau_{\sigma m}$	$10.2 \pm 0.3 \times 10^{-6}\,\mathrm{s}$
Photoluminescence intensity ratio R_B	2.02 ± 0.03
Analysis	
Bulk lifetime τ	$20.4 \pm 0.7 \times 10^{-6}\,\mathrm{s}$
Surface recombination velocity S	$2.35(+0.97, -0.56) \times 10^5\,\mathrm{cm\,s}^{-1}$
Validation via PLIR	
Lower bound of bulk lifetime $\tau\|_{S,D_a}$	$13.5 \times 10^{-6}\,\mathrm{s}$
Upper bound of bulk lifetime $\tau\|_{S,D_a}$	$28.0 \times 10^{-6}\,\mathrm{s}$

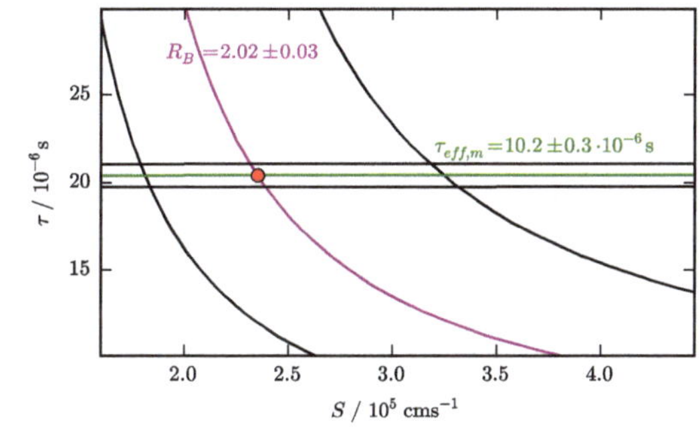

Fig. 8.22 Depiction of the uncertainty of the experimental results given in Table 8.4. The uncertainty of measurements of harmonically modulated lifetime $\tau_{\sigma m}$ and a luminescence intensity ratio R_B together with the uncertainty of the diffusion coefficient D_a define limiting curves $\tau(S)|_{D_a}$, whose intercepts yield the uncertainty of the measurement result. Adapted after [80]

$n_\# = 10^4$ modulations was performed in order to achieve a sufficient signal-to-noise ratio for the specified measurement precision. As stated above, lasers with inverse penetration depths α_γ of 947 cm^{-1} and 417 cm^{-1} in crystalline silicon at $T = 300\,\mathrm{K}$ were used, while measurements conducted with the latter were used for the self-consistent lifetime determination.

Measurement results and relevant data can be found in Table 8.4. The resistivity and the minority carrier diffusion coefficient were inferred from (four-point-probe calibrated) eddy current measurements [87, 94, 96]. The specified uncertainty of the measured effective lifetime $\tau_{\sigma m}$ sums up the standard deviation of the mean effective lifetime of ten successive measurements and the uncertainty of the synchronization measurement, as detailed in Sect. 7.2.1. The specified uncertainty of the measured amplitude ratio R_B accounts for the fluctuation of measured amplitudes. The depic-

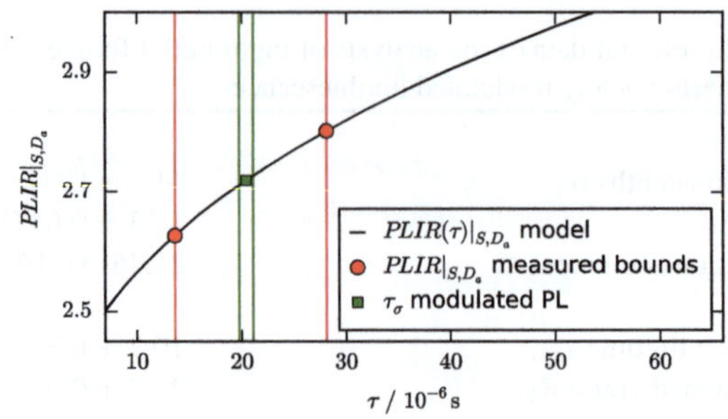

Fig. 8.23 Modeled and measured spectral PLIR according to Mitchell et al. [85]. The modeled curve represents the expected ratio of steady-state photoluminescence intensity with and without a 1,050 nm short pass filter. Corresponding depth-dependent luminescence detection probabilities $p_\Phi(z)$ are shown in Fig. 4.2. Lower and upper bounds of measured PLIR are explained in the discussion of results. Adapted after [80]

tion of the curve intercept in Fig. 8.22 illustrates how the measurement uncertainty propagates to the uncertainty of the resulting recombination properties.

8.4.3.1 Validation and Discussion

In order to validate the precise measurement results provided in Table 8.4, bulk lifetime was alternatively approximated via the approach based on steady-state spectral luminescence intensity ratios by Mitchell et al. [68, 85, 88] (PLIR). For this purpose, steady-state photoluminescence images with and without an optical short pass filter with an edge wavelength of 1,050 nm (with accurately determined spectral transmission) were divided, and the resulting ratio was compared to a calculation of the expected photoluminescence intensity ratio $PLIR\,(\tau)\,|_{S,D_a}$ as a function of bulk lifetime τ. As previously pointed out by Mitchell et al., this method is highly sensitive to spectral information: Most prominently, this involves the emitted luminescence spectrum, which is very sensitive to the band-to-band absorption coefficient α_γ of crystalline silicon. Further, the detector quantum efficiency and the transmission spectra of optics and filters must be accurately known. In a previous PLIR implementation at Fraunhofer ISE [97], all relevant spectra had been accurately determined. The resulting parametrization $PLIR\,(\tau)\,|_{S,D_a}$ is depicted in Fig. 8.23. It was calculated on the basis of Eqs. 3.72 and 4.31, assuming the same diffusion coefficient D_a as for the harmonically modulated photoluminescence analysis, the surface recombination velocity S resulting from the harmonically modulated lifetime analysis (cf. Table 8.4), and an inverse light penetration depth $\alpha_\gamma = 947\,\mathrm{cm}^{-1}$ (corresponding to the used laser light source).

Figure 8.23 indicates an upper and a lower bound to a best bulk lifetime according to the PLIR analysis. These bounds are due to an uncertainty of the measured net photoluminescence intensities. Although the detection of irradiation light in the photoluminescence detector was suppressed by more than 10 orders of magnitude by a 250 μm thick polished monocrystalline silicon wafer, and by another 6 orders of magnitude with a dielectric long pass filter, there is clear evidence of a small but non-negligible proportion of non-luminescent light in the detector. It is likely to originate from a faint background spectrum of the laser light source within the spectral range of silicon band-to-band luminescence, but parts of it could also be caused by optical transitions in filters. The lower bound of bulk lifetime $\tau > 13.5\,\mu s$ results from assuming negligible non-luminescent light in the detector. The upper bound of $\tau < 28.0\,\mu s$ accounts for the non-luminescent light detected when covering the silicon ingot with a highly reflective aluminum foil and additionally considering the difference in reflectance between the ingot interface and the aluminum foil. The latter bound is to be perceived as an upper bound, because the aluminum interface is definitely less specular than the silicon ingot, thereby causing more (non-specular) scattering of non-luminescent light into the detector. Thus, the silicon ingot image is expected to feature a substantially lower non-luminescent contribution than the aluminum foil image—even after a correction of reflectance R_γ.

Looking at the cautious upper and lower bounds of bulk lifetime according to the PLIR approach, and taking into account that, with the above reasoning, true lifetime must be substantially higher than the lower bound and substantially lower than the upper bound, the dynamic method is well confirmed by the PLIR reference measurement. Notably, the arithmetic average of the two specified bounds even coincides with the precise harmonically modulated measurement result.

The dynamic lifetime approach elaborated here is not entirely unaffected by spectral assumptions. This is due to the fact that both, the excess carrier decay time in a silicon ingot with a dominant surface recombination, as well as the luminescence detection probability $p_\phi(z)$ (cf. Eq. 4.31) feature a pronounced dependence on substrate depth z. However, the sensitivity of the harmonically modulated (dynamic) approach, with respect to spectral assumptions, is much weaker than for (steady-state) techniques which infer bulk lifetime entirely from spectral information [85, 88, 97, 98]. Such techniques are largely affected by minor uncertainties of temperature and absorption data. Harmonically modulated luminescence is very robust in this respect. The reported bulk lifetime varies less than 0.5 % upon exchange of the near-band-gap absorption data as tabulated by Green and Keevers [31] by data as measured by Daub [99]. In contrast, the lifetime result of the PLIR analysis varies more than 30 % in light of the latter (very faint) uncertainty of absorption data.[38]

References

1. T. Trupke, R.A. Bardos, M.D. Abbott, Self-consistent calibration of photoluminescence and photoconductance lifetime leasurements. Appl. Phys. Lett. **87**, 184102 (2005)

[38] Note that the PLIR analysis was not optimized in terms of signal contrast for this reference measurement. Yet, even if slightly higher signal contrast could be obtained, the gap in relative measurement accuracy between the dynamic analysis as elaborated here and the PLIR approach is likely to remain significantly above an order of magnitude.

2. J.A. Giesecke, M.C. Schubert, W. Warta, Self-sufficient minority carrier lifetime in silicon from quasi-steady-state photoluminescence. Phys. Stat. Sol. A **209**, 2286–2290 (2012)
3. P.P. Altermatt, F. Geelhaar, T. Trupke, X. Dai, A. Neisser, E. Daub, Injection dependence of spontaneous radiative recombination in crystalline silicon: experimental verification and theoretical analysis. Appl. Phys. Lett. **88**, 261901 (2006)
4. W. van Roosbroeck, W. Shockley, Photon-radiative recombination of electrons and holes in germanium. Phys. Rev. **94**, 1558–1560 (1954)
5. A.M. Emel'yanov, The mechanism of radiative recombination in the region of interband transitions in single crystal silicon. Tech. Phys. Lett. **30**, 964–966 (2004)
6. J.A. Giesecke, M.C. Schubert, D. Walter, W. Warta, Minority carrier lifetime in silicon wafers from quasi-steady-state photoluminescence. Appl. Phys. Lett. **97**, 092109 (2010)
7. R.A. Sinton, A. Cuevas, M. Stuckings, Quasi-Steady-State Photoconductance, a New Method for Solar Cell Material and Device Characterization, in *Proceedings of the 25th IEEE PVSC* (Washington, D.C., 1996), pp. 457–460
8. R.A. Sinton, A. Cuevas, Contactless determination of current-voltage characteristics and minority-carrier lifetimes in semiconductors from quasi-steady-state photoconductance data. Appl. Phys. Lett. **69**, 2510–2512 (1996)
9. J.A. Giesecke, M. Kasemann, W. Warta, Determination of local minority carrier diffusion lengths in crystalline silicon from luminescence images. J. Appl. Phys. **106**, 014907 (2009)
10. D. Macdonald, T. Roth, P.N.K. Deenapanray, K. Bothe, P. Pohl, J. Schmidt, Formation rates of iron-acceptor pairs in crystalline silicon. J. Appl. Phys. **98**, 083509 (2005)
11. K. Bothe, J. Schmidt, Electronically activated boron-oxygen-related recombination centers in crystalline silicon. J. Appl. Phys. **99**, 013701 (2006)
12. D. Macdonald, F. Rougieux, A. Cuevas, B. Lim, J. Schmidt, M. Di Sabatino, L.J. Geerligs, Light-induced boron-oxygen defect generation in compensated p-type Czochralski silicon. J. Appl. Phys. **105**, 093704 (2009)
13. D. Macdonald, A. Cuevas, L.J. Geerligs, Measuring dopant concentrations in compensated p-type crystalline silicon via iron-acceptor pairing. Appl. Phys. Lett. **92**, 202119 (2008)
14. J.A. Giesecke, M.C. Schubert, B. Michl, F. Schindler, W. Warta, Minority carrier lifetime imaging of silicon wafers calibrated by quasi-steady-state photoluminescence. Sol. Energy Mater. Sol. Cells **95**, 1011–1018 (2011)
15. J.A. Giesecke, S.W. Glunz, W. Warta, Determination of actual carrier lifetime from differential measurements. Energy Procedia **38**, 59–65 (2013)
16. F.M. Schuurmans, A. Schönecker, A.R. Burgers, W.C. Sinke, Simplified evaluation method for light-biased effective lifetime measurements. Appl. Phys. Lett. **71**, 1795–1797 (1997)
17. J. Schmidt, Measurement of differential and actual recombination parameters on crystalline silicon wafers. IEEE Trans. Electron. Dev. **46**, 2018–2025 (1999)
18. S. Bowden, R.A. Sinton, Determining lifetime in silicon blocks and wafers with accurate expressions for carrier density. J. Appl. Phys. **102**, 124501 (2007)
19. J.A. Giesecke, T. Niewelt, M. Rüdiger, M. Rauer, M.C. Schubert, W. Warta, Broad range injection-dependent minority carrier lifetime from photoluminescence. Sol. Energy Mater. Sol. Cells **102**, 220–224 (2012)
20. J.A. Giesecke, M.C. Schubert, W. Warta, Carrier lifetime from dynamic electroluminescence. IEEE J. Photovoltaics **3**, 1012–1015 (2013)
21. R.N. Hall, Germanium rectifier characteristics. Phys. Rev. **83**, 228 (1951)
22. W. Shockley, W.T. Read, Statistics of the recombinations of holes and electrons. Phys. Rev. **87**, 835–842 (1952)
23. A. Cuevas, F. Recart, Capacitive effects in quasi-steady-state voltage and lifetime measurements of silicon devices. J. Appl. Phys. **98**, 074507 (2005)
24. T. Trupke, R.A. Bardos, F. Hudert, P. Würfel, J. Zhao, A. Wang, M.A. Green, Effective excess carrier lifetimes exceeding 100 milliseconds in float zone silicon determined from photoluminescence, in *Proceedings of the 19th EU PVSEC* (Paris, 2004), pp. 758–761
25. T. Trupke, R.A. Bardos, Photoluminescence: a surprisingly sensitive lifetime technique, in *Proceedings of the 31st IEEE PVSC* (Orlando, 2005), pp. 903–906

26. T. Trupke, Influence of photon reabsorption on quasi-steady-state photoluminescence measurements on crystalline silicon. J. Appl. Phys. **100**, 063531 (2006)
27. M. Rüdiger, T. Trupke, P. Würfel, T. Roth, S.W. Glunz, Influence of photon reabsorption on temperature dependent quasi-steady-state photoluminescence lifetime measurements on crystalline silicon. Appl. Phys. Lett. **92**, 222112 (2008)
28. D.B.M. Klaassen, A unified mobility model for device simulation–I. Model equations and concentration dependence. Solid-State Electron. **35**, 953–959 (1992)
29. D.B.M. Klaassen, A unified mobility model for device simulation–II. Temperature dependence of carrier mobility and lifetime. Solid-State Electron. **35**, 961–967 (1992)
30. A.B. Sproul, Dimensionless solution of the equation describing the effect of surface recombination on carrier decay in semiconductors. J. Appl. Phys. **76**, 2851–2854 (1994)
31. M.A. Green, M.J. Keevers, Optical properties of intrinsic silicon at 300 k. Prog. Photovoltaics: Res. Appl. **3**, 189–192 (1995)
32. A.G. Aberle, T. Lauinger, J. Schmidt, R. Hezel, Injection-level dependent surface recombination velocities at the silicon-plasma silicon nitride interface. Appl. Phys. Lett. **66**, 2828–2830 (1995)
33. C. Leguijt, P. Lölgen, J.A. Eikelboom, A.W. Weeber, F.M. Schuurmans, W.C. Sinke, P.F.A. Alkemade, P.M. Sarro, C.H.M. Maree, L.A. Verhoef, Low temperature surface passivation for silicon solar cells. Sol. Energy Mater. Sol. Cells **40**, 297–345 (1996)
34. J.R. Elmiger, R. Schieck, M. Kunst, Recombination at the silicon nitride/silicon interface. J. Vac. Sci. Tech. A **15**, 2418–2425 (1997)
35. M.J. Kerr, A. Cuevas, Recombination at the interface between silicon and stoichiometric plasma silicon nitride. Semicond. Sci. Tech. **17**, 166–172 (2002)
36. S. Steingrube, P.P. Altermatt, D.S. Steingrube, J. Schmidt, R. Brendel, Interpretation of recombination at c $- si/sin_x$ interfaces by surface damage. J. Appl. Phys. **108**, 014506 (2010)
37. B. Steuer, Aufbau eines Photolumineszenz-Messplatzes zur Charakterisierung von Wafern und Solarzellen aus kristallinem Silizium, Master thesis, Universität Konstanz (2011)
38. J.A. Giesecke, B. Michl, F. Schindler, M.C. Schubert, W. Warta, Minority carrier lifetime of silicon solar cells from quasi-steady-state photoluminescence. Sol. Energy Mater. Sol. Cells **95**, 1979–1982 (2011)
39. J. Isenberg, J. Dicker, W. Warta, Averaging of laterally inhomogeneous lifetimes for one-dimensional modeling of solar cells. J. Appl. Phys. **94**, 4122–4130 (2003)
40. R.A. Sinton, Predicting multi-crystalline solar cell efficiency from lifetime measured during cell fabrication, in *Proceedings of the 3rd WC PVEC* (Osaka, 2003), pp. 1028–1031
41. T.I. Kamins, Hall mobility in chemically deposited polycrystalline silicon. J. Appl. Phys. **42**, 4357–4365 (1971)
42. J.Y.W. Seto, The electrical properties of polycrystalline silicon films. J. Appl. Phys. **46**, 5247–5254 (1975)
43. G. Baccarani, B. Ricco, G. Spadini, Transport properties of polycrystalline silicon films. J. Appl. Phys. **49**, 5565–5570 (1978)
44. G.J. Korsh, R.S. Muller, Conduction properties of lightly doped, polycrystalline silicon. Solid-State Electron. **21**, 1045–1051 (1978)
45. C.H. Seager, T.G. Castner, Zero-bias resistance of grain boundaries in neutron-transmutation-doped polycrystalline silicon. J. Appl. Phys. **49**, 3879–3889 (1978)
46. A.K. Ghosh, A. Rose, H.P. Maruska, D.J. Eustace, T. Feng, Hall measurements and grain-size effects in polycrystalline silicon. Appl. Phys. Lett. **37**, 544–546 (1980)
47. H.P. Maruska, A.K. Ghosh, A. Rose, T. Feng, Hall mobility of polycrystalline silicon. Appl. Phys. Lett. **36**, 381–383 (1980)
48. N.C.C. Lu, L. Gerzberg, C.Y. Lu, J.D. Meindl, Modeling and optimization of monolithic polycrystalline silicon resistors. IEEE Trans. Electron Dev. **28**, 818–830 (1981)
49. N.C.C. Lu, L. Gerzberg, C.Y. Lu, J.D. Meindl, A conduction model for semiconductor-grain-boundary-semiconductor barriers in polycrystalline-silicon films. IEEE Trans. Electron Dev. **30**, 137–149 (1983)

50. M. Spitz, S. Rein, Impact of Potential Barriers at Grain Boundaries of Multicrystalline Silicon Wafers on Inductively Coupled Resistivity Measurements, in *Proceedings of the 26th EU PVSEC* (Hamburg, 2011), pp. 338–343
51. A. Cuevas, The Paradox of Compensated Silicon, in *Proceedings of the IEEE Conference on Optoelectronic and Microelectronic Materials and Devices* (Sydney, 2008), pp. 238–241
52. T. Fuyuki, H. Kondo, T. Yamazaki, Y. Takahashi, Y. Uraoka, Photographic surveying of minority carrier diffusion length in polycrystalline silicon solar cells by electroluminescence. Appl. Phys. Lett. **86**, 262108 (2005)
53. T. Trupke, R.A. Bardos, M.C. Schubert, W. Warta, Photoluminescence imaging of silicon wafers. Appl. Phys. Lett. **89**, 044107 (2006)
54. S. Herlufsen, J. Schmidt, D. Hinken, K. Bothe, R. Brendel, Photoconductance-calibrated photoluminescence lifetime imaging of crystalline silicon. Phys. Stat. Sol. (RRL) **2**, 245–247 (2008)
55. D. Macdonald, A. Cuevas, Trapping of minority carriers in multicrystalline silicon. Appl. Phys. Lett. **74**, 1710–1712 (1999)
56. R.A. Bardos, T. Trupke, M.C. Schubert, T. Roth, Trapping artifacts in quasi-steady-state photoluminescence and photoconductance lifetime measurements on silicon wafers. Appl. Phys. Lett. **88**, 053504 (2006)
57. S. Herlufsen, K. Ramspeck, D. Hinken, A. Schmidt, J. Müller, K. Bothe, J. Schmidt, R. Brendel, Dynamic photoluminescence lifetime imaging for the characterisation of silicon wafers. Phys. Stat. Sol. (RRL) **5**, 25–27 (2011)
58. D. Kiliani, G. Micard, B. Steuer, B. Raabe, A. Herguth, G. Hahn, Minority charge carrier lifetime mapping of crystalline silicon wafers by time-resolved photoluminescence imaging. J. Appl. Phys. **110**, 054508 (2011)
59. F. Dannhäuser, Die abhängigkeit der trägerbeweglichkeit in silicium von der konzentration der freien ladungsträger - i. Solid-State Electron. **15**, 1371–1375 (1972)
60. J. Krausse, Die abhängigkeit der trägerbeweglichkeit in silizium von der konzentration der freien ladungsträger - ii. Solid-State Electron. **15**, 1377–1381 (1972)
61. D.M. Caughey, R.E. Thomas, Carrier mobilities in silicon empirically related to doping and field. Proc. IEEE **55**, 2192–2193 (1967)
62. W.R. Thurber, R.L. Mattis, Y.M. Liu, J.J. Filliben, Resistivity-dopant density relationship for boron-doped silicon. J. Electrochem. Soc. **127**, 2291–2294 (1980)
63. W.R. Thurber, R.L. Mattis, Y.M. Liu, J.J. Filliben, Resistivity-dopant density relationship for phosphorus-doped silicon. J. Electrochem. Soc. **127**, 1807–1812 (1980)
64. F. Schindler, M.C. Schubert, A. Kimmerle, J. Broisch, S. Rein, W. Kwapil, W. Warta, Modeling majority carrier mobility in compensated crystalline silicon for solar cells. Sol. Energy Mater. Sol. Cells **106**, 31–36 (2012)
65. F. Schindler, J. Geilker, W. Kwapil, J. Giesecke, M.C. Schubert, W. Warta, Conductivity mobility and hall mobility in compensated multicrystalline silicon, in *Proceedings of the 25th EU PVSEC* (Valencia, 2010), pp. 2364–2368
66. F. Schindler, J. Geilker, W. Kwapil, W. Warta, M.C. Schubert, Hall mobility in multicrystalline silicon. J. Appl. Phys. **110**, 043722 (2011)
67. D. Walter, A. Liu, E. Franklin, D. Macdonald, B. Mitchell, T. Trupke, Contrast enhancement of luminescence images via point spread deconvolution, in *Proceedings of the 38th IEEE PVSC* (Austin, 2012), pp. 307–312
68. B. Mitchell, J.W. Weber, D. Walter, D. Macdonald, T. Trupke, On the method of photoluminescence spectral intensity ratio imaging of silicon bricks: advances and limitations. J. Appl. Phys. **112**, 063116 (2012)
69. M.C. Schubert, S. Pingel, M. The, W. Warta, Quantitative carrier lifetime images optically measured on rough silicon wafers. J. Appl. Phys. **101**, 124907 (2007)
70. B. Michl, M. Rüdiger, J.A. Giesecke, M. Hermle, W. Warta, M.C. Schubert, Efficiency limiting bulk recombination in multicrystalline silicon solar cells. Sol. Energy Mater. Sol. Cells **98**, 441–447 (2012)

71. M. C. Schubert, Detektion von infraroter Strahlung zur Beurteilung der Materialqualität von Solar-Silizium, Ph.D. thesis, Universität Konstanz (2008)
72. S. Rein, W. Kwapil, J. Geilker, G. Emanuel, M. Spitz, I. Reis, A. Weil, D. Biro, M. Glatthaar, A.K. Soiland, E. Enebakk, R. Tronstad, Impact of compensated solar-grade silicon on Czochralski silicon wafers and solar cells, in *Proceedings of the 24th EU PVSEC* (Hamburg, 2009), pp. 1140–1147
73. J. Geilker, W. Kwapil, I. Reis, S. Rein, Doping concentration and mobility in compensated material: comparison of different determination methods, in *Proceedings of the 25th EU PVSEC* (Valencia, 2010), pp. 1322–1327
74. P.P. Altermatt, A. Schenk, F. Geelhaar, G. Heiser, Reassessment of the intrinsic carrier density in crystalline silicon in view of band-gap narrowing. J. Appl. Phys. **93**, 1598–1604 (2003)
75. A.B. Sproul, M.A. Green, Intrinsic carrier concentration and minority-carrier mobility of silicon from 77 to 300 k. J. Appl. Phys. **73**, 1214–1225 (1993)
76. S.M. Sze, *Semiconductor Devices: Physics and Technology* (Wiley, New York, 2001)
77. A. Schenk, Finite-temperature full random-phase approximation model of band gap narrowing for silicon device simulation. J. Appl. Phys. **84**, 3684–3695 (1998)
78. K. Ramspeck, K. Bothe, J. Schmidt, R. Brendel, Combined dynamic and steady-state infrared camera based carrier lifetime imaging of silicon wafers. J. Appl. Phys. **106**, 114506 (2009)
79. D. Kiliani, A. Herguth, G. Micard, J. Ebser, G. Hahn, Time-resolved photoluminescence imaging with electronic shuttering using an image intensifier unit. Sol. Energy Mater. Sol. Cells **106**, 55–59 (2012)
80. J.A. Giesecke, R.A. Sinton, M.C. Schubert, S. Riepe, W. Warta, Determination of bulk lifetime and surface recombination velocity of silicon ingots from dynamic photoluminescence. IEEE J. Photovoltaics **3**, 1311–1318 (2013)
81. R.A. Sinton, T. Mankad, S. Bowden, N. Enjalbert, Evaluating silicon blocks and ingots with quasi-steady-state lifetime measurements, in *Proceedings of the 19th EU PVSEC* (Paris, 2004), pp. 520–523
82. J.S. Swirhun, R.A. Sinton, M.K. Forsyth, T. Mankad, Contactless measurement of minority carrier lifetime in silicon ingots and bricks. Prog. Photovoltaics: Res. Appl. **19**, 313–319 (2011)
83. D. Klein, F. Wuensch, M. Kunst, The determination of charge-carrier lifetime in silicon. Phys. Stat. Sol B **245**, 1865–1876 (2008)
84. N. Schüler, D. Mittelstrass, K. Dornich, J.R. Niklas, H. Neuhaus, Next generation inline minority carrier lifetime metrology on multicrystalline silicon bricks for PV, in *Proceedings of the 35th IEEE PVSC* (Honolulu, 2010), pp. 852–857
85. B. Mitchell, T. Trupke, J.W. Weber, J. Nyhus, Bulk minority carrier lifetimes and doping of silicon bricks from photoluminescence intensity ratios. J. Appl. Phys. **109**, 083111 (2011)
86. S. Herlufsen, K. Bothe, J. Schmidt, R. Brendel, S. Siegmund, Dynamic photoluminescence lifetime imaging of multicrystalline silicon bricks. Sol. Energy Mater. Sol. Cells **106**, 42–46 (2012)
87. A.B. Sproul, M.A. Green, A.W. Stephens, Accurate determination of minority carrier- and lattice scattering-mobility in silicon from photoconductance decay. J. Appl. Phys. **72**, 4161–4171 (1992)
88. P. Würfel, T. Trupke, T. Puzzer, E. Schäffer, W. Warta, S.W. Glunz, Diffusion lengths of silicon solar cells from luminescence images. J. Appl. Phys. **101**, 123110 (2007)
89. F.E. Rougieux, D. Macdonald, A. Cuevas, S. Ruffell, J. Schmidt, B. Lim, A.P. Knights, Electron and hole mobility reduction and hall factor in phosphorus-compensated p-type silicon. J. Appl. Phys. **108**, 013706 (2010)
90. E. Fourmond, M. Forster, R. Einhaus, H. Lauvray, J. Kraiem, M. Lemiti, Electrical properties of boron, phosphorus and gallium co-doped silicon. Energy Procedia **8**, 349–354 (2011)
91. B. Lim, M. Wolf, J. Schmidt, Carrier mobilities in multicrystalline silicon wafers made from umg-si. Phys. Stat. Sol. C **8**, 835–838 (2011)
92. M. Forster, F.E. Rougieux, A. Cuevas, B. Dehestru, A. Thomas, E. Fourmond, M. Lemiti, Incomplete ionization and carrier mobility in compensated p-type and n-type silicon. IEEE J. Photovoltaics **3**, 108–113 (2013)

93. D. Baek, S. Rouvimov, B. Kim, T.C. Jo, D.K. Schroder, Surface recombination velocity of silicon wafers by photoluminescence. Appl. Phys. Lett. **86**, 112110 (2005)
94. W.R. Thurber, R.L. Mattis, Y.M. Liu, J.J. Filliben, *The Relationship between Resistivity and Dopant Density for Phosphorus- and Boron-Doped Silicon* (US Department of Commerce, National Bureau of Standards, Washington, D.C., 1981)
95. B. Mitchell, J. Greulich, T. Trupke, Quantifying the effect of minority carrier diffusion and free carrier absorption on photoluminescence bulk lifetime imaging of silicon bricks. Sol. Energy Mater. Sol. Cells **107**, 75–80 (2012)
96. L.B. Valdes, Resistivity measurements on germanium for transistors. Proc. IRE **42**, 420–427 (1954)
97. J.A. Giesecke, M. Kasemann, M.C. Schubert, P. Würfel, W. Warta, Separation of local bulk and surface recombination in crystalline silicon from luminescence reabsorption. Prog. Photovoltaics: Res. Appl. **18**, 10–19 (2010)
98. M.P. Peloso, J.S. Lew, T. Trupke, M. Peters, R. Utama, A.G. Aberle, Evaluating the electrical properties of silicon wafer solar cells using hyperspectral imaging of luminescence. Appl. Phys. Lett. **99**, 221915 (2011)
99. E. Daub, Photolumineszenz von Silizium, Ph.D. thesis, University of Karlsruhe (1995)

Chapter 9
Conceptual Advances: Transport Properties

This chapter details the novel techniques and applications to resolve and quantify transport properties via modulated luminescence arising from this work. These involve a class of modulated photoluminescence techniques sensitive to net dopant concentration, which could be combined with four-point-probe measurements for studies of majority carrier mobility, and a transit time approach to accurately determine minority carrier mobility.

9.1 Net Dopant Concentration

This section elaborates a class of methods to determine net dopant concentration of silicon wafers via quasi-steady-state harmonically modulated luminescence. In contrast to resistivity measurements, this approach is independent of assumptions about majority carrier mobility. The latter fact particularly suggests application to compensated silicon material classes (e.g. upgraded metallurgical grade material), for which conventional mobility models involving only one dopant species have been reported to fail. This approach also allows access to majority carrier mobility via combination of net dopant concentration as inferred from photoluminescence with a resistivity measurement.[1]

9.1.1 Review of Relevant Facts and Literature

The concentration of dopant atoms (acceptors or donors) in silicon is an essential quantity for the device performance not only of solar cells, but of almost any technical application of silicon as a semiconductor, because it critically affects electronic and transport properties such as carrier lifetime, carrier mobilities, and resistivity. In photovoltaics, accurate knowledge of dopant concentration is of particular relevance for several reasons: First, together with carrier lifetime, it substantially accounts for

[1] The account given here is adapted from the corresponding original publications [1, 2].

J. Giesecke, *Quantitative Recombination and Transport Properties in Silicon from Dynamic Luminescence*, Springer Theses, DOI: 10.1007/978-3-319-06157-3_9,
© Springer International Publishing Switzerland 2014

the voltage of a solar cell. Therefore, its precise optimization can be a prerequisite for achieving optimal efficiencies. Second, assumptions about dopant concentration are often required for measurement techniques of other relevant quantities such as recombination properties or carrier mobilities. These facts corroborate the need for precise and accurate quantitative measurement methods of dopant concentration—and particularly of net dopant concentration (or dark majority carrier concentration), which shall be perceived here as the absolute value of the difference between acceptor and donor concentrations.

There is a broad amount of literature covering measurement techniques aiming at a determination of dopant concentration. Therefore, the brief survey of literature given here does not claim comprehensiveness. Rather, it focuses on the most practically relevant as well as on recent luminescence-based techniques: one of the most established measurement techniques of net dopant concentration is the four-point-probe resistivity measurement technique, which is sensitive to the product of majority carrier mobility and net dopant concentration [3, 4]. Measurements of either of these two quantities require knowledge of the other. As majority carrier mobility is again a function of dopant concentration, the latter is usually determined iteratively from four-point-probe measurements via a mobility model [5–11]. However, this measurement technique is restricted to silicon material classes which only feature one dopant species. This excludes compensated silicon, because carrier mobilities depend not only on net dopant concentration, but on both acceptor and donor concentration [12–14], which are not simultaneously accessible via the four-point -probe technique. Inductively coupled eddy current techniques [4] can be calibrated via four-point-probe measurements of base resistivity, as also exploited for most established photoconductance based carrier lifetime measurements [15, 16]. In spite of the advantage of contactless measurements, the eddy current technique suffers from the same inherent limitations as the four-point-probe technique. Beyond, its application to multicrystalline wafers is possibly compromised due to crystal defects (e.g. grain boundaries) acting as electrical potential barriers for eddy currents [17].

An alternative method to access dark majority carrier concentration—electrochemical capacitance-voltage—has long been known and used [18, 19]. It features a very broad range of accessible dopant concentrations, but for wafers it requires a physical substrate contact with an electrolyte. On textured interfaces, it additionally suffers from the uncertainty of the effective measurement area [20]. Recently published contactless measurement techniques of net dopant concentration involve free carrier absorption and Fourier transform infrared spectroscopy [21–23], as well as photoluminescence [2, 24–27].

Original photoluminescence applications to access dopant concentration were based on height ratios of phonon-assisted exciton lines and bands [28, 29]. As such, they were restricted to low temperatures (4–15 K) and to dopant concentrations less than typical for photovoltaic applications ($n_0 < 10^{15}\,\mathrm{cm}^{-3}$). Yet, with the deployment of luminescence-based metrology in photovoltaics, several more broadly applicable approaches working at room temperature have been published: Mitchell et al. introduced a method to simultaneously image both carrier lifetime and the relative course of net dopant concentration of multicrystalline silicon ingots [24], based on a methodology which was originally introduced by Würfel et al. [30]. Macdonald et al. introduced a technique to determine the total acceptor concentration via carrier lifetime measurements during re-pairing of optically dissociated acceptor-iron pairs [31]. This formalism was extended to photoluminescence measurements [25] and photoluminescence imaging [26] of total acceptor concentration by Lim et al. Also, Lim et al. [26] determined maps of relative dopant concentration from photoluminescence images of wafers featuring strongly surface limited recombination properties. Recently, Hameiri et al. combined photoluminescence and photoconductance measurements carried out at both high- and low-level injection conditions in order to access dopant concentration [27]—as originally suggested by Trupke and Bardos [32].

9.1.2 Net Dopant Concentration via Modulated Luminescence

The class of methods to be derived in the following provide quantitative access to net dopant concentration via quasi-steady-state harmonically modulated carrier lifetime measurements. This class of methods is distinguished by the fact that it exploits injection-level dependent carrier lifetime measurements in combination with the relation between photoluminescence intensity, excess carrier density, and net dopant concentration [33] (cf. Eqs. 4.36 or 8.10). A quantitative determination of net dopant concentration without a priori knowledge of carrier mobilities or the dopant type is thereby enabled.

9.1.2.1 Peak Curvature Analysis and Phase-Sensitive Lifetime

Under quasi-steady-state measurement conditions, a relative deformation of the photoluminescence intensity curve $\Phi(t)$ with respect to excess carrier generation rate $G(t)$ may have two causes: one possible cause is an injection dependence of carrier lifetime. Regardless of the injection level, a positive derivative of lifetime with respect to excess carrier density $\frac{d\tau_{eff}}{d\Delta n} > 0$ leads to a narrowing of the width of $\Phi(t)$, whereas a negative derivative $\frac{d\tau_{eff}}{d\Delta n} < 0$ leads to a broadening of the width of $\Phi(t)$. The second cause is related to the nonlinearity of photoluminescence intensity as a function of excess carrier density under intermediate- to high-level injection conditions (cf. Eq. 4.36), which leads to a narrowed width of $\Phi(t)$ with respect to $G(t)$. These two effects are illustrated in Fig. 9.1.

Rigorously speaking, there is also a third cause due to the injection dependence of the integral coefficient of radiative recombination [34, 35] (cf. Sect. 4.2.2 and Appendix A.5), but this can be taken into account via an iterative transformation of photoluminescence intensity $\Phi(t)$ to be expected if the coefficient of radiative recombination was constant. Both of the above mentioned causes of relative deformations of $\Phi(t)$ with respect to $G(t)$ have an impact on the peak curvature ratio of these signals. Yet, it is demonstrated in Sect. 8.1 that a phase-sensitive

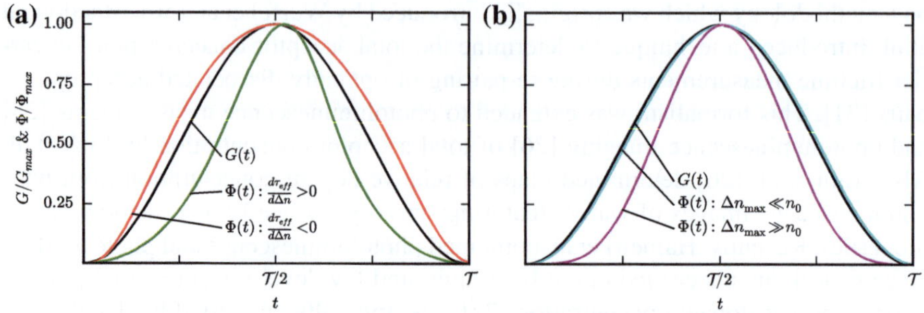

Fig. 9.1 The major mechanisms affecting the relative curve shape of photoluminescence intensity $\Phi(t)$—and hence the ratio between peak curvatures of photoluminescence intensity and generation rate $G(t)$: **a** the injection dependence of carrier lifetime largely affects the broadening or narrowing of the photoluminescence curve $\Phi(t)$. The modeled curves shown here represent true low-level injection conditions. **b** The peak curvature of $\Phi(t)$ is also influenced by the ratio between excess carrier density and net dopant concentration. In this model, true low- and high-level injection conditions are compared, while carrier lifetime is injection-independent. Adapted after [1]

(self-sufficient or *d2a*) injection-dependent determination of effective lifetime is possible without a priori assumptions about dopant concentration. Thereby obtaining an *actual* effective lifetime $\tau_{eff,0}$ as well as its local derivative $\gamma_\tau = \frac{d\tau_{eff}}{d\Delta n}\big|_{\Delta n=G_0\tau_{eff,0}}$ at a peak excess carrier generation rate G_0, one may recall quasi-steady-state harmonically modulated excess carrier generation rate of Eq. 3.83 and excess carrier density of Eq. 3.87. With an injection-dependent lifetime $\tau_{eff}(\Delta n)$ and accounting for the fact that excess carrier density is shifted with respect to generation rate by *differential* lifetime τ_m rather than by *actual* lifetime τ_{eff} (cf. Sect. 6.2), $G(t)$ and $\Delta n(t)$ read[2]

$$G(t) = G_0 \sin^2 \frac{\pi t}{T} \tag{9.1}$$

$$\Delta n(t) = G_0 \tau_{eff}(\Delta n(t)) \sin^2\left(\frac{\pi}{T}(t - \tau_m)\right). \tag{9.2}$$

The linear approximation of effective lifetime in the proximity of $\Delta n = G_0\tau_{eff}$ is

$$\tau_{eff} = \tau_{eff,0} + \gamma_\tau\left(\Delta n - G_0\tau_{eff,0}\right). \tag{9.3}$$

This allows accurate expression of excess carrier density of Eq. 9.2 in the proximity of its maximum as

$$\Delta n(t) = G_0\tau_{eff,0}\frac{1-\gamma_\tau G_0}{\sin^2\left(\frac{\pi}{T}(t-\tau_m)\right) - \gamma_\tau G_0}. \tag{9.4}$$

[2] Disregarding the injection dependence of the peak shift τ_m, which however does not affect the peak curvature sought here.

Recalling the relation between luminescence intensity and excess carrier density as of Eq. 4.36,

$$\Phi(t) = c_\Phi \Delta n(t)(n_0 + \Delta n(t)), \tag{9.5}$$

one may evaluate the second derivative of normalized luminescence intensity with respect to time at its maximum at $t = \frac{\mathcal{T}}{2} + \tau_m$, yielding

$$\mathcal{C}_\Phi = \frac{1}{\Phi}\frac{d^2\Phi}{dt^2}\bigg|_{t=\frac{\mathcal{T}}{2}+\tau_m} = -\frac{2\pi^2}{\mathcal{T}^2}\frac{2G_0\tau_{eff,0}+n_0}{G_0\tau_{eff,0}+n_0}\frac{1}{1-\gamma_\tau G_0}. \tag{9.6}$$

The second derivative of normalized excess carrier generation rate with respect to time at its maximum $t = \frac{\mathcal{T}}{2}$ is

$$\mathcal{C}_G = \frac{1}{G}\frac{d^2G}{dt^2}\bigg|_{t=\frac{\mathcal{T}}{2}} = -\frac{2\pi^2}{\mathcal{T}^2}. \tag{9.7}$$

Therefore, the ratio $\mathcal{R}_\Phi = \frac{\mathcal{C}_\Phi}{\mathcal{C}_G}$ of normalized peak curvatures is

$$\mathcal{R}_\Phi = \frac{2G_0\tau_{eff,0}+n_0}{G_0\tau_{eff,0}+n_0}\frac{1}{1-\gamma_\tau G_0}, \tag{9.8}$$

thereby yielding net dopant concentration n_0 as a function of the peak curvature ratio $\mathcal{R}_\Phi$ as

$$n_0 = G_0\tau_{eff,0}\frac{\mathcal{R}_\Phi(1-\gamma_\tau G_0)-2}{1-\mathcal{R}_\Phi(1-\gamma_\tau G_0)}. \tag{9.9}$$

The phase-sensitive lifetime analyses elaborated in Sect. 8.1 accurately provide access to both, $\tau_{eff,0}$ and γ_τ at a given peak excess carrier generation rate G_0. Further evaluation of the peak curvature ratio $\mathcal{R}_\Phi$ of a harmonically modulated luminescence measurement carried out under intermediate injection conditions thereby allows access to net dopant concentration n_0.[3]

9.1.2.2 Combined Self-Consistent and Phase-Sensitive Analysis

As addressed in Sect. 5.2.3, under intermediate- and high-level injection conditions, the self-consistent determination of carrier lifetime from a quasi-steady-state harmonically modulated luminescence measurement depends on the choice of net dopant concentration. This drawback turns into an advantage when combining the

[3] For the determination of $\mathcal{R}_\Phi$, the effect of the injection dependence of the radiative recombination coefficient B on luminescence intensity $\Phi(t)$ is to be taken into account.

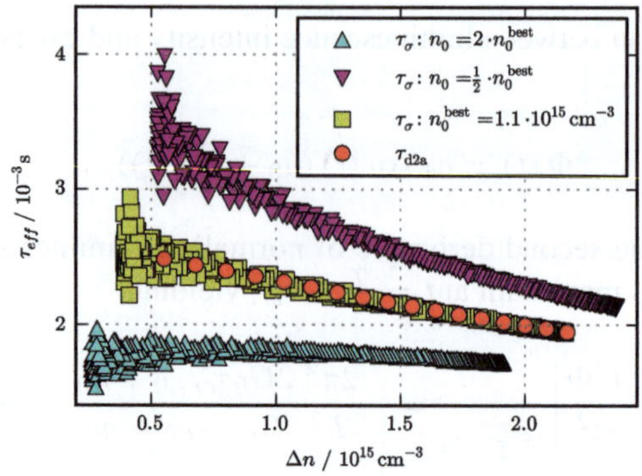

Fig. 9.2 Depiction of the working principle of the combined self-consistent (τ_σ) and phase-sensitive (τ_{d2a}) analysis of net dopant concentration. The correct choice of net dopant concentration yields coinciding *d2a* and self-consistent carrier lifetimes. If net dopant concentration is chosen too high, self-consistent carrier lifetime underestimates true lifetime. If net dopant concentration is chosen too low, self-consistent carrier lifetime overestimates true lifetime. Measurement data are taken from one out of five measurements evaluated in Table 9.1. Adapted after [1]

self-consistent approach with one of the phase-sensitive lifetime approaches elaborated in Sect. 8.1 (preferrably the *d2a* approach) in order to access net dopant concentration n_0.

For this purpose, n_0 is adapted in the self-consistent analysis such that injection-dependent carrier lifetime curves $\tau_\sigma\,(\Delta n)$ and $\tau_{d2a}\,(\Delta n)$ coincide. The working principle of this method is depicted on the basis of real experimental data in Fig. 9.2.

9.1.2.3 Entirely Self-Consistent Analysis

From a theoretical standpoint, one might argue that true self-consistency of the continuity equation could only be satisfied if both, injection-dependent carrier lifetime and net dopant concentration were correctly chosen. Therefore, it should be possible to simultaneously determine net dopant concentration and carrier lifetime from a self-consistent analysis of one modulated luminescence measurement, conducted at intermediate injection conditions. Such an advanced analysis would involve not only the adaption of the scaling factor c_Φ of Eq. 9.5 (detailed in Sect. 5.2) but also the adaption of n_0 as an additional degree of freedom. Even though such an analysis would be numerically feasible, it fails in practice.

As detailed in Sect. 5.2, self-consistency is inferred from a numerical process of hysteresis minimization. In spite of the impact of net dopant concentration on the self-consistent lifetime analysis under intermediate-level injection conditions, it must be stated that the contrast of hysteresis with respect to net dopant concentration is only very faint. Thus, very minor inaccuracies may cause the failure of the self-consistent approach, and it is therefore only possible to roughly determine the order

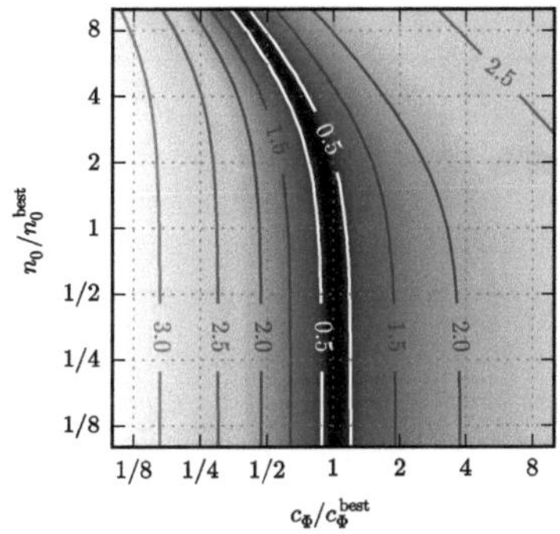

Fig. 9.3 Contour plot of hysteresis—being a numerical measure of self-consistency of the continuity equation—as function of both, the scaling factor c_Φ and net dopant concentration n_0. Here, hysteresis is depicted as the logarithmized sum of squares of residuals of lifetime expressions, subtracted by its global minimum. Theoretically, self-consistency is equivalent to a global minimum of hysteresis. This plot clearly demonstrates the very faint contrast of hysteresis with respect to relative variations of n_0, while the contrast with respect to relative variations of c_Φ (and thereby of self-consistent lifetime τ_σ) is quite pronounced. Measured data and best values of n_0 and c_Φ correspond to the analysis of lifetime and net dopant concentration depicted in Fig. 9.2 (cf. Table 9.1). Adapted after [1]

of magnitude of net dopant concentration through the self-consistent analysis of only one harmonically modulated measurement. This limitation is illustrated in Fig. 9.3.

In spite of the above limitation, it is possible to obtain net dopant concentration n_0 from an entirely self-consistent harmonically modulated luminescence lifetime analysis. This is accomplished by conducting two measurements, one at low-level injection and another one at intermediate- to high-level injection, respectively. As outlined in Sect. 5.2.3, under low-level injection, the self-consistent carrier lifetime analysis is not affected by the choice of net dopant concentration. Under intermediate- to high-level injection conditions, this does not apply any longer. The resulting carrier lifetime τ_σ then also depends on the choice of n_0. Yet, the scaling factor c_Φ should be the same for both measurements—except for a quantifiable and corrigible effect of the injection dependence of the coefficient of radiative recombination, and except for possibly different photocurrent amplification settings, which are likewise well-known.

The fact that scaling factors of low- and intermediate- to high-level injection measurements must be identical can be exploited for an entirely self-consistent determination of net dopant concentration. For this purpose, the low-level injection measurement is evaluated in two ways: First, lifetime is evaluated self-consistently. Second, apparent lifetime is evaluated via the scaling factor and net dopant concentration taken from the self-consistent lifetime analysis of the intermediate- to high-level injection measurement. Accordingly, net dopant concentration is to be adapted such that the lifetime expressions resulting from both analyses coincide.

Figure 9.4 visualizes the working principle of the entirely self-consistent approach: in three equally scaled plots, net dopant concentration is either mismatched by a

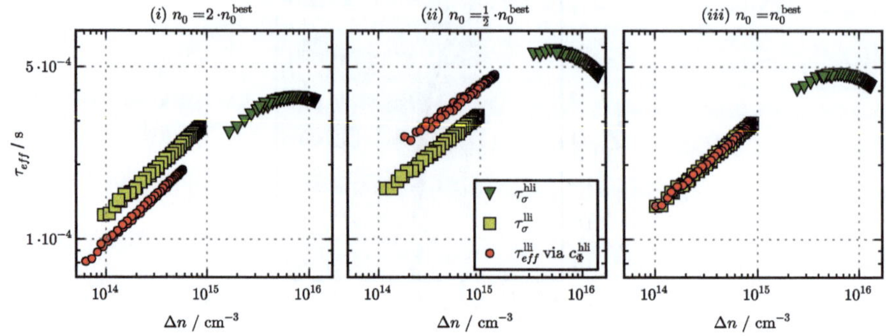

Fig. 9.4 The working principle of the entirely self-consistent approach to determine net dopant concentration n_0: at low-level injection, self-consistent carrier lifetime is only very weakly affected by the choice of n_0 for the lifetime analysis. The aim of this method is to adapt n_0 such that the lifetime evaluation of the low-level injection measurement τ_{eff}^{lli} with the scaling factor c_Φ^{hli} of the self-consistent lifetime analysis τ_σ^{hli} of an intermediate- to high-level injection measurement coincides with the self-consistent lifetime analysis τ_σ^{lli} of the low-level injection measurement. Three cases are shown: (i) overestimation of n_0; (ii) underestimation of n_0; (iii) correct choice of n_0. Adapted after [1]

factor of two (i, ii) or chosen correctly (iii). The impact of the choice of net dopant concentration on the self-consistent carrier lifetime analysis of the low-level injection measurement (yellow squares) is very faint. It is only nonzero because low-level injection conditions are not rigorously satisfied. Note that this does not compromise the described method. In contrast, the impact of the choice of net dopant concentration on the self-consistent carrier lifetime analysis of the intermediate- to high-level injection measurement (green triangles) is quite pronounced. Application of the scaling factor of the latter measurement to a lifetime analysis of the low-level injection measurement (red circles) yields too low carrier lifetime if net dopant concentration is chosen higher than the true value, and vice versa. This illustrates the high contrast of the described method with respect to net dopant concentration.

9.1.3 Experimental Results

Experimental results are presented as follows: an experimental proof of principle is given for the methods involving a phase-sensitive (e.g. *d2a*) approach. The entirely self-consistent approach is then used for an investigation of net dopant concentration of several silicon material classes, also comprising upgraded metallurgical grade silicon. The latter experimental data are combined with resistivity measurements in order to determine majority carrier mobility.

9.1.3.1 A Note on the Measurement Configuration

In order to keep temperature variations sufficiently small ($\Delta T < 2\,\mathrm{K}$) between low- and high-level injection measurements, all measurements were conducted in the *reflective* measurement configuration of the luminescence detector. This enabled effective dissipation of heat through positioning of the investigated substrates directly on a solid aluminum block. This was particularly important for the entirely self-consistent analysis due to the temperature dependence of the coefficient of radiative recombination (cf. Appendix A.5). An additional advantage of the reflection geometry is the option to measure carrier lifetime (and net dopant concentration) on metalized wafers (cf. Sect. 7.3).

9.1.3.2 Phase-Sensitive Approaches

Net dopant concentration on a surface-passivated $4\,\Omega$ cm n-type float zone wafer was measured in three different ways: an inductive eddy current technique calibrated by four-point-probe measurements, a phase-sensitive *d2a* lifetime analysis [36–38] (cf. Sect. 8.1) in combination with a peak curvature analysis [2] of a quasi-steady-state harmonically modulated photoluminescence measurement conducted at intermediate- to high-level injection conditions, and a combination of *d2a* and self-consistent lifetime analyses of that same measurement. The measurement was repeated five times in order to quantify the measurement precision of n_0. The result is shown in Table 9.1. Good agreement of both dynamic photoluminescence methods with the inductive reference measurement can be stated.

9.1.3.3 Entirely Self-Consistent Approach

For the entirely self-consistent analysis, a set of (undiffused) wafers covering three different material classes was used, comprising monocrystalline float zone, standard multicrystalline (gas-phase-cleaned and uncompensated), and multicrystalline upgraded metallurgical grade (compensated) silicon. Measurements were carried out on laterally uniform spots (homogeneous lifetime) of a size of $1 \times 1\,\mathrm{cm}^2$. Lateral uniformity was verified via photoluminescence images.

Figure 9.5 depicts resulting net dopant concentrations as well as relative deviations from net dopant concentration as inferred from four-point-probe [and inductive, but four-point-probe calibrated (float zone wafers only)] resistivity measurements. For float zone wafers, all results coincide with four-point-probe reference measurements within the specified measurement precision. The measurement precision as specified represents the standard deviation upon repetition of measurements (five measurements per investigated wafer spot). For uncompensated multicrystalline silicon, results of the majority of samples coincide with four-point-probe reference measurements within the specified measurement precision. There are only a few samples featuring significant deviations in both directions. This could be related to

Table 9.1 Comparison between phase-sensitive harmonically modulated photoluminescence methods to determine net dopant concentration, and an inductive eddy current technique (four-point-probe calibrated). Photoluminescence techniques involve a combination of the phase-sensitive lifetime τ_{d2a} with both, an analysis of the peak curvature ratio $\mathcal{R}_{\Phi}$, and a self consistent lifetime analysis τ_{σ}. Net dopant concentration was measured on a surface-passivated 4 Ω cm n-type float zone wafer. The uncertainty of the luminescence measurements relates to the encountered measurement precision of five successive measurements

Method	$n_0/10^{15}\,\text{cm}^{-3}$
Inductive (eddy current) [3, 9]	1.1
τ_{d2a} & $\mathcal{R}_{\Phi}$	1.01 ± 0.13
τ_{d2a} & τ_{σ}	1.15 ± 0.10

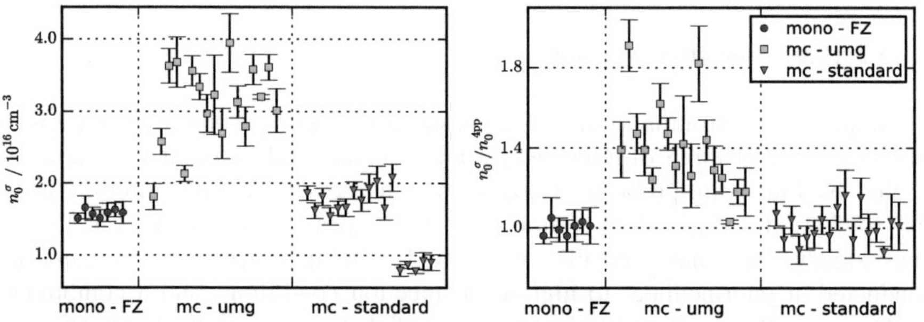

Fig. 9.5 Determination of net dopant concentration via an entirely self-consistent quasi-steady-state harmonically modulated photoluminescence approach: the *left plot* shows net dopant concentrations n_0^σ (with a superscript σ indicating the entierely self-consistent determination) as determined on 41 electronically homogeneous spots ($1 \times 1\,\text{cm}^2$) of boron-doped wafers of three different material classes, involving monocrystalline float zone material (mono–FZ), uncompensated standard gas-phase-cleaned multicrystalline material (mc–standard), and multicrystalline upgraded metallurgical grade material (mc–umg), which was partially compensated with phosphorus, but still p-type. The *right plot* shows the relative deviation of the same measurements from a determination of n_0^{4pp} as inferred from resistivity measurements via the four-point-probe technique. The latter technique iteratively determines net dopant concentration via a conventional mobility model, involving only one dopant species [10]. Adapted after [1]

specific characteristics of multicrystalline material, such as the impact of crystal defects like grain boundaries on the distribution of the electrical field in a four-point-probe geometry [17]. Note that four-point-probe measurements in this study were never carried out across grain boundaries. Nevertheless, the resulting resistivities cannot be claimed to be totally artifact-free, because the grain boundary configuration did not in any case allow sufficient spacing between grain boundaries and the four-point-probe electrical field.

In spite of the reasonable agreement between net dopant concentration as inferred from modulated luminescence and four-point-probe reference measurements on uncompensated silicon, there is a systematic deviation (of invariable direction) between the two methods throughout all investigated compensated wafers. This was to be expected, as the implied majority carrier mobility model of the four-point-probe

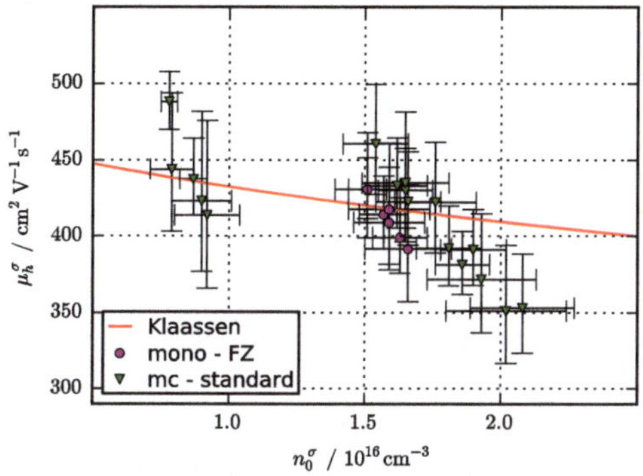

Fig. 9.6 Calculation of majority carrier (hole) mobility μ_h^σ for the uncompensated wafers used in this study (cf. Fig. 9.5): For this purpose, net dopant concentration n_0^σ as determined via entirely self-consistent quasi-steady-state harmonically modulated photoluminescence was combined with four-point-probe measurements. The expected majority carrier mobility according to the unified mobility model of Klaassen [12, 13] is also plotted. Adapted after [1]

analysis must fail in the case of coexisting dopant species. This finding confirms the necessity and relevance of measurement methods of net dopant concentration that do not require information about carrier mobility.

9.1.3.4 Access to Majority Carrier Mobility

The combination of a measurement of wafer base resistivity ρ with a measurement of net dopant concentration n_0 independent of majority carrier mobility μ provides access to the latter via

$$\mu = \frac{1}{q n_0 \rho} \tag{9.10}$$

This was used for a determination of majority carrier mobility from both, harmonically modulated luminescence and four-point-probe measurements. The resulting hole mobilities in the uncompensated p-type samples of the self-consistent study presented here are shown in Fig. 9.6 as a function of net dopant concentration. Comparison to the unified mobility model by Klaassen [12, 13], which can be regarded as largely equivalent to the state-of-the-art one-dopant-species mobility model [10] for uncompensated material, reveals reasonable agreement for most samples.

9.1.4 Discussion of Limitations

Light-induced changes of carrier lifetime [39–41] during measurements do not affect the methods to determine net dopant concentration presented here. Even significant lifetime changes between low- and high-level injection measurements for a self-consistent analysis of net dopant concentration would leave the resulting net dopant concentration unaffected, because such lifetime changes usually have no impact on the scaling factor c_Φ of a self-consistent lifetime analysis (cf. Eq. 9.5). If the scaling factor remains invariant throughout measurements, net dopant concentration can be determined.

The scaling factor c_Φ relates photoluminescence intensity to measured effective carrier lifetime τ_σ. Due to the fact that harmonically modulated luminescence is a dynamic method, measured effective carrier lifetime never corresponds to the arithmetic average over local lifetimes in the case of lateral lifetime inhomogeneities. Rather, it is much dominated by comparatively high local lifetimes (cf. Eq. 8.24). This fact imposes a limitation on the entirely self-consistent approach to determine net dopant concentration: optical generation rates may well differ by more than an order of magnitude between low- and intermediate-level injection measurements conducted for the above-shown self-consistent analyses (cf. Fig. 9.4). If the active measurement area was electronically inhomogeneous—and particularly if it contained spots of significantly differing injection dependence—this might affect the determination of the scaling factor c_Φ. Such a variation may perturb the analysis of net dopant concentration. Therefore, the entirely self-consistent analysis is to be carefully restricted to spots of laterally uniform electronic quality for any of the investigated material classes. Such spots were identified via photoluminescence images.

Any of the methods discussed here require at least one carrier lifetime measurement conducted under intermediate- to high-level injection conditions. This implies a targeted excess carrier density Δn in the range of net dopant concentration n_0, which again imposes a limitation on the range of applicability of the class of methods described herein: With quasi-steady-state excess carrier density corresponding to the product of generation rate and lifetime, the question whether $\Delta n \sim n_0$ can be reached depends to a large extent on effective carrier lifetime at that injection-level. To exemplify this, reaching an injection level of $\Delta n \approx 1.5 \times 10^{16}\,\text{cm}^{-3}$ with an irradiance equivalent to 50 suns on a $d = 250\,\mu\text{m}$ thick wafer requires a minimum effective lifetime of $\tau_{eff} > 3 \times 10^{-5}$ s. Also note, that at net dopant concentrations typically in the range and above $\Delta n \sim 10^{16}\,\text{cm}^{-3}$, Auger recombination may dominate at the targeted injection-level. A further boost of laser irradiance would be only of very limited benefit. This is due to both, the onset of Auger recombination and to the fact that at even higher irradiances specimen may substantially heat up even in spite of an efficient heat sink (dissipation limited by their inherent heat conduction capability).

The fact that monochromatic irradiation with a mean free path much less than typical specimen width (in silicon at room temperature) is used here brings about a systematic error for very small carrier lifetimes on the order of several microseconds. All the methods outlined here relate net dopant concentration to measured excess carrier density. Yet, excess carrier density is inferred as the product of carrier lifetime and generation rate, while the latter is usually a depth-averaged quantity. Thus, for low carrier lifetimes corresponding to excess carrier densities featuring a pronounced depth profile, Δn (and thereby n_0) are significantly underestimated. This effect could be corrected through an estimate of an effective (rather than a depth-averaged) excess carrier density [42] ($\Delta n_{eff} = \varkappa \Delta n$ cf. Sect. 8.1.3).

It must be stated that any of the methods to determine n_0 which are described above make use of the dependence of the coefficient of radiative recombination B [35] on temperature and free carrier density. Any systematic error of Altermatt's model of B (cf. Appendix A.5) would also propagate to a systematic error of n_0 as determined via quasi-steady-state harmonically modulated luminescence.

9.2 Minority Carrier Mobility

This section describes the determination of minority carrier mobility from self-consistent quasi-steady-state harmonically modulated photoluminescence measurements of effective carrier lifetime.[4] The measurement design is distinguished by a limitation of carrier recombination through minority carrier transport—with excess carrier generation and recombination confined to opposite interfaces, respectively. Minority carrier mobility is inferred from the minority carrier diffusion coefficient via the Einstein relation. An experimental proof of concept on monocrystalline p-type material is provided, showing good agreement with state-of-the-art data and models. Considerations for the applicability of the method to compensated and multicrystalline silicon materials are discussed.

9.2.1 Review of Relevant Facts and Literature

Carrier mobility is a crucial semiconductor transport property, which substantially affects the performance of devices such as silicon solar cells. Accurate knowledge of both, majority and minority carrier mobility is therefore essential for device simulation and characterization. With strong interest in multicrystalline and particularly in partially compensated material in silicon photovoltaics in recent years, investigations of majority carrier mobility have seen a revival [22, 44–50]. Despite a similarly high interest in the minority carrier mobility of the above mentioned materials, experimental investigations of minority carrier mobility have remained relatively scarce [51–55]. This could be attributed to the fact that measurement techniques to accurately and reliably resolve minority carrier mobility are demanding, especially if their application is to be extended away from highest quality single crystals obtained via float zone crystal growth.

Originally, Haynes and Shockley [56] suggested an experimental design to determine minority carrier mobility (or drift mobility) in germanium from measurements of the minority carrier transit time in the presence of an electrical field. This design was used and adapted in numerous works, among others by Prince [57], by Ludwig and Watters [58], by Zerbst and Heywang [59], and by Tang et al. [60]. Several alternative techniques have also been published: Dziewior and Silber [61] calculated

[4] The account given here is adapted from the corresponding original publication [43].

minority carrier diffusion coefficients from measurements of the complex diffusion length in a radio frequency time-modulated experiment, Neugroschel [62] calculated minority carrier diffusion coefficients from the minority-carrier transit time across the base of a pn-junction device, Misiakos and Neugroschel [63] performed a frequency-domain analysis of the transient free carrier response following optical generation, Susi et al. [64] and Sontag et al. [65] inferred minority carrier diffusion coefficients from a combination of measurements of minority carrier lifetime and diffusion length, and Sproul et al. [66] determined minority carrier diffusion coefficients from photoconductance decay measurements on wafers with intentionally enhanced surface recombination velocity.

A comprehensive survey of formerly published minority carrier mobility data on monocrystalline material by Sproul et al. [66] reveals a broad scattering. This scattering has been attributed to majority carrier drag effects in Haynes-Shockley-type drift experiments [61], insufficient crystal purity [59, 62, 66], or compensation of dopant species [57, 59]. Other reasons for this scattering could be deviations from rigorous low-level injection resulting in a measurement of ambipolar mobility rather than minority carrier mobility, or a discrepancy between differential carrier lifetime and actual carrier lifetime leading to minor measurement artifacts in published data based on light-biased photoconductance decay [37, 67–70]. While the origin of the observed scattering has not yet been completely resolved, mobility models [10, 12–14, 71] are today used for device simulation and characterization at an increasing demand of accuracy, and often on challenging materials such as multicrystalline and compensated silicon. The accurate minority carrier mobility data by Sproul et al. are based on the use of highest quality float zone crystals. A reliable and at the same time broadly applicable measurement technique of minority carrier mobility is thus deemed of considerable interest in order to validate and to extend the existing knowledge about minority carrier mobility.

In the following account, a technique to determine minority carrier mobility from carrier lifetime measurements via quasi-steady-state harmonically modulated photoluminescence is presented. Like the aforementioned technique by Sproul et al. [66], it infers minority carrier mobility via the Einstein relation from the minority carrier diffusion coefficient obtained in a state of predominant surface recombination. In contrast to Sproul's approach, the technique introduced here is based on a heavily asymmetrical surface passivation scheme—featuring optimal surface passivation on an irradiated wafer interface and predominant surface recombination on the opposite wafer interface with respect to irradiation. In this configuration, carrier recombination is limited by minority carrier transport from the irradiated wafer interface (where electron–hole-pair generation takes place) to the opposite interface (where carrier recombination takes place). Therefore, as in Sproul's approach [66], predominantly the minority carrier diffusion coefficient and the wafer thickness affect effective carrier lifetime. Beyond its capability to independently validate both state-of-the-art experimental data [66] and theoretical models [12, 13] of minority carrier mobility, the presented method further takes into account finite surface recombination velocities and bulk lifetimes for the determination of the minority carrier diffusion coefficient. This is particularly relevant for wafers featuring either surface recombination velocities well below the thermal carrier drift velocity (at the interface featuring high recombination), high dopant or impurity concentrations, or substantial crystal defects. In contrast to Sproul's approach [66], the measurement design described in the following is suitable for an accurate determination of minority carrier mobility even under such conditions.

9.2.2 Experimental Design

Measurement techniques of minority carrier mobility $\mu_{e/h}$ can be classified into two categories, comprising either measurements of minority carrier drift in the presence of an electric field [56–60] or measurements of the minority carrier diffusion coefficient [61–66]. The latter is related to minority carrier mobility through the Einstein relation (cf. Eq. 3.15), which follows from the requirement that in thermodynamic equilibrium the field induced local drift current must balance the local diffusion current. In the following, an approach to access the minority carrier diffusion coefficient D_a via a self-consistent quasi-steady-state harmonically modulated photoluminescence lifetime analysis is described.

9.2.2.1 A Transit Time Approach

The measurement design of a determination of the minority carrier diffusion coefficient via modulated photoluminescence is distinguished by illumination of an optimally surface-passivated wafer interface ($S_0 \rightarrow 0$), while the opposite wafer interface features an intentionally enhanced high surface recombination velocity ($S_d \rightarrow \infty$). Note that at a room temperature inverse penetration depth of irradiation light of $\alpha_\gamma = 417\,\text{cm}^{-1}$ and a typical wafer thickness of $d = 200\,\mu\text{m}$, excess carrier generation is much confined to the illuminated wafer interface. Therefore, if recombination rates associated with both, the wafer bulk and the illuminated wafer interface, are negligible compared to the recombination rate associated with the opposite wafer interface with respect to illumination, there is essentially only one nontrivial property left to affect the depth profile of excess carrier density $\Delta n(z)$—and therefore steady-state effective carrier lifetime $\tau_{eff,s}$, which is the ratio between the depth integrals of $\Delta n(z)$ and $G(z)$: This is the ambipolar minority carrier diffusion coefficient D_a. Figure 9.7 illustrates the high contrast of the depth profile $\Delta n(z)$ (and thereby also of its depth integral) with respect to D_a. This high contrast results from a limitation of rear surface recombination by carrier transport.

A straightforward analytical representation of minority carrier transit time through the wafer bulk can be derived from the steady-state solution of the continuity equation, as given in Sect. 3.2.2. For a negligible surface recombination velocity at an irradiated interface, a quasi-infinite surface recombination velocity at the opposite interface, and negligible bulk recombination, a derivation given in Appendix A.1.3 yields a straightforward expression of transit time

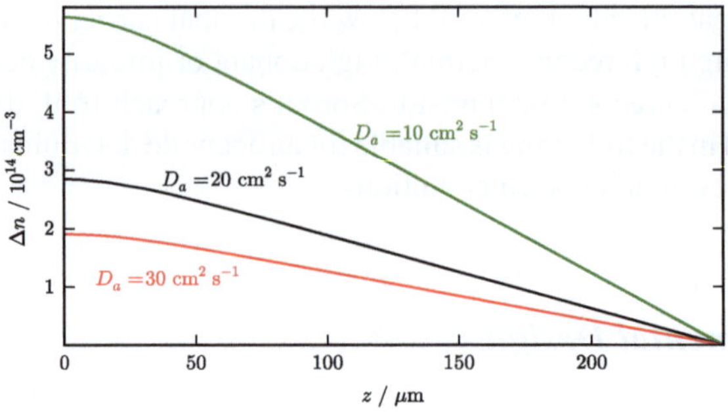

Fig. 9.7 Depth distribution of excess carrier density $\Delta n(z)$ for different ambipolar diffusion coefficients D_a, according to the steady-state solution of the continuity equation (cf. Sect. 3.2.2). Parameters: wafer thickness $d = 250\,\mu\,\text{m}$, surface recombination velocities $S_0 = 1\,\text{cm}\,\text{s}^{-1}$ and $S_d = 10^6\,\text{cm}\,\text{s}^{-1}$, bulk lifetime $\tau = 10^{-3}\,\text{s}$, absorbed photon current density $j_\gamma = 2.55 \times 10^{17}\,\text{cm}^{-2}\,\text{s}^{-1}$, and inverse light penetration depth $\alpha_\gamma = 417\,\text{cm}^{-1}$. Adapted after [43]

$$\tau_{tr} = \frac{1}{D_a}\left(\frac{d^2}{2} - \alpha_\gamma^{-2}\right), \tag{9.11}$$

which essentially depends on the thickness d, the penetration depth α_γ^{-1}, and the (ambipolar) minority carrier diffusion coefficient D_a. This idealized representation of minority carrier transit time is plotted as a limiting curve in the depiction of the analysis of p-type electron diffusion coefficients in Fig. 9.11. Among the wafers investigated therein, those featuring very high bulk lifetime (cf. Table 9.2) approach the idealized expression given in Eq. 9.11 very closely.

Despite the very faint contrast of this transport-limited measurement design with respect to sufficiently high bulk lifetimes τ and surface recombination velocities S_d, these properties could be additionally estimated for a most accurate determination of D_a—even in nonideal material—rather than to assume these properties as quasi-infinite. The estimation of these recombination properties shall be detailed among other relevant considerations later (cf. Sect. 9.2.4).

9.2.2.2 The Measurement Procedure

In a first optional step, injection-dependent bulk lifetime of a symmetrically surface-passivated silicon wafer of interest is determined through a broad range injection-dependent photoluminescence lifetime measurement (as detailed in Sect. 8.2). This measurement must cover a sufficiently broad injection range, including the injection level of interest of an asymmetrically surface-passivated lifetime measurement on the same wafer active measurement area (cf. Fig. 9.8)—aiming at an approximation of bulk lifetime for the following harmonically modulated measurement and analysis of the minority carrier diffusion coefficient D_a.

Subsequently, the surface passivation is removed at one wafer surface—preferably mechanically [66] in order to ensure a sufficiently high S_d. Optionally, the thereby

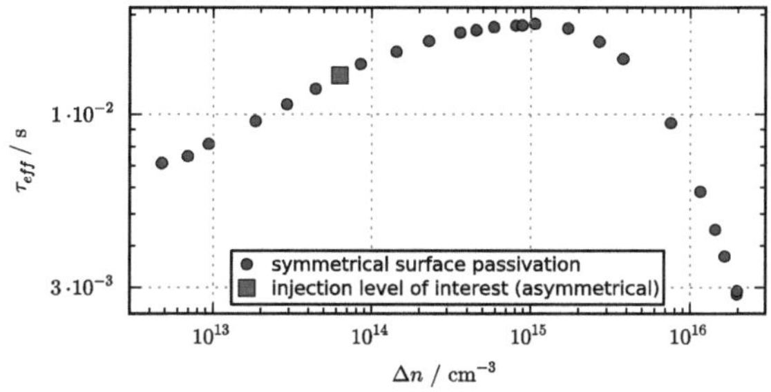

Fig. 9.8 Example plot of a broad range injection-dependent measurement of effective carrier lifetime τ_{eff} (cf. Sect. 8.2) in a symmetrically surface-passivated state: effective lifetime is evaluated at the excess carrier density of interest of a carrier lifetime measurement in the asymmetrically surface-passivated state, in order to estimate bulk carrier lifetime $\tau|_{S_0}$ of the harmonically modulated lifetime analysis of the ambipolar diffusion coefficient D_a. Adapted after [43]

intentionally deteriorated surface recombination velocity can be determined for highest accuracy. S_d can be approximated via a modified photoluminescence intensity ratio approach, as originally proposed by Baek et al. [72] (cf. Sect. 9.2.4).

Finally, a self-consistent quasi-steady-state harmonically modulated photoluminescence carrier lifetime measurement is to be carried out on the wafer featuring asymmetrical surface recombination velocities—under illumination of the wafer interface featuring optimal surface passivation. The resulting effective carrier lifetime τ_σ is evaluated in terms of the (ambipolar) minority carrier diffusion coefficient D_a according to the steady-state solution of the continuity equation, while a dynamic lifetime correction to be detailed in Sect. 9.2.3 accounts for the mismatch between self-consistent and steady-state effective lifetime expressions. As the abovementioned estimation of both bulk lifetime and surface recombination velocity also depends on D_a, the analysis of D_a ought to be carried out iteratively.

9.2.3 Modulated Lifetime Versus Steady-State Lifetime

Under quasi-steady-state conditions, the depth profile of excess carrier density $\Delta n(z)$ is at any time appropriately represented by the steady-state solution of the continuity equation (cf. Sect. 3.2.2). However, this fact does not allow the conclusion that steady-state effective lifetime $\tau_{eff,s}$ as the ratio between cumulative excess carrier density and generation rate [73, 74] would accurately coincide with a quasi-steady-state harmonically modulated self-consistent lifetime interpretation τ_σ. In fact, the injective assignment $\tau_\sigma \overset{ss}{\mapsto} D_a$ solely via the steady-state solution of the continuity equation in the above-described experimental design would give rise to substantial inaccuracy.

In Sect. 6.1.2, the cause to a mismatch between quasi-steady-state harmonically modulated effective lifetime and steady-state effective lifetime in the presence of very high surface recombination velocities is derived: High surface recombination

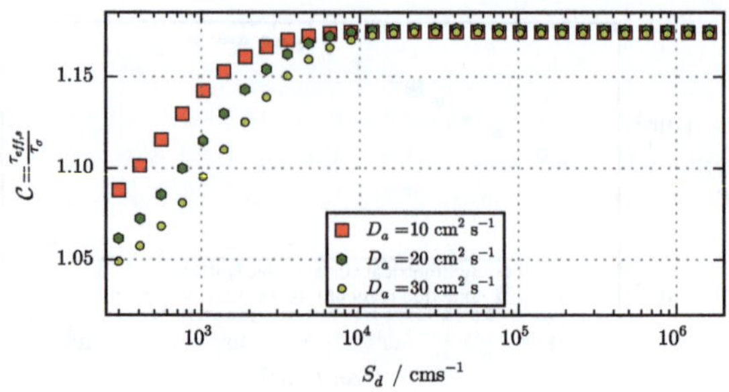

Fig. 9.9 Depiction of the ratio $\mathcal{C}$ between effective carrier lifetime $\tau_{eff,s}$ inferred from the analytic steady-state solution of the continuity equation [ratio of cumulative $\Delta n(z)$ and $G(z)$], and effective self-consistent carrier lifetime τ_σ calculated from the distribution of weights of excess carrier decay modes of a harmonically modulated Green's function solution of the continuity equation (cf. Sects. 6.1.2 and 3.2.4). $\mathcal{C}$ is modeled as a function of the surface recombination velocity S_d of the opposite wafer interface with respect to illumination for varying ambipolar diffusion coefficients D_a. Model parameters: wafer thickness $d = 250\,\mu$ m, illuminated surface recombination velocity $S_0 = 1\,\text{cm}\,\text{s}^{-1}$, bulk lifetime $\tau = 10^{-4}$ s, and inverse light penetration depth $\alpha_\gamma = 417\,\text{cm}^{-1}$. Adapted after [43]

velocities bring about significant higher order modes of excess carrier decay. Harmonically modulated effective lifetime is found to feature a fundamentally different (nonlinear) weighting of the different modes of excess carrier decay than steady-state effective lifetime. Therefore, harmonically modulated effective lifetime only coincides with steady-state effective lifetime, if all excess carrier decay modes except the fundamental mode are negligible. Notably, the mismatch between quasi-steady -state effective lifetime τ_σ and steady-state effective lifetime $\tau_{eff,s}$ is independent of modulation time $\mathcal{T}$, if a quasi-steady-state condition is fulfilled. Therefore, the mismatch between τ_σ and $\tau_{eff,s}$ sustains even in the limit $\mathcal{T} \rightarrow \infty$.

In the above-described experimental design of a transit time measurement, the surface recombination velocity S_d of the opposite interface with respect to illumination is enhanced. As illustrated in Fig. 6.2, this configuration of recombination properties leads to a harmonically modulated effective lifetime τ_σ less than steady-state effective lifetime $\tau_{eff,s} > \tau_\sigma$. This phenomenon—introduced as the paradox of steady-state effective lifetime exceeding a dynamic effective lifetime expression in Sect. 6.1.2—can be quantitatively explained via weighting of excess carrier decay modes. Consequently, it is also possible to accurately and straightforwardly correct the respective mismatch via a factor $\mathcal{C}$ such that the corresponding injective assignment would be

$$\mathcal{C}\tau_\sigma = \tau_{eff,s} \overset{ss}{\mapsto} D_a. \tag{9.12}$$

The mismatch factor $\mathcal{C} = \frac{\tau_{eff,s}}{\tau_\sigma}$ is a function of all the relevant parameters involved in the steady-state solution of the continuity equation, comprising the wafer thickness d, the bulk lifetime τ, surface recombination velocities S_0 and S_d, the diffusion coefficient D_a, and the inverse light penetration depth α_γ. A most accurate calculation

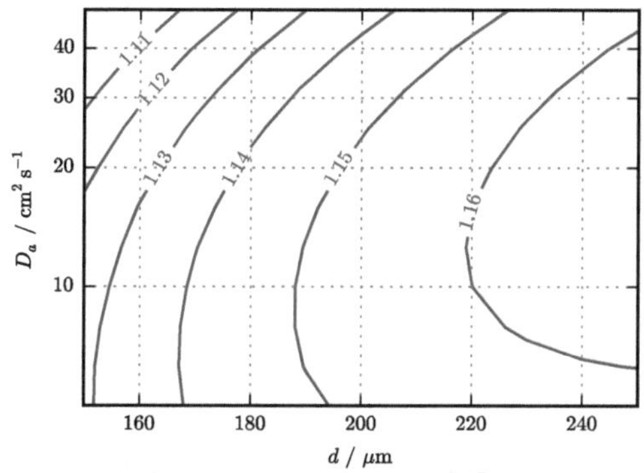

Fig. 9.10 Contour plot of the ratio $\mathcal{C}$ (cf. Fig. 9.9) as a function of both, the ambipolar diffusion coefficient D_a and the wafer thickness d. Model parameters: surface recombination velocities $S_0 = 1\,\mathrm{cm\,s^{-1}}$ and $S_d = 10^6\,\mathrm{cm\,s^{-1}}$, bulk lifetime $\tau = 10^{-4}$ s, and inverse light penetration depth $\alpha_\gamma = 417\,\mathrm{cm^{-1}}$. Adapted after [43]

of $\mathcal{C}$ compares steady-state effective lifetime at a given set of parameters with harmonically modulated effective lifetime according to the distribution of weights of excess carrier decay modes (cf. Sect. 6.1.2) inferred from a harmonically modulated Green's function solution of the continuity equation (cf. Sect. 3.2.4).

At present, the fastest and most reliable implementation to determine minority carrier mobility from quasi-steady-state harmonically modulated lifetime measurments exploits the steady-state solution of the continuity equation in combination with a precalculated array of mismatch factors $\mathcal{C}\,(D_a, \tau, d)\,|_{\alpha_\gamma, S_d}$—to be interpolated for an individual measurement analysis $\mathcal{C}\tau_\sigma \overset{ss}{\mapsto} D_a$, accordingly. Modeling of $\mathcal{C}\,(S_d)\,|_{D_a}$ in Fig. 9.9 suggests that $\mathcal{C}$ does not have to be calculated as a function of varying surface recombination velocity S_d, as typically encountered values of S_d significantly exceed $10^4\,\mathrm{cm\,s^{-1}}$. Therefore, it suffices to precalculate $\mathcal{C}\,(D_a, \tau, d)\,|_{\alpha_\gamma, S_d}$ at the constant inverse light penetration depth α_γ of the used light source and at a sufficiently high surface recombination velocity S_d. Figure 9.10 shows the modeled mismatch factor as a function of wafer thickness d and of the diffusion coefficient D_a.

9.2.4 Considerations for Broad Applicability and Accuracy

The measurement procedure outlined in Sect. 9.2.2 comprises two optional measurements, aiming at an estimation of both bulk lifetime τ and of the intentionally deteriorated surface recombination velocity S_d. These optional measurements become relevant if the assumption of quasi-infinity of the respective recombination properties is not fulfilled.

9.2.4.1 Approximation of Enhanced Surface Recombination Velocity

A sufficiently high intentionally enhanced surface recombination velocity S_d is essential to the method described here. S_d can be estimated according to an adaption of a steady-state approach originally introduced by Baek et al. [72]. This technique relates the ratio R_B of low-level-injection photoluminescence intensities ($\propto \Delta n$) obtained with different irradiation wavelengths (but identical excess carrier generation rates) to surface recombination velocity S.[5]

Provided that the minority carrier diffusion coefficient D_a is known and that the investigated interface (which is in this case the illuminated interface) features the predominant recombination rate of the entire wafer, surface recombination velocity is unambiguously related to the ratio $R_B(S)|_{D_a}$ of excess carrier densities (according to the steady-state solution of the continuity equation) resulting from irradiation with two different laser wavelengths. Here, two laser wavelengths are used—performing a laterally uniform wafer illumination of the active measurement area at room temperature inverse light penetration depths α_γ of 417 and 947 cm^{-1}, respectively (cf. Appendix D).

Application of this method requires several precautionary measures: As mentioned above, the ratio R_B depends on both surface recombination velocity and the (ambipolar) minority carrier diffusion coefficient D_a. Thus, S_d is to be determined iteratively.

9.2.4.2 Approximation of Bulk Lifetime

In contrast to an intentionally enhanced rear surface recombination velocity S_d, both front surface recombination velocity S_0 and bulk lifetime τ are to feature relatively small recombination rates in the experimental design described in Sect. 9.2.2. In order to check this, and to be able to use most realistic (rather than quasi-infinite) recombination properties for the assignment $C\tau_\sigma \overset{SS}{\mapsto} D_a$ in the steady-state solution of the continutiy equation, broad range injection-dependent photoluminescence carrier lifetime of a symmetrically surface-passivated wafer (as detailed in Sect. 8.2) can be determined (cf. Fig. 9.8). This optional measurement is particularly important in the case of high dopant concentrations (cf. the 0.1 Ω cm wafer shown in Fig. 9.11) or high impurity concentrations, because bulk recombination is then non-negligible compared to surface recombination. The injection-dependent lifetime measured on the symmetrically surface-passivated wafer is later evaluated at the excess carrier density of interest of the quasi-steady-state harmonically modulated self-consistent lifetime measurement in the asymmetrically surface-passivated state. The corresponding effective lifetime value is used for an approximation of bulk lifetime $\tau|_{S_0}$ according

[5] The modified Baek approach to surface recombination velocity, as used here, is detailed for silicon ingots in Sect. 8.4.2, and model curves $R_B(S)|_{D_a}$ are shown in Fig. 8.19. Such curves can likewise be calculated for wafers on the basis of the steady-state solution of the continuity equation for wafers (rather than ingots) detailed in Sect. 3.2.2.

to the steady-state solution of the continuity equation—assuming a very low surface recombination velocity S_0. The resulting values $\tau|_{S_0}$ and S_0 may subsequently be used for a most appropriate assignment $\mathcal{C}\tau_\sigma \overset{ss}{\mapsto} D_a$ accordig to the steady-state solution of the continuity equation (cf. Sect. 3.2.2).

For wafers featuring relatively low bulk lifetime, this procedure accurately yields bulk lifetime. For wafers featuring very high bulk lifetime, substantial uncertainty remains concerning the exact allocation of recombination rates to S_0 and $\tau|_{S_0}$. However, in the latter case, resulting errors only negligibly affect the assignment $\mathcal{C}\tau_\sigma \overset{ss}{\mapsto} D_a$, because recombination rates associated with S_0 and τ are then negligible compared to the recombination rate associated with the predominant rear surface recombination velocity S_d. The relation between effective lifetime and the diffusion coefficient is then accurately approximated by the analytical transit time expression of Eq. 9.11.

Note that the consideration of bulk lifetime—if relevant—must proceed according to the steady-state solution of the continuity equation. A transit time correction according to the widespread simplistic approach to effective lifetime via the sum of inverse lifetime expressions

$$\tau_{tr}^{-1} = (\mathcal{C}\tau_\sigma)^{-1} - \tau^{-1} \tag{9.13}$$

would fail here. This is because Eq. 9.13 is based on the assumption of equal or uniform depth distributions of all recombination rates involved. In fact, a non-negligible bulk recombination rate ($\propto \tau^{-1}$) has a higher impact on effective lifetime than assumed in this simplistic treatment, because it gives rise to recombination of excess carriers already at the very location of their generation (as opposed to an intentionally deteriorated rear surface recombination velocity S_d).

9.2.4.3 Consideration of Nonuniform Excess Carrier Density

Note that the depth profile of excess carrier density $\Delta n(z)$ is far from uniform in the asymmetrical case of interest. Rather, it typically approaches a triangular shape (cf. Fig. 9.7). This effect is not accounted for in the self-consistent harmonically modulated lifetime analysis, which is based on arithmetically averaged generation rate and excess carrier density. The determination of self-consistent carrier lifetime is unaffected by this effect. Yet, the assigned excess carrier density is systematically underestimated in this case. It is possible to correct this mismatch with a weighted average Δn_{eff}, as given in Eq. 8.14 [42]. With a triangularly shaped approximation of excess carrier density (cf. Fig. 9.7)

$$\Delta n(z) \approx \left(1 - \frac{z}{d}\right) \Delta n(z = 0), \tag{9.14}$$

the resulting effective excess carrier density Δn_{eff} corresponds to the arithmetically averaged self-consistent excess carrier density $\frac{\Delta n(z=0)}{2}$ multiplied by a factor of $\frac{4}{3}$.

Table 9.2 Summary of relevant measurement results of the p-type wafers investigated here. These are the wafer base resistivity ρ, the estimated electron bulk lifetime $\tau|_{S_0=1\,\mathrm{cm\,s^{-1}}}$, the intentionally deteriorated back surface recombination velocity S_d, the self-consistent effective carrier lifetime τ_σ (measurement uncertainties specified in Fig. 9.11), the mismatch factor between dynamic and steady-state effective lifetime $\mathcal{C} = \frac{\tau_{eff,s}}{\tau_\sigma}$, and the depth averaged excess carrier density $\langle \Delta n \rangle$

$\rho/\Omega\,\mathrm{cm}$	$\tau\|_{S_0}/\mathrm{s}$	$S_d/\mathrm{cm\,s^{-1}}$	$\tau_\sigma/10^{-6}\,\mathrm{s}$	$\mathcal{C}$	$\langle \Delta n \rangle/\mathrm{cm^{-3}}$
0.1	2.8×10^{-5}	1×10^{7}	11.32	1.152	8.81×10^{13}
1.0	5.9×10^{-4}	9×10^{5}	9.78	1.174	7.75×10^{13}
8.0	1.3×10^{-2}	1×10^{7}	7.73	1.172	6.12×10^{13}

Thus, for injection-dependent carrier lifetime (which is usually encountered) the above mentioned excess carrier density of interest for the determination of $\tau|_{S_0}$ corresponds to the self-consistently determined excess carrier density of interest of the asymmetrical modulated photoluminescence measurement multiplied by $\frac{4}{3}$.

9.2.5 Experimental Results for p-Type Silicon

As an experimental proof of concept, a determination of minority carrier mobility was carried out on three $d = 253\,\mu\mathrm{m}$ thick boron-doped float zone wafers with dopant concentrations n_0 of 1.7×10^{15}, 1.5×10^{16}, and $2.3 \times 10^{17}\,\mathrm{cm^{-3}}$. Wafer interfaces were passivated with a 10–30 nm aluminum oxide layer, featuring a reflectivity of 33 % at the laser irradiation wavelength of 880 nm ($\alpha_\gamma = 417\,\mathrm{cm^{-1}}$). The corresponding surface recombination velocity was estimated to amount to $S_0 = 1\,\mathrm{cm\,s^{-1}}$ in the analysis of minority carrier mobility. Photoluminescence images ensured lateral homogeneity of carrier lifetime. The minority carrier diffusion coefficient was determined according to the measurement procedure outlined in Sect. 9.2.2. Relevant measurement results are listed in Table 9.2, the resulting diffusion coefficients D_a are shown in Fig. 9.11, corresponding minority carrier mobilities are depicted in Fig. 9.12, and the exact measurement conditions are detailed in the following.

9.2.5.1 Measurement Conditions

Measurements on asymmetrically surface-passivated wafers were carried out at a maximum irradiance equivalent to one sun, at a temperature of $T = 298 \pm 1\,\mathrm{K}$ and at a modulation time of $\mathcal{T} = 1.25 \times 10^{-2}\,\mathrm{s}$. Modulated 880 nm ($\alpha_\gamma = 417\,\mathrm{cm^{-1}}$) and 791 nm ($\alpha_\gamma = 947\,\mathrm{cm^{-1}}$, cf. Appendix D) diode lasers were used at a maximal impinging photon current density equivalent to one sun (cf. Appendix A.6). Wafers were uniformly illuminated on a $2 \times 2\,\mathrm{cm^2}$ quadratic measurement area. The specified uncertainty of asymmetrical lifetime measurements is a superposition of the uncertainty of the individually performed synchronization measurement

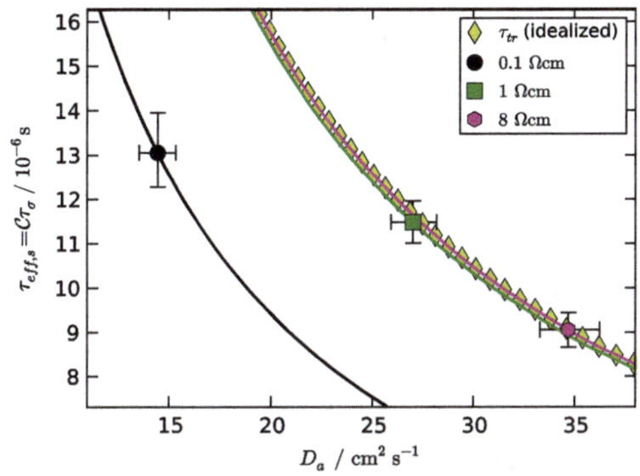

Fig. 9.11 Parametrization of steady-state effective carrier lifetime $\tau_{eff,s} = C\tau_\sigma$ as a function of the ambipolar electron diffusion coefficient according to the steady-state solution of the continuity equation (cf. Sect. 3.2.2). The *solid lines* represent $\tau_{eff,s}\,(D_a)\,|_{\tau|_{S_0},S_d}$ according to the encountered recombination properties S_d and $\tau|_{S_0}$ of the investigated wafers as given in Table 9.2. *Symbols with indicated uncertainties* represent measured self-consistent effective lifetimes τ_σ (multiplied by the corresponding dynamic mismatch factor C as described in Sect. 9.2.3), which are related to resulting values of D_a through the modeled (*solid*) curves $\tau_{eff,s}\,(D_a)\,|_{\tau|_{S_0},S_d}$. *Diamond symbols* represent the idealized analytic expression of minority carrier transit time according to Eq. 9.11. Adapted after [43]

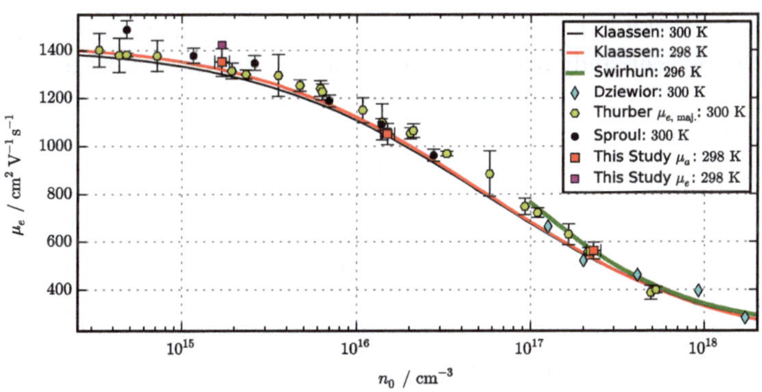

Fig. 9.12 Minority electron mobility μ_e as a function of dopant concentration n_0, determined from the electron diffusion coefficients shown in Fig. 9.11. The results of the present study are compared to previously published data and carrier mobility models from [10, 12, 13, 61, 66, 71]. The *small purple squares* indicate a correction of an ambipolar diffusion effect in the data of the present study. This effect is significant for the lightly doped wafer featuring a dopant concentration of $n_0 = 1.7 \times 10^{15}$ cm^{-3}. Adapted after [43]

(detailed in Sect. 7.2.1) and of the standard deviation of five subsequently carried out harmonically modulated carrier lifetime measurements, respectively. Together with the uncertainty of temperature, it propagates to the specified uncertainty of minority carrier mobility. Wafer thickness was precisely measured with a gauge, interface reflectivities at the irradiation wavelengths of interest were measured with a spectrophotometer and subsequently used to determine the transmitted photon current density from a measurement of the impinging photon current density. The symmetrically surface-passivated lifetime measurement was estimated to feature an uncertainty of well below 10 %.

9.2.5.2 Discussion of Results

Figure 9.12 depicts minority carrier mobilities determined from diffusion coefficients according to the Einstein relation (Eq. 3.15), and relates these to a selection of previously published data. Within the specified measurement precision, the results confirm those obtained by Sproul et al. [66], as well as those obtained by Dziewior et al. for very high dopant concentrations [61]. Majority electron mobility data by Thurber et al. are also plotted as a reference, as it has been argued that majority carrier mobility does not significantly differ from minority carrier mobility at dopant concentrations $n_0 < 10^{17}\,\mathrm{cm}^{-3}$ (e.g. in [66]).

Toward low dopant concentrations, some of the precise data published by Sproul et al. [66] significantly exceed the unified carrier mobility model by Klaassen [12, 13] as well as the majority electron mobility data by Thurber et al. [10]. Concerning this remarkable discrepancy, at first glance the data shown in Fig. 9.12 suggest that the 298 K harmonically modulated photoluminescence measurement result of the present study accords better with the Klaassen model than the 300 K measurement results by Sproul et al. However, further evaluation reveals that the $8\,\Omega\,\mathrm{cm}$ data point (at a dopant concentration of $n_0 = 1.7 \times 10^{15}\,\mathrm{cm}^{-3}$) turns out to be affected by ambipolar diffusion (cf. Sect. 3.1.5) at the encountered excess carrier density (cf. Table 9.2). This is a result of differing minority and majority carrier mobilities, which may lead to local charges in the case of excess carrier density gradients upon departure from rigorous low-level injection. An electrical field caused by the above mentioned space charge leads to an acceleration of the relatively inert holes and a deceleration of the relatively mobile electrons. An exact expression of ambipolar minority carrier mobility is given in Sect. 3.1.5. Ambipolar minority carrier mobility directly follows from an expression for the ambipolar minority carrier diffusion coefficient as given in Eq. 3.41, yielding

$$\mu_a = \frac{\mu_e \mu_h \left(n_e + n_h\right)}{\mu_e n_e + \mu_h n_h}, \tag{9.15}$$

with free electron and hole densities n_e and n_h. Notably, this expression suggests a significant onset of ambipolar diffusion already at an injection level far below dopant concentration.

Here, Eq. 9.15 is used for an estimate of a best value of low-level injection minority electron mobility from the measurement results, thereby taking advantage of accurate knowledge of the injection level of the conducted photoluminescence measurements. Accounting for the depth profiles of excess carrier density and excess carrier generation rate, and using Klaassen's unified mobility model for an approximation of

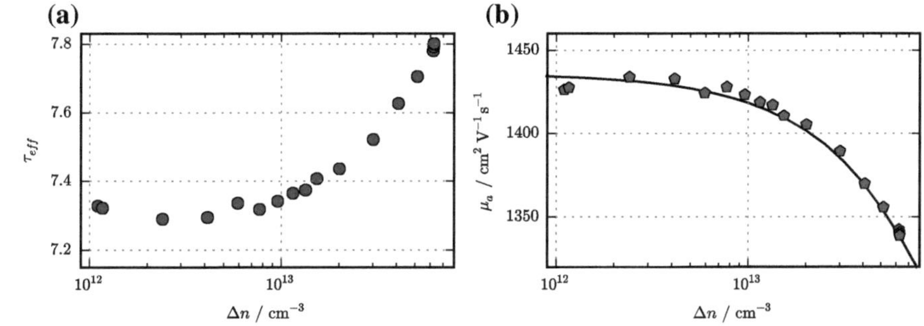

Fig. 9.13 Experimental evidence of an ambipolar effect in the minority carrier mobility measurement of the 8 Ω cm wafer shown in Figs. 9.11 and 9.12 and in Table 9.2: *on the left*, part of a broad range injection-dependent effective lifetime scan (cf. Sect. 8.2) is depicted. *On the right*, the minority carrier mobility analysis of effective lifetimes as depicted *on the left* is shown and compared to modeled ambipolar mobility according to Eq. 9.15. For this purpose, an estimate of $\mu_h \approx 460\,\text{cm}^2\,\text{V}^{-1}\,\text{s}^{-1}$ was taken from the Klaassen model, while μ_e was fitted to the experimental result. As can be seen, the injection dependence of μ_a agrees well with the ambipolar model

minority and majority carrier mobilities at the dopant concentrations of interest (in Eq. 9.15), the ratio between the measured ambipolar mobility μ_a and true minority electron mobility μ_e is calculated. As expected, this ratio proves to be very close to unity in the case of the highly doped 0.1 Ω cm wafer ($\frac{\mu_a}{\mu_e} = 0.9996$), and close to unity in the case of the moderately doped 1 Ω cm wafer ($\frac{\mu_a}{\mu_e} = 0.993$). However, in the case of the lightly doped 8 Ω cm wafer, this ratio turns out to be $\frac{\mu_a}{\mu_e} = 0.95$, such that a corresponding mobility correction amounts to $\mu_e - \mu_a \approx 70\,\text{cm}^2\,\text{V}^{-1}\,\text{s}^{-1}$ here.

It was possible to experimentally confirm this ambipolar effect quantitatively via broad range injection-dependent effective lifetime measurements on the investigated 1 Ω cm wafer down to an excess carrier density of $\Delta n \approx 10^{12}\,\text{cm}^{-3}$ (cf. Sect. 8.2), as shown in Fig. 9.13. A correction of the ambipolar effect is also shown in Fig. 9.12. With virtually the same precision as the uncorrected data point, the 8 Ω cm result now significantly deviates from the Klaassen model, while confirming the experimental data by Sproul et al. When taking into account the difference in temperature between the photoluminescence measurements of the present study and the measurements by Sproul et al., the agreement between the measurement results is convincing. Therefore, the results of the present study independently and entirely confirm the findings of Sproul et al. [66], in contrast to previous results of numerous groups yielding substantially lower minority carrier mobilities in p-type material (cf. data review in [66]).

References

1. J.A. Giesecke, M.C. Schubert, W. Warta, Measurement of net dopant concentration via dynamic photoluminescence. J. Appl. Phys. **112**, 063704 (2012)
2. J.A. Giesecke, D. Walter, F. Kopp, P. Rosenits, M.C. Schubert, W. Warta, Simultaneous determination of carrier lifetime and net dopant concentration of silicon wafers from photoluminescence, in *Proceedings of the 35th IEEE PVSC* (Honolulu, 2010), pp. 847–851
3. L.B. Valdes, Resistivity measurements on germanium for transistors. Proc. IRE **42**, 420–427 (1954)
4. D.K. Schroder, *Semiconductor Material and Device Characterization* (Wiley, New York, 2006)
5. D.M. Caughey, R.E. Thomas, Carrier mobilities in silicon empirically related to doping and field. Proc. IEEE **55**, 2192–2193 (1967)
6. F. Dannhäuser, Die Abhängigkeit der Trägerbeweglichkeit in Silicium von der Konzentration der freien Ladungsträger—I. Solid-State Electron. **15**, 1371–1375 (1972)
7. J. Krausse, Die Abhängigkeit der Trägerbeweglichkeit in Silizium von der Konzentration der freien Ladungsträger—II. Solid-State Electron. **15**, 1377–1381 (1972)
8. W.R. Thurber, R.L. Mattis, Y.M. Liu, J.J. Filliben, Resistivity-dopant density relationship for boron-doped silicon. J. Electrochem. Soc. **127**, 2291–2294 (1980)
9. W.R. Thurber, R.L. Mattis, Y.M. Liu, J.J. Filliben, Resistivity-dopant density relationship for phosphorus-doped silicon. J. Electrochem. Soc. **127**, 1807–1812 (1980)
10. W.R. Thurber, R.L. Mattis, Y.M. Liu, J.J. Filliben, *The Relationship Between Resistivity and Dopant Density for Phosphorus- and Boron-Doped Silicon* (US Department of Commerce, National Bureau of Standards, Washington, D.C., 1981)
11. G. Masetti, M. Severi, S. Solmi, Modeling of carrier mobility against carrier concentration in arsenic-, phosphorus-, and boron-doped silicon. IEEE Trans. Electron Dev. **30**, 764–769 (1983)
12. D.B.M. Klaassen, A unified mobility model for device simulation—I. Model equations and concentration dependence. Solid-State Electron. **35**, 953–959 (1992)
13. D.B.M. Klaassen, A unified mobility model for device simulation—II. Temperature dependence of carrier mobility and lifetime. Solid-State Electron. **35**, 961–967 (1992)
14. F. Schindler, M.C. Schubert, A. Kimmerle, J. Broisch, S. Rein, W. Kwapil, W. Warta, Modeling majority carrier mobility in compensated crystalline silicon for solar cells. Sol. Energy Mater. Sol. Cells **106**, 31–36 (2012)
15. R.A. Sinton, A. Cuevas, M. Stuckings, Quasi-steady-state photoconductance, a new method for solar cell material and device characterization, in *Proceedings of the 25th IEEE PVSC* (Washington, D.C., 1996), pp. 457–460
16. R.A. Sinton, A. Cuevas, Contactless determination of current-voltage characteristics and minority-carrier lifetimes in semiconductors from quasi-steady-state photoconductance data. Appl. Phys. Lett. **69**, 2510–2512 (1996)
17. M. Spitz, S. Rein, Impact of potential barriers at grain boundaries of multicrystalline silicon wafers on inductively coupled resistivity measurements, in *Proceedings of the 26th EU PVSEC* (Hamburg, 2011), pp. 338–343
18. T. Ambridge, C.R. Elliott, M.M. Faktor, The electrochemical characterization of n-type gallium arsenide. J. Appl. Electroch. **3**, 1–15 (1973)
19. E. Peiner, A. Schlachetzki, D. Krüger, Doping profile analysis in Si by electrochemical capacitance-voltage measurements. J. Electrochem. Soc. **142**, 576–580 (1995)
20. D. Hinken, A. Milsted, R. Bock, B. Fischer, K. Bothe, M. Schütze, J. Isenberg, A. Schulze, M. Wagner, Determination of the base-dopant concentration of large-area crystalline silicon solar cells. IEEE Trans. Electron Dev. **57**, 2831–2837 (2010)
21. S. Rein, W. Kwapil, J. Geilker, G. Emanuel, M. Spitz, I. Reis, A. Weil, D. Biro, M. Glatthaar, A.K. Soiland, E. Enebakk, R. Tronstad, Impact of compensated solar-grade silicon on Czochralski silicon wafers and solar cells, in *Proceedings of the 24th EU PVSEC* (Hamburg, 2009), pp. 1140–1147

22. J. Geilker, W. Kwapil, I. Reis, S. Rein, Doping concentration and mobility in compensated material: comparison of different determination methods, in *Proceedings of the 25th EU PVSEC* (Valencia, 2010), pp. 1322–1327
23. R. Zierer, G. Gärtner, H.J. Möller, Determination of dopant concentrations in UMG-silicon by FTIR spectroscopy, hall- and resistivity measurements, in *Proceedings of the 25th EU PVSEC* (Valencia, 2010), pp. 1318–1321
24. B. Mitchell, T. Trupke, J.W. Weber, J. Nyhus, Bulk minority carrier lifetimes and doping of silicon bricks from photoluminescence intensity ratios. J. Appl. Phys. **109**, 083111 (2011)
25. S.Y. Lim, D. Macdonald, Measuring dopant concentrations in p-type silicon using iron-acceptor pairing monitored by band-to-band photoluminescence. Sol. Energy Mater. Sol. Cells **95**, 2485–2489 (2011)
26. S.Y. Lim, S.P. Phang, T. Trupke, A. Cuevas, D. Macdonald, Dopant concentration imaging in crystalline silicon wafers by band-to-band photoluminescence. J. Appl. Phys. **110**, 113712 (2011)
27. Z. Hameiri, T. Trupke, N. Gao, R.A. Sinton, J.W. Weber, Effective bulk doping concentration of diffused and undiffused silicon wafers obtained from combined photoconductance and photoluminescence measurements. Prog. Photovoltaics: Res. Appl. **21**, 942–949 (2013)
28. M. Tajima, Determination of boron and phosphorus concentration in silicon by photoluminescence analysis. Appl. Phys. Lett. **32**, 719–721 (1978)
29. P.M. Colley, E.C. Lightowlers, Calibration of the photoluminescence technique for measuring B, P and Al concentrations in Si in the range 10^{12} to 10^{15} cm^{-3} using Fourier transform spectroscopy. Semicond. Sci. Technol. **2**, 157 (1987)
30. P. Würfel, T. Trupke, T. Puzzer, E. Schäffer, W. Warta, S.W. Glunz, Diffusion lengths of silicon solar cells from luminescence images. J. Appl. Phys. **101**, 123110 (2007)
31. D. Macdonald, A. Cuevas, L.J. Geerligs, Measuring dopant concentrations in compensated p-type crystalline silicon via iron-acceptor pairing. Appl. Phys. Lett. **92**, 202119 (2008)
32. T. Trupke, R.A. Bardos, Photoluminescence: a surprisingly sensitive lifetime technique, in *Proceedings of the 31st IEEE PVSC* (Orlando, 2005), pp. 903–906
33. W. van Roosbroeck, W. Shockley, Photon-radiative recombination of electrons and holes in germanium. Phys. Rev. **94**, 1558–1560 (1954)
34. P.P. Altermatt, F. Geelhaar, T. Trupke, X. Dai, A. Neisser, E. Daub, Injection dependence of spontaneous radiative recombination in c-Si: experiment, theoretical analysis, and simulation, in *Proceedings of the 5th International Conference on Numerical Simulation and Optoelectronic Devices*, IEEE (Berlin, 2005), pp. 47–48
35. P.P. Altermatt, F. Geelhaar, T. Trupke, X. Dai, A. Neisser, E. Daub, Injection dependence of spontaneous radiative recombination in crystalline silicon: experimental verification and theoretical analysis. Appl. Phys. Lett. **88**, 261901 (2006)
36. J.A. Giesecke, M.C. Schubert, W. Warta, Self-sufficient minority carrier lifetime in silicon from quasi-steady-state photoluminescence. Phys. Stat. Sol. A **209**, 2286–2290 (2012)
37. J.A. Giesecke, S.W. Glunz, W. Warta, Understanding and resolving the discrepancy between differential and actual minority carrier lifetime. J. Appl. Phys. **113**, 073706 (2013)
38. J.A. Giesecke, S.W. Glunz, W. Warta, Determination of actual carrier lifetime from differential measurements. Energy Procedia **38**, 59–65 (2013)
39. D. Macdonald, T. Roth, P.N.K. Deenapanray, K. Bothe, P. Pohl, J. Schmidt, Formation rates of iron-acceptor pairs in crystalline silicon. J. Appl. Phys. **98**, 083509 (2005)
40. K. Bothe, J. Schmidt, Electronically activated boron-oxygen-related recombination centers in crystalline silicon. J. Appl. Phys. **99**, 013701 (2006)
41. D. Macdonald, F. Rougieux, A. Cuevas, B. Lim, J. Schmidt, M. Di Sabatino, L.J. Geerligs, Light-induced boron-oxygen defect generation in compensated p-type Czochralski silicon. J. Appl. Phys. **105**, 093704 (2009)
42. S. Bowden, R.A. Sinton, Determining lifetime in silicon blocks and wafers with accurate expressions for carrier density. J. Appl. Phys. **102**, 124501 (2007)
43. J.A. Giesecke, F. Schindler, M. Bühler, M.C. Schubert, W. Warta, Accurate determination of minority carrier mobility in silicon from quasi-steady-state photoluminescence. J. Appl. Phys. **113**, 213705 (2013)

44. F. Schindler, J. Geilker, W. Kwapil, W. Warta, M.C. Schubert, Hall mobility in multicrystalline silicon. J. Appl. Phys. **110**, 043722 (2011)
45. J. Libal, S. Novaglia, M. Acciarri, S. Binetti, R. Petres, J. Arumughan, R. Kopecek, A. Prokopenko, Effect of compensation and of metallic impurities on the electrical properties of Cz-grown solar grade silicon. J. Appl. Phys. **104**, 104507 (2008)
46. F. Schindler, J. Geilker, W. Kwapil, J. Giesecke, M.C. Schubert, W. Warta, Conductivity mobility and hall mobility in compensated multicrystalline silicon, in *Proceedings of the 25th EU PVSEC* (Valencia, 2010), pp. 2364–2368
47. J. Veirman, S. Dubois, N. Enjalbert, J.P. Garandet, D.R. Heslinga, M. Lemiti, Hall mobility reduction in single-crystalline silicon gradually compensated by thermal donors activation. Solid-State Electron. **54**, 671–674 (2010)
48. S. Dubois, J. Veirman, N. Enjalbert, F. Tanay, G. Raymond, Studies on compensated and UMG materials and solar cells, in *Proceedings of the 21st Workshop on Crystalline Siicon Solar Cells and Modules* (Beckenridge, 2011), pp. 19–26
49. M. Forster, E. Fourmond, R. Einhaus, H. Lauvray, J. Kraiem, M. Lemiti, Ga co-doping in Cz-grown silicon ingots to overcome limitations of B and P compensated silicon feedstock for PV applications. Phys. Stat. Sol. C **8**, 678–681 (2011)
50. J. Veirman, S. Dubois, N. Enjalbert, J.P. Garandet, M. Lemiti, Electronic properties of highly-doped and compensated solar-grade silicon wafers and solar cells. J. Appl. Phys. **109**, 103711 (2011)
51. B. Lim, F.E. Rougieux, D. Macdonald, K. Bothe, J. Schmidt, Generation and annihilation of boron-oxygen-related recombination centers in compensated p-and n-type silicon. J. Appl. Phys. **108**, 103722 (2010)
52. F.E. Rougieux, D. Macdonald, A. Cuevas, S. Ruffell, J. Schmidt, B. Lim, A.P. Knights, Electron and hole mobility reduction and hall factor in phosphorus-compensated p-type silicon. J. Appl. Phys. **108**, 013706 (2010)
53. E. Fourmond, M. Forster, R. Einhaus, H. Lauvray, J. Kraiem, M. Lemiti, Electrical properties of boron, phosphorus and gallium co-doped silicon. Energy Procedia **8**, 349–354 (2011)
54. B. Lim, M. Wolf, J. Schmidt, Carrier mobilities in multicrystalline silicon wafers made from UMG-Si. Phys. Stat. Sol. C **8**, 835–838 (2011)
55. M. Forster, F.E. Rougieux, A. Cuevas, B. Dehestru, A. Thomas, E. Fourmond, M. Lemiti, Incomplete ionization and carrier mobility in compensated p-type and n-type silicon. IEEE J. Photovoltaics **3**, 108–113 (2013)
56. J.R. Haynes, W. Shockley, The mobility and life of injected holes and electrons in germanium. Phys. Rev. **81**, 835–843 (1951)
57. M.B. Prince, Drift mobilities in semiconductors—II. Silicon. Phys. Rev. **93**, 1204–1206 (1954)
58. G.W. Ludwig, R.L. Watters, Drift and conductivity mobility in silicon. Phys. Rev. **101**, 1699–1701 (1956)
59. M. Zerbst, W. Heywang, Zur Driftbeweglichkeit der Ladungsträger in hochreinem Silicium. Zeitschrift für Naturforschung A **11**, 608–609 (1956)
60. D.D. Tang, F.F. Feng, M. Scheuermann, T.C. Chen, G. Sai-Halasz, Minority carrier transport in silicon, in *Proceedings of the IEEE International Electron Devices Meeting* (Los Angeles, 1986), pp. 20–23
61. J. Dziewior, D. Silber, Minority-carrier diffusion coefficients in highly doped silicon. Appl. Phys. Lett. **35**, 170–172 (1979)
62. A. Neugroschel, Minority-carrier diffusion coefficients and mobilities in silicon. IEEE Electron Dev. Lett. **6**, 425–427 (1985)
63. K. Misiakos, A. Neugroschel, Lifetime and diffusivity determination from frequency-domain transient analysis. IEEE Electron Dev. Lett. **8**, 358–360 (1987)
64. E. Susi, L. Passari, M. Merli, M.C. Carotta, Determination of minority carrier mobility in silicon from stationary and transient lifetime measurements. Phys. Stat. Sol. A **106**, 583–587 (1988)
65. D. Sontag, G. Hahn, P. Geiger, P. Fath, E. Bucher, Two-dimensional resolution of minority carrier diffusion constants in different silicon materials. Sol. Energy Mater. Sol. Cells **72**, 533–539 (2002)

66. A.B. Sproul, M.A. Green, A.W. Stephens, Accurate determination of minority carrier- and lattice scattering-mobility in silicon from photoconductance decay. J. Appl. Phys. **72**, 4161–4171 (1992)
67. R. Brendel, Note on the interpretation of injection-level-dependent surface recombination velocities. Appl. Phys. A **60**, 523–524 (1995)
68. A.G. Aberle, J. Schmidt, R. Brendel, On the data analysis of light-biased photoconductance decay measurements. J. Appl. Phys. **79**, 1491–1496 (1996)
69. F.M. Schuurmans, A. Schönecker, A.R. Burgers, W.C. Sinke, Simplified evaluation method for light-biased effective lifetime measurements. Appl. Phys. Lett. **71**, 1795–1797 (1997)
70. J. Schmidt, Measurement of differential and actual recombination parameters on crystalline silicon wafers. IEEE Trans. Electron Dev. **46**, 2018–2025 (1999)
71. S.E. Swirhun, Y.H. Kwark, R.M. Swanson, Measurement of electron lifetime, electron mobility and band-gap narrowing in heavily doped p-type silicon, in *Proceedings of the IEEE International Electron Devices Meeting* (Los Angeles, 1986), pp. 24–27
72. D. Baek, S. Rouvimov, B. Kim, T.C. Jo, D.K. Schroder, Surface recombination velocity of silicon wafers by photoluminescence. Appl. Phys. Lett. **86**, 112110 (2005)
73. R.A. Sinton, T. Trupke, Limitations on dynamic excess carrier lifetime calibration methods. Prog. Photovoltaics: Res. Appl. **20**, 246–249 (2012)
74. M. Turek, Interplay of bulk and surface properties for steady-state measurements of minority carrier lifetimes. J. Appl. Phys. **111**, 123703 (2012)

Chapter 10
Summary and Outlook

10.1 Summary of Advances

The present work adds to the versatility of harmonically modulated luminescence lifetime techniques via miscellaneous theoretical, conceptual, and experimental advances.

10.1.1 Theoretical Advances

Dynamic effective lifetime may exceed steady-state effective lifetime by several orders of magnitude—particularly if investigated substrates are thick and excess carrier injection is much confined to highly recombinative interfaces. This also applies to quasi-steady-state harmonically modulated effective lifetime. Yet, configurations of steady-state effective lifetime exceeding harmonically modulated lifetime exist. Straightforward expressions of effective lifetime via weighting of modes of excess carrier decay were found to fully explain the mismatch between quasi-steady-state harmonically modulated lifetime and steady-state effective lifetime.

Light-biased decay time measurements have long been known to yield *differential* rather than *actual* lifetime [1–3]. Harmonically modulated lifetime techniques were shown herein to represent a most affected limiting case of *differential* lifetime measurements—most relevant for all phase-sensitive lock-in lifetime approaches. A quantitative theory of light-biased decay time now allows an immediate determination of *actual* carrier lifetimes from *differential* lifetime measurements.

10.1.2 Conceptual Advances

A class of phase-sensitive injection-dependent lifetime analyses of quasi-steady-state harmonically modulated luminescence measurements was developed—exceeding

J. Giesecke, *Quantitative Recombination and Transport Properties in Silicon from Dynamic Luminescence*, Springer Theses, DOI: 10.1007/978-3-319-06157-3_10,
© Springer International Publishing Switzerland 2014

the application range of a self-consistent lifetime approach [4] to accurate high-level injection lifetime analyses without a priori knowledge of net dopant concentration.

Virtually trapping-free access to carrier lifetime under extremely low injection conditions is enabled via combination of a harmonically modulated luminescence lifetime measurement with an (automated) injection-dependent scan of steady-state luminescence measurements—thereby reaching excess carrier densities of 10^8 cm^{-3}.

Considerations on how laterally varying local lifetimes contribute to a measured harmonically modulated lifetime enabled a highly sensitive lifetime calibration of luminescence images, which requires no assumptions on other material properties such as dopant concentration or carrier mobilities.

Based on a Green's function solution of the continuity equation, a most accurate technique to determine both bulk lifetime and surface recombination velocity of silicon ingots via harmonically modulated photoluminescence was introduced.

Harmonically modulated luminescence was exploited to determine net dopant concentration without a priori knowledge of minority carrier mobility. This approach exploits the nonlinear relation between luminescent emission and excess carrier density. Combination with resistivity measurements allows access to majority carrier mobility.

A transit time approach was exploited for the determination of minority carrier mobility via harmonically modulated photoluminescence lifetime measurements, thereby confirming previous controversial results [5]. Minority carrier mobility analyses represent the most rigorous benchmark of the experimental setup in terms of lifetime accuracy.

10.1.3 Experimental Advances

An essential boost of harmonically modulated luminescence lifetime accuracy at low lifetimes was achieved as a result of a systematic error analysis—thereby tackling the range of microsecond effective lifetimes.

A *reflective* luminescence detection configuration now enables harmonically modulated luminescence lifetime measurements on metalized wafers and solar cells—suggesting effective lifetime as one consistent and comparable recombination quality control parameter in any stage of the solar cell production process.

Harmonically modulated electroluminescence features advantages over photoluminescence mainly in terms of the cost of instrumentation and the redundance of optical filtering of irradiation. A prototype harmonically modulated electroluminescence experimental setup was designed and successfully tested.

10.2 Outlook

Numerous derivative applications of substantial innovation capacity could evolve from the theoretical advances, the novel concepts, and the experimental innovations of the present work. Some shall be addressed. Yet, some observations also call for refined investigations.

10.2.1 Pending Investigations

The optical optimization of the experimental setup is a task which remains to be perfected. With silicon absorption filters and an arrangement of optics far from hemispherical capture of luminescent emission, there is still room for improvement.

As for lifetime measurements on substrates featuring efficient lateral transport mechanisms (highly doped emitters or metalization) potential lifetime measurement artifacts (e.g. due to local recombination centers or local illumination) need to be further investigated.

With regard to apparent deviations between harmonically modulated and QSSPC lifetime calibrations of luminescence images of multicrystalline silicon wafers, the potential causes call for further investigation. Beyond the rather unquestionable cause of an insufficient mobility model of QSSPC in the case of compensating dopant species, the comprehensive statistical analysis of Appendix C suggests further causal connections, which could not yet be unambiguously identified. Most importantly, the possible impact of crystal defects acting as electrical potential barriers, and the impact of lateral optical blurring of luminescence images in combination with the non-arithmetical averaging procedure must be analyzed. State-of-the-art measures to suppress lateral optical blurring of luminescence images are to be implemented.

Concerning signal synchronization, the response times of the detection system (consisting of photodetectors and amplifiers) were found to not only depend on the discrete amplification settings, but also on signal amplitudes—particularly at the upper end of the dynamic range. In order to avoid a tedious individual synchronization procedure for each low-lifetime measurement, either the dependencies of response times should be unambiguously identified (and corrections implemented), or an automated individual synchronization measurement should be pursued.

In some cases, a weak but systematic correlation between quasi-steady-state harmonically modulated lifetime and modulation time has been observed. At present, the only known explanation for such an artifact appears to be trapping. However, such observations have occasionally sustained at an injection range which allows a priori exclusion of trapping. The cause to this artifact is yet to be found.

10.2.2 Derivative Applications and Novel Pathways

The combination of phase-sensitive harmonically modulated lock-in lifetime techniques with the *d2a* technique suggests itself. In terms of noise, a lock-in approach is superior to the synchronized detection of amplified time-dependent photocurrents, but it yields *differential* rather than *actual* lifetime. Therefore, the combination of the concept of lock-in phase detection with the findings of the theory of light-biased decay time could open up novel pathways in lifetime metrology.

An integrated experimental setup combining luminescence imaging and harmonically modulated lifetime measurements could fully be implemented on the basis of concepts elaborated in this work. Beyond the prototype and proof of concept studies of dynamic electroluminescence given herein, it would be worthwhile to also take advantage of the benefits of electrical injection in such a setup. A stand-alone harmonically modulated imaging application could be imagined as well.

Among the miscellaneous novel applications of harmonically modulated luminescence in materials science, further investigations of carrier mobility should be explicitly mentioned here. Investigations could involve n-type, multicrystalline, and compensated material.

References

1. R. Brendel, Note on the interpretation of injection-level-dependent surface recombination velocities. Appl. Phys. A **60**, 523–524 (1995)
2. A.G. Aberle, J. Schmidt, R. Brendel, On the data analysis of light-biased photoconductance decay measurements. J. Appl. Phys. **79**, 1491–1496 (1996)
3. J. Schmidt, Measurement of differential and actual recombination parameters on crystalline silicon wafers. IEEE Trans. Electron Dev. **46**, 2018–2025 (1999)
4. T. Trupke, R.A. Bardos, M.D. Abbott, Self-consistent calibration of photoluminescence and photoconductance lifetime measurements. Appl. Phys. Lett. **87**, 184102 (2005)
5. A.B. Sproul, M.A. Green, A.W. Stephens, Accurate determination of minority carrier- and lattice scattering-mobility in silicon from photoconductance decay. J. Appl. Phys. **72**, 4161–4171 (1992)

Appendix A
Miscellaneous

A.1 Special Cases of Steady-State Effective Lifetime

In Sect. 3.1.4, steady-state effective carrier lifetime of a silicon substrate of thickness d was defined as the ratio of cumulative excess carrier density and generation rate [1]

$$\tau_{eff,s} = \frac{\int_0^d dz\, \Delta n(z)}{\int_0^d dz\, G(z)}. \tag{A.1}$$

Assuming a vanishing internal reflectivity (without loss of generality with regard to the conclusions of the derivations given here), cumulative excess carrier generation rate reads

$$\int_0^d dz\, G(z) = j_\gamma \left(1 - e^{-\alpha_\gamma d}\right). \tag{A.2}$$

According to the steady-state solution of the continuity equation $\Delta n(z)$ as given in Sect. 3.2.2, the effective lifetime expression to be calculated in the following has the general form

$$\tau_{eff,s} = \frac{A \cdot L\left(e^{\frac{d}{L}} - 1\right) + B \cdot L\left(1 - e^{-\frac{d}{L}}\right) - \frac{j_\gamma}{D_a\left(\alpha_\gamma^2 - L^{-2}\right)}\left(1 - e^{-\alpha_\gamma d}\right)}{j_\gamma\left(1 - e^{-\alpha_\gamma d}\right)}. \tag{A.3}$$

Here, several special cases are derived—distinguished by the structure of the coefficients A and B of the steady-state solution of the continuity equation.

J. Giesecke, *Quantitative Recombination and Transport Properties in Silicon from Dynamic Luminescence*, Springer Theses, DOI: 10.1007/978-3-319-06157-3,
© Springer International Publishing Switzerland 2014

A.1.1 Surface Limited Lifetime—High SRV

For infinite surface recombination velocitities, the steady-state effective lifetime of a wafer for uniform excess carrier generation G (independent of z) is

$$\tau_{eff,s} = \frac{d^2}{12D_a}. \tag{A.4}$$

This is to be shown here on the basis of the steady-state solution of the continuity equation given in Sect. 3.2.2. For $S_0 = S_d \to \infty$ and in the limit $\alpha_\gamma \to 0$ corresponding to a depth-independent excess carrier generation, the matrix elements accounting for the coefficients A and B of the steady-state solution of the continuity equation simplify such that

$$\begin{pmatrix} 1 & 1 \\ e^{\frac{d}{L}} & e^{-\frac{d}{L}} \end{pmatrix} \begin{pmatrix} A \\ B \end{pmatrix} = -\frac{\alpha_\gamma j_\gamma L^2}{D_a} \begin{pmatrix} 1 \\ 1 \end{pmatrix}. \tag{A.5}$$

With this, the coefficients A and B read

$$A = -\frac{\alpha_\gamma j_\gamma L^2}{D_a} \cdot \frac{e^{-\frac{d}{L}} - 1}{e^{-\frac{d}{L}} - e^{\frac{d}{L}}} \tag{A.6}$$

$$B = -\frac{\alpha_\gamma j_\gamma L^2}{D_a} \cdot \frac{e^{\frac{d}{L}} - 1}{e^{\frac{d}{L}} - e^{-\frac{d}{L}}} \tag{A.7}$$

For uniform excess carrier generation ($\alpha_\gamma d \ll 1$), the first order Taylor series expansion of $e^{-\alpha_\gamma d}$ is accurate, yielding $1 - e^{-\alpha_\gamma d} = \alpha_\gamma d$, such that in the limit $\alpha_\gamma \to 0$ effective lifetime becomes

$$\tau_{eff,s} = -\frac{L^3}{dD_a}\left(2\tanh\frac{d}{2L} - \frac{d}{L}\right). \tag{A.8}$$

This expression is to be considered for large minority carrier diffusion lengths L, so the domain of interest of $\tanh\frac{d}{2L}$ are small values $\frac{d}{2L} \ll 1$. Third order Taylor series expansion leads to

$$\tanh\frac{d}{2L} = \frac{d}{2L} - \frac{1}{3}\left(\frac{d}{2L}\right)^3 + O\left(\left(\frac{d}{2L}\right)^4\right), \tag{A.9}$$

and insertion into Eq. A.8 finally yields the sought expression Eq. A.4.

A.1.2 Surface Limited Lifetime—Low SRV

Here, it is shown that in the case of a very large diffusion length of minority carriers L and dominant surface recombination velocities S (identical on both wafer interfaces), which are yet sufficiently small in order to satisfy the condition $S \ll \alpha_\gamma D_a$, steady-state effective lifetime approaches the form

$$\tau_{eff,s} = \frac{d}{2S}. \tag{A.10}$$

This relation is particularly relevant for the investigation of surface recombination velocities of miscellaneous semiconductor interfaces on wafers featuring very high bulk lifetimes. Numerous works imply that this relation—originally derived from a dynamic perception of effective lifetime—would also hold under steady-state conditions even in the light of possibly shallow excess carrier generation $\alpha_\gamma d \gg 1$ (this is no condition in the following). The following derivation confirms this assumption. With $LS \gg D_a$, $L \gg \alpha_\gamma^{-1}$, and $S \ll \alpha_\gamma D_a$, the matrix elements accounting for the coefficients A and B of the steady-state solution of the continuity equation simplify such that[1]

$$\begin{pmatrix} 1 & 1 \\ e^{\frac{d}{L}} & e^{-\frac{d}{L}} \end{pmatrix} \begin{pmatrix} A \\ B \end{pmatrix} = \frac{j_\gamma}{S} \begin{pmatrix} 1 \\ e^{-\alpha_\gamma d} \end{pmatrix}. \tag{A.11}$$

Therefore, the coefficients A and B read

$$A = \frac{j_\gamma}{S} \cdot \frac{e^{-\frac{d}{L}} + e^{-\alpha_\gamma d}}{e^{-\frac{d}{L}} - e^{\frac{d}{L}}} \tag{A.12}$$

$$B = \frac{j_\gamma}{S} \cdot \frac{e^{\frac{d}{L}} + e^{-\alpha_\gamma d}}{e^{\frac{d}{L}} - e^{-\frac{d}{L}}} \tag{A.13}$$

With excess carrier generation rate as of Eq. A.2, the factors $1 - e^{-\alpha_\gamma d}$ in both the numerator and the denominator of the steady-state effective lifetime (cf. Eq. A.3) cancel out such that it can be simplified to

$$\tau_{eff,s} = \frac{L}{S} \tanh \frac{d}{2L} - \frac{1}{D_a \alpha_\gamma^2} \tag{A.14}$$

accordingly. The first order Taylor series expansion of $\tanh \frac{d}{2L}$, which is accurate if the condition $d \ll 2L$ is satisfied, then yields

[1] L must approach very high values in order for all three conditions to be satisfied at once. Particularly the combination of $LS \gg D_a$ and $S \ll \alpha_\gamma D_a$ is limiting in this context, as this is seldom encountered in practice. Yet, the point of this derivation is to demonstrate that dynamic and steady-state expressions of effective lifetime coincide in the very limit $L \rightarrow \infty$. In practice, the typically used relation $\tau_{eff,d} = \tau^{-1} + 2S/d$ accounts for finite bulk lifetime τ.

$$\tau_{eff,s} = \frac{d}{2S} - \frac{1}{D_a \alpha_\gamma^2}, \tag{A.15}$$

and with the above condition of $S \ll \alpha_\gamma D_a$ and $\alpha_\gamma^{-1} \lesssim d$, the second term is negligible such that one obtains the sought expression of Eq. A.10.

A.1.3 Transit Time

Transit time denotes the average duration for excess minority carriers to reach an opposite wafer interface with respect to the interface close to which they were injected. In Sect. 9.2 the ambipolar minority carrier diffusion coefficient D_a is calculated from an advanced transit time approach. Here, an analytic expression of transit time shall be derived.

Let us assume a situation of shallow excess carrier generation $\alpha_\gamma d \gg 1$ leading to

$$\int_0^d dz\, G(z) = j_\gamma. \tag{A.16}$$

The illuminated wafer interface shall exhibit a negligible surface recombination velocity compared to the opposite wafer interface. This is reflected in the set of conditions $S_0 \ll \alpha_\gamma D_a$, $S_0 L \ll D_a$, and $S_d \to \infty$. Further, bulk recombination rate shall be negligible compared to back surface recombination rate. This is accounted for by the additional condition $L \gg d$ (which also implies $L \ggg \alpha_\gamma^{-1}$). With these conditions, the matrix elements accounting for the coefficients A and B of the steady-state solution of the continuity equation simplify such that

$$\begin{pmatrix} -\frac{D_a}{S_0 L} & \frac{D_a}{S_0 L} \\ e^{\frac{d}{L}} & e^{-\frac{d}{L}} \end{pmatrix} \begin{pmatrix} A \\ B \end{pmatrix} = \frac{j_\gamma}{S_0} \begin{pmatrix} 1 \\ 0 \end{pmatrix}. \tag{A.17}$$

Thus, the coefficients A and B read

$$A = -\frac{j_\gamma L}{D_a} \cdot \frac{e^{-\frac{d}{L}}}{e^{\frac{d}{L}} + e^{-\frac{d}{L}}} \tag{A.18}$$

$$B = \frac{j_\gamma L}{D_a} \cdot \frac{e^{\frac{d}{L}}}{e^{\frac{d}{L}} + e^{-\frac{d}{L}}} \tag{A.19}$$

Transit time τ_{tr} is then represented by steady-state effective lifetime $\tau_{tr} = \tau_{eff,s}$, as excess carriers only recombine at the opposite interface with respect to the location of their generation, and as steady-state effective lifetime is defined as the arithmetic average of the duration of excitation of an ensemble of excess carriers. With the above coefficients, transit time is

$$\tau_{tr} = \frac{1}{D_a}\left(L^2 \frac{\cosh \frac{d}{L} - 1}{\cosh \frac{d}{L}} - \frac{1}{\alpha_\gamma^2}\right) \tag{A.20}$$

Second order Taylor series expansion of the $\cosh \frac{d}{L}$ fraction for small arguments $\frac{d}{L}$ (cf. condition $L \gg d$) yields

$$\frac{\cosh \frac{d}{L} - 1}{\cosh \frac{d}{L}} = \frac{1}{2}\left(\frac{d}{L}\right)^2 + O\left(\left(\frac{d}{L}\right)^3\right), \tag{A.21}$$

leading to the transit time expression

$$\tau_{tr} = \frac{1}{D_a}\left(\frac{d^2}{2} - \frac{1}{\alpha_\gamma^2}\right). \tag{A.22}$$

A.2 Extended Integration of Gaussians

In Sect. 3.2.4, integrals of the type

$$\int_0^\infty d\zeta \, e^{-\alpha_\gamma \zeta - \frac{(\zeta \pm z)^2}{4D_a t}} \tag{A.23}$$

occur. Such integrals can be expressed in terms of complementary error functions, as outlined in the following.

The error function is a non-elementary function defined as the integral

$$\mathrm{erf}(x) = \frac{2}{\sqrt{\pi}} \int_0^x dt \, e^{-t^2} \tag{A.24}$$

With the Gaussian integral

$$\frac{2}{\sqrt{\pi}} \int_0^\infty dt \, e^{-t^2} = 1 \tag{A.25}$$

a complementary error function is defined as

$$\mathrm{erfc}(x) = 1 - \mathrm{erf}(x) = \frac{2}{\sqrt{\pi}} \int_x^{\infty} dt\, e^{-t^2}. \tag{A.26}$$

An essential property of error functions, which directly follows from their definition, is

$$\frac{d}{dx}\,\mathrm{erf}(x) = -\frac{d}{dx}\,\mathrm{erfc}(x) = \frac{2}{\sqrt{\pi}} e^{-x^2}. \tag{A.27}$$

A derivative property of the complementary error function is

$$\int_0^{\infty} d\zeta\, e^{-\left(\frac{\zeta-b}{a}\right)^2} = \frac{\sqrt{\pi}}{2} a\, \mathrm{erfc}\left(-\frac{b}{a}\right). \tag{A.28}$$

With this property of the complementary error function, the integrals A.23 are to be expressed as

$$\int_0^{\infty} d\zeta\, e^{-\alpha_\gamma \zeta - \frac{(\zeta \pm z)^2}{4D_a t}} = \sqrt{D_a t \pi} e^{\pm \alpha_\gamma (z \pm D_a t \alpha_\gamma)}\, \mathrm{erfc}\left(\pm \frac{z}{2\sqrt{D_a t}} + \alpha_\gamma \sqrt{D_a t}\right). \tag{A.29}$$

A.3 Numerical Solution of the Continuity Equation

A.3.1 Time-Dependent Continuity Equation

The discrete representation of the time-dependent continuity equation (Eq. 3.75) for depth-averaged excess carrier density $\Delta n(t_i) = \Delta n_i$ at times $t_i = \Delta t, 2\Delta t, ..., T$, depth-averaged excess carrier generation rate $G(t_i) = G_i$, and effective lifetime $\tau_{eff}(\Delta n_i)$ is

$$\frac{\Delta n_{i+1} - \Delta n_i}{\Delta t} = G_i - \frac{\Delta n_i}{\tau_{eff}(\Delta n_i)}. \tag{A.30}$$

With an initial value $\Delta n(t_0 = 0) = \Delta n_0$, and with known excess carrier generation rate G_i and lifetime $\tau_{eff}(\Delta n_i)$, the finite element solution for times t_k with $k > 0$ can be iteratively determined via

$$\Delta n_{i+1} = \Delta n_i + \Delta t \left(G_i - \frac{\Delta n_i}{\tau_{eff}(\Delta n_i)}\right). \tag{A.31}$$

There are two conditions for an adequate discretization Δt. First, the timescale of significant changes of excess carrier generation rate Δt_G must be much greater than $\Delta t \ll \Delta t_G$. Second, $\Delta t \ll \tau_{eff}$ must hold. For quasi-steady-state harmonic time modulation of excess carrier generation, if the condition $\Delta t \ll \tau_{eff}$ is satisfied, then the weaker condition $\Delta t \ll \Delta t_G$ holds as well. Finite element modeling of the time-dependent continuity equation is numerically straightforward and requires few computational effort.

A.3.2 Time- and Depth-Dependent Continuity Equation

The discrete representation of the time- and depth-dependent continuity equation (Eq. 3.19) for excess carrier density $\Delta n(t_i, z_j) = \Delta n_{i,j}$ at times $t_i = \Delta t, 2\Delta t, ..., \mathcal{T}$ and at a substrate depth $z_j = \Delta z\left(\frac{1}{2}, \frac{3}{2}, \frac{5}{2}, ..., \frac{d}{\Delta z} - \frac{1}{2}\right)$, excess carrier generation rate $G\left(t_i, z_j\right) = G_{i,j}$, and bulk lifetime $\tau_j\left(\Delta n_{i,j}\right)$ is

$$\frac{\Delta n_{i+1,j} - \Delta n_{i,j}}{\Delta t} = G_{i,j} - \frac{\Delta n_{i,j}}{\tau_j\left(\Delta n_{i,j}\right)} - D_a\frac{2\Delta n_{i,j} - \Delta n_{i,j+1} - \Delta n_{i,j-1}}{\Delta z^2}. \quad (A.32)$$

With $\#_z = \frac{d}{\Delta z}$ the discrete representation of boundary conditions (cf. Eqs. 3.21 and 3.22) is

$$D_a\frac{\Delta n_{i,2} - \Delta n_{i,1}}{\Delta z} = S_0\Delta n_{i,1} \quad (A.33)$$

$$D_a\frac{\Delta n_{i,\#_z} - \Delta n_{i,\#_z-1}}{\Delta z} = -S_d\Delta n_{i,\#_z}. \quad (A.34)$$

With an initial value $\Delta n\left(t_0 = 0, z_j\right) = \Delta n_{0,j}$ and with known excess carrier generation rate $G_{i,j}$, lifetime $\tau_j\left(\Delta n_{i,j}\right)$, surface recombination velocities $S_{0,d}$ and diffusion coefficient D_a, the finite element solution $\Delta n_{k,j}$ for times t_k with $k > 0$ can be iteratively determined via

$$\Delta n_{i+1,j} = \mathcal{M}_S\left[\left(\infty\left(1 - \frac{\Delta t}{\tau_j\left(\Delta n_{i,j}\right)}\right) + \mathcal{M}_D\right)\Delta n_{i,j} + \infty\Delta t G_{i,j}\right]. \quad (A.35)$$

Here, time propagation is represented by the discrete operators or matrices acting on the function or vector $\Delta n_{i,j}$. With $\#_z = \frac{d}{\Delta z}$, the vectors $\Delta n_{i,j}$ and $G_{i,j}$ are of dimension $\mathbb{R}^{\#_z}$, and the matrices acting on $\Delta n_{i,j}$ are of dimension $\mathbb{R}^{\#_z \times \#_z}$. By acting on $\Delta n_{i,j}$, these matrices impose excess carrier generation and recombination rates, a surface recombination rate ($\mathcal{M}_S$), and a redistribution rate ($\mathcal{M}_D$) of excess carriers due to diffusion during a discrete time interval Δt.

Diffusion Matrix: According to the discrete representation of the continuity equation (Eq. A.32), the matrix $\mathcal{M}_D$ must contain the matrix elements

$$d_{1,1} = d_{\#_z,\#_z} = -\frac{D_a \Delta t}{\Delta z^2}, \quad d_{i,i} = -\frac{2D_a \Delta t}{\Delta z^2} \quad \forall \;\; i \in \left[2, \#_z - 1\right], \tag{A.36}$$

along its diagonal. Off-diagonal elements are

$$d_{i,j} = \frac{D_a \Delta t}{\Delta z^2} \quad \forall \;\; j = i \pm 1, \quad d_{i,j} = 0 \;\; \forall \;\; j > i+1, j < i-1. \tag{A.37}$$

In this form, the diffusion matrix satisfies the condition of a vanishing change of total excess carrier density (redistribution only)

$$\sum_{j=1}^{\#_z} \mathcal{M}_D \Delta n_{i,j} = 0. \tag{A.38}$$

Surface Recombination Matrix: According to the discrete representation of the boundary conditions of the continuity equation (Eqs. A.33 and A.34), the diagonal matrix elements of $\mathcal{M}_S$ are

$$s_{1,1} = 0, \quad s_{\#_z,\#_z} = 0, \quad s_{i,i} = 1 \;\; \forall \;\; i \in \left[2, \#_z - 1\right], \tag{A.39}$$

and the only non-zero off-diagonal matrix elements are

$$s_{1,2} = \frac{1}{\frac{S_0 \Delta z}{D_a} + 1} \tag{A.40}$$

and

$$s_{\#_z,\#_z-1} = \frac{1}{\frac{S_d \Delta z}{D_a} + 1}. \tag{A.41}$$

Discretization Conditions: As for the time-dependent continuity equation, the discretization conditions $\Delta t \ll \tau$ and $\Delta t \ll \Delta t_G$ must be satisfied. However, these conditions proved not to be limiting for the time and depth-dependent numerical solution of the continuity equation. The more limiting conditions are

$$\frac{D_a \Delta t}{\Delta z^2} \ll 1 \tag{A.42}$$

and—particularly for highly localized recombination, e.g. for a high surface recombination velocity of the illuminated interface

$$\Delta z \ll \alpha_\gamma^{-1}. \tag{A.43}$$

While the condition A.42 implies redistribution rates of excess carrier density much less than differences between adjacent locations $\Delta n_{i,j+1} - \Delta n_{i,j}$, the condition A.43 ensures accurate excess carrier generation rates and negligible surface artifacts in boundary elements in the case of very high surface recombination rates. The combination of the latter two conditions represents the bottleneck of the finite element approach. For quasi-steady-state harmonic modulation of excess carrier density, the dynamics of excess carriers near the most relevant peak domain could also be accurately investigated by performing a time propagation only near the times of maximal excess carrier density and generation rate—starting with an initial value corresponding to the steady-state solution of the continuity equation at the corresponding initial excess carrier generation rate.
An Example: Assume a wafer with an effective lifetime of $\tau_{eff} = 10\,\mu\text{s}$, and a harmonic time modulation with a periodic time of $\mathcal{T} = 10^{-2}\,\text{s}$. Accepting only one order of magnitude in order for the "$\ll$" operator to be justified in the above conditions, the condition $\Delta t \ll \tau_{eff}$ would be satisfied with 10^4 time steps of $\Delta t = 10^{-6}\,\text{s}$. However, with a substrate thickness of $d = 200\,\mu\text{m}$ and a light penetration depth of $\alpha_\gamma^{-1} = 10\,\mu\text{m}$, a depth interval of $\Delta z = 1\,\mu\text{m}$ would yield $\Delta t \lesssim 3 \times 10^{-11}\,\text{s}$ with a diffusion coefficient of $D_a = 30\,\text{cm}^2\text{s}^{-1}$.

A.4 Photon Density of States

The following derivation aims at an analytic expression of the density of states of photons in a cubic cavity of sidelength l with periodic boundary conditions. This expression shall be given as a function of photon energy ε_γ. Periodic boundary conditions imply the formation of standing waves—equivalent to nodes (vanishing fields or light amplitudes) at boundaries. With a wavevector $\mathbf{k}$ of light featuring a norm of $|\mathbf{k}| = k = \frac{2\pi}{\lambda}$ (with wavelength λ), periodic boundary conditions impose constraints upon the components k_i of $\mathbf{k}$ along the coordinates $i = x, y, z$ (aligned with cube axes) such that with any positive natural number n_i the equation

$$k_i = \frac{\pi n_i}{l} \tag{A.44}$$

must be satisfied. Therefore, allowed states are equally distributed throughout positive $\mathbf{k}$-space, with a $\mathbf{k}$-space volume per state of

$$V_{\mathbf{k},0} = \left(\frac{\pi}{l}\right)^3. \tag{A.45}$$

With photon energy $\varepsilon_\gamma = \frac{hc'}{\lambda}$ (with the velocity of light c' in a matter of interest), ε_γ and k can be expressed as injective functions of each other:

$$\varepsilon_\gamma = \hbar c' k \tag{A.46}$$

With this, the infinitesimal **k**-space volume associated with the photon energy interval $[\varepsilon_\gamma, \varepsilon_\gamma + d\varepsilon_\gamma]$ is the volume confined by the concentric spheres distinguished by $k\left(\varepsilon_\gamma\right)$ and $k\left(\varepsilon_\gamma + d\varepsilon_\gamma\right)$. Thus, the infinitesimal volume in positive **k**-space to be associated with the photon energy interval $[\varepsilon_\gamma, \varepsilon_\gamma + d\varepsilon_\gamma]$ is

$$dV_{\mathbf{k},\varepsilon_\gamma} = \frac{1}{8} \int\limits_{\mathbf{k}(\varepsilon_\gamma)}^{\mathbf{k}(\varepsilon_\gamma + d\varepsilon_\gamma)} d\mathbf{k} = \frac{1}{8} 4\pi k^2 \left(\varepsilon_\gamma\right) \frac{dk}{d\varepsilon_\gamma} d\varepsilon_\gamma. \tag{A.47}$$

Here, the factor $\frac{1}{8}$ accounts for integration solely over positive **k**-space volume. Note that every state may be occupied by two photons of different spin quantum number. This is accounted for in an additional factor of 2 in the density of states. With this, the number of states per volume in the energy interval $[\varepsilon_\gamma, \varepsilon_\gamma + d\varepsilon_\gamma]$ is

$$D_\gamma \left(\varepsilon_\gamma\right) d\varepsilon_\gamma = \frac{2}{l^3} \frac{dV_{\mathbf{k},\varepsilon_\gamma}}{V_{\mathbf{k},0}} = \frac{\varepsilon_\gamma^2}{\pi^2 \hbar^3 c'^3} d\varepsilon_\gamma. \tag{A.48}$$

A.5 On the Coefficient of Radiative Recombination

The dependence of the relative integral coefficient of radiative recombination $\frac{B}{B_0}$ (with a low-level injection value B_0) was parametrized by Altermatt et al. [2]. This parametrization (with modifications initiated by Roth, to be found in [3]) shall be given here.

With coefficients, $b_2 = 0.54$, $b_4 = 1.25$, $r_1 = 320$, $r_2 = 2.5$, $r_{max} = 0.2$, $s_1 = 550$, $s_2 = 3$, $s_{max} = 1.5 \times 10^{18}$, $s_{min} = 10^7$, $w_1 = 365$, $w_2 = 3.54$, $w_{max} = 4 \times 10^{18}$, $w_{min} = 10^9$, and derivative temperature-dependent coefficients

$$b_{min}(T) = r_{max} \left(1 - \frac{1}{1 + \left(\frac{T}{r_1}\right)^{r_2}}\right),$$

$$b_1(T) = 2 \left(s_{max} + \frac{s_{min} - s_{max}}{1 + \left(\frac{T}{s_1}\right)^{s_2}}\right),$$

$$b_3(T) = 2 \left(w_{max} + \frac{w_{min} - w_{max}}{1 + \left(\frac{T}{w_1}\right)^{w_2}}\right),$$

with $B_0 = B(T = 0, n_0 = 0, \Delta n = 0)$ the relative integral coefficient of radiative recombination is deemed to be best parametrized as the following function of temperature T, dopant concentration n_0, and excess carrier density Δn:

$$B(T, n_0, \Delta n) = B_0 \left(b_{min}(T) + \frac{1 - b_{min}(T)}{1 + \left(\frac{2\Delta n + n_0}{b_1(T)}\right)^{b_2} + \left(\frac{2\Delta n + n_0}{b_3(T)}\right)^{b_4}} \right) \tag{A.49}$$

A.6 The Equivalent of Solar Irradiance

With the absorption coefficient of intrinsic (band-to-band) crystalline silicon $\alpha_\gamma(\varepsilon_\gamma, T)$ as a function of photon energy ε_γ and temperature T [4, 5], the absorptance a of a (vertically irradiated) silicon substrate of thickness d is

$$a(\varepsilon_\gamma, d, T) = 1 - e^{-\alpha_\gamma(\varepsilon_\gamma, T)d}. \tag{A.50}$$

With the spectral photon current density of the sun $\frac{dj_{sun}}{d\varepsilon_\gamma}$ taken from the direct hemispherical solar spectral irradiance on a 37° tilted surface (AM1.5 spectrum) [6], the monochromatic equivalent of the electron-hole pair generating photon current density of the sun in crystalline silicon j_γ^{1sun} may be defined as the product of the inverse fraction a^{-1} of monochromatic light of energy ε_γ absorbed by a substrate of thickness d and of the sun's total photon current density absorbed by this substrate:[2]

$$j_\gamma^{1sun}(\varepsilon_\gamma, d, T) = \frac{1}{a(\varepsilon_\gamma, d, T)} \int_0^\infty d\varepsilon_\gamma \frac{dj_{sun}}{d\varepsilon_\gamma} a(\varepsilon_\gamma, d, T) \tag{A.51}$$

Figure A.1 depicts the monochromatic equivalent photon current density of one sun at a temperature of $T = 300\,\text{K}$ as a function of monochromatic irradiation wavelength $\lambda_\gamma = \frac{hc}{\varepsilon_\gamma}$ and of substrate thickness d. Therefore, at a substrate thickness $d \approx 250\,\mu\text{m}$ and irradiation wavelengths of $\lambda_\gamma \leq 900\,\text{nm}$, the monochromatic sun's equivalent amounts to $j_\gamma^{1sun} \approx 2.5 \times 10^{17}\,\text{cm}^{-2}\text{s}^{-1}$.

A.7 Validity of Theory of Light-Biased Decay Time

In the following, a proof of the general validity of Eq. 6.40 in the limit of a bias light parameter $\xi \rightarrow 1$ is given:[3]

Statement The curvature of any nonlinear effective lifetime $\tau_{eff}(\Delta n)$—represented by its second derivative with respect to excess carrier density—does not affect

[2] Note that more complex perceptions of the equivalent of solar irradiance exist—incorporating interface reflectivity and light trapping. The definition given herein aspires independence of substrate optical properties.

[3] This proof proceeds along the lines of its original publication in [7].

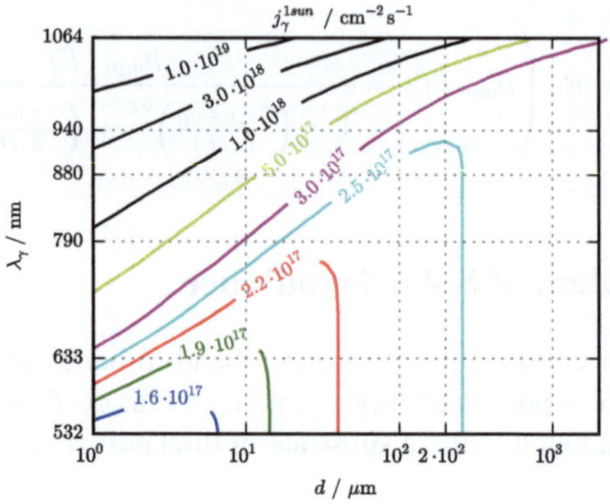

Fig. A.1 Plot of the monochromatic equivalent of the photon current density of the sun j_γ^{1sun} at a temperature of $T = 300\,\mathrm{K}$ as a function of monochromatic irradiation wavelength $\lambda_\gamma = \frac{hc}{\varepsilon_\gamma}$ and of substrate absorber thickness d. Data on the 300 K intrinsic absorption coefficient of crystalline silicon were taken from [4], data on the hemispherical AM1.5 spectral solar irradiance accord with the standard ASTM G173-03 [6]. Note that the absorptance of the solar spectrum largely depends on the absorber thickness due to weakly absorbed light near and below the band gap

$\Delta n_\infty = \Delta n\,(t \to \infty)$ in the limit $\xi \to 1$, e.g. represented by quasi-steady-state harmonically modulated excess carrier generation.

Proof With $\Delta n_0 = \Delta n\,(t = 0) = \tau_{eff,0} G_0$, $d_1 = \frac{d\tau_{eff}}{d\Delta n}\big|_{\Delta n_0}$, and $d_2 = \frac{d^2\tau_{eff}}{d\Delta n^2}\big|_{\Delta n_0}$, the second order Taylor series expansion of lifetime as a function of excess carrier density Δn_∞ reads

$$\tau_{eff}\,(\Delta n_\infty) = \tau_{eff,0} + (\Delta n_\infty - \Delta n_0)\,d_1 + \frac{1}{2}\,(\Delta n_\infty - \Delta n_0)^2\,d_2 + O\left((\Delta n_\infty - \Delta n_0)^3\right). \tag{A.52}$$

Insertion of this expression into Eq. 6.39 yields

$$\Delta n_\infty = \xi G_0 \left(\tau_{eff,0} + (\Delta n_\infty - \Delta n_0)\,d_1 + \frac{1}{2}\,(\Delta n_\infty - \Delta n_0)^2\,d_2\right). \tag{A.53}$$

The positive solution Δn_∞ of this quadratic equation is

$$\Delta n_\infty = \frac{1 - \xi G_0 d_1 + \xi G_0 \Delta n_0 d_2 + \sqrt{(\xi G_0 d_1 - 1)^2 + 2\xi G_0 \Delta n_0 d_2\,(1 - \xi)}}{\xi G_0 d_2}. \tag{A.54}$$

The first order Taylor series expansion of the square root term (here denoted as g) of the latter equation

$$g(1-\xi) = \xi G_0 d_1 - 1 + \frac{\xi G_0 \Delta n_0 d_2}{\xi G_0 d_1 - 1}(1-\xi) + O\left((1-\xi)^2\right). \tag{A.55}$$

is exact ($O = 0$) in the limit $\xi \to 1$, yielding

$$\lim_{\xi \to 1} \Delta n_\infty = \xi G_0 \frac{1 - G_0 d_1}{1 - \xi G_0 d_1}. \tag{A.56}$$

This expression is independent of the curvature d_2 and identical to Eq. 6.40 □.

References

1. R.A. Sinton, T. Trupke, Limitations on dynamic excess carrier lifetime calibration methods. Prog. Photovoltaics Res. Appl. **20**, 246–249 (2012)
2. P. P. Altermatt, F. Geelhaar, T. Trupke, X. Dai, A. Neisser, E. Daub, Injection dependence of spontaneous radiative recombination in c-Si: experiment, theoretical analysis, and simulation, in *Proceedings of the 5th International Conference on Numerical Simulation and Optoelectronic Devices, IEEE* (Berlin, 2005), pp. 47–48
3. P. Rosenits, Electrical characterisation of crystalline silicon thin-film material, Ph.D. thesis, Universität Konstanz (2010)
4. M.A. Green, *Silicon Solar Cells: Advanced Principles & Practice* (University of New South Wales, Sydney, 1995)
5. M.A. Green, M.J. Keevers, Optical properties of intrinsic silicon at 300K. Prog. Photovoltaics: Res. Appl. **3**, 189–192 (1995)
6. Standard tables for reference solar spectral irradiances: direct normal and hemispherical on 37° tilted surface, ASTM G173–03, ASTM International (2012), http://www.astm.org/Standards/G173.htm
7. J.A. Giesecke, S.W. Glunz, W. Warta, Understanding and resolving the discrepancy between differential and actual minority carrier lifetime. J. Appl. Phys. **113**, 073706 (2013)

Appendix B
Reassessment of the Radial Sensitivity of QSSPC

Quasi-steady-state photoconductance measurements are a prevalent measurement technique of minority carrier lifetime in silicon photovoltaics. As this technique is also used for wafers with a laterally inhomogeneous distribution of electronic material quality, its lateral sensitivity distribution is of interest for reliable lifetime measurements. Two independent investigations of the lateral sensitivity distribution of quasi steady-state photoconductance have recently been reported. In spite of the reasonable agreement between the results, there were systematic deviations calling for a reassessment—as presented here.

B.1 Essentials of QSSPC

Quasi-steady-state photoconductance (QSSPC) [1, 2] is an eddy current measurement technique of injection-dependent minority carrier lifetime. A flash lamp featuring a characteristic decay time much greater than expected effective carrier lifetime illuminates a silicon wafer positioned on top of a radio frequency coil. A calibrated sensor coil is sensitive to an enhanced inductance due to both dark majority carrier concentration (net dopant concentration) and light induced excess carrier density. Via a carrier mobility model [3, 4], the measured light induced conductance

$$\Delta\sigma = q\,(\mu_e + \mu_h)\,\Delta n \tag{B.1}$$

is related to excess carrier density $\Delta n(t)$. Simultaneous measurement of the excess carrier generation rate $G(t)$ finally enables the determination of (steady-state) effective carrier lifetime $\tau_{\mathit{eff}}\,(\Delta n(t)) = \frac{\Delta n(t)}{G(t)}$.

J. Giesecke, *Quantitative Recombination and Transport Properties in Silicon from Dynamic Luminescence*, Springer Theses, DOI: 10.1007/978-3-319-06157-3,
© Springer International Publishing Switzerland 2014

B.2 Consideration of Local Sensitivity of QSSPC

The magnetic flux through a radio frequency coil is usually not homogeneously distributed throughout its cross section (designated active measurement area of QSSPC). Therefore, lifetime measurement results on wafers featuring a laterally inhomogeneous distribution of carrier lifetime critically depend not only on the relative position of the wafer with respect to the sensor region, but also on the lateral lifetime distribution within the sensor region. This is particularly relevant for the carrier lifetime calibration of photoluminescence images of multicrystalline wafers via quasi-steady-state or steady-state photoconductance [5], as also implemented in commercially available quantitative luminescence imaging applications. From the results of recent investigations of the lateral sensitivity distribution of QSSPC [6, 7], and given the radial symmetry of the measurement setup, a radial symmetry of sensitivity can be assumed. One may assume a radial sensitivity function $\mathcal{S}_R(r)$ normalized such that

$$2\pi \int_0^R rdr\, \mathcal{S}_R(r) \stackrel{!}{=} \pi R^2, \tag{B.2}$$

with a radius R chosen such that sensitivity would only significantly differ from null for $r < R$. In this form, a steady-state photoluminescence image (cf. Eq. 8.15) comprising all n_φ pixels $r_i \leq R$ corresponding to the active QSSPC measurement area can be calibrated in terms of effective carrier lifetime via a QSSPC lifetime measurement $\tau_{\mathrm{QSSPC}}(G_0)$ at an excess carrier generation rate G_0:

$$\tau_{\mathrm{QSSPC}}(G_0) \stackrel{!}{=} n_\varphi^{-1} \sum_{r_i \leq R} \tau_{\mathit{eff},i}(G_0) \mathcal{S}_R(r_i). \tag{B.3}$$

Through Eq. B.3, local carrier lifetimes $\tau_{\mathit{eff},i}$ can be calculated from local photoluminescence intensities according to Eq. 8.15 (as e.g. measured with a CCD or CMOS camera) [6]. Comparing this sensitivity-weighted average to an arithmetic average over the designated QSSPC sensor region, systematic deviations are likely for wafers featuring a laterally inhomogeneous distribution of carrier lifetime, if the characteristic dimension of laterally homogeneous substrate areas does not substantially exceed the QSSPC sensor region.

B.3 Previous Investigations

Two independent investigations of the radial sensitivity $\mathcal{S}_R(r)$ of QSSPC (*Sinton Instruments, WCT-120*) were released lately: In [6], measurements of $\mathcal{S}_R(r)$ were carried out in order to get a reliable reference for quasi-steady-state harmonically modulated photoluminescence lifetime measurements. For this purpose, a 4″ float

Table B.1 Summary of fit parameters for a fit of the radial sensitivity function $\mathcal{S}_R(r)$ according to Eq. B.4

Investigation	r_0/mm	σ_S/mm
Giesecke et al., SOLMAT 95 (2011) 1011	8.34	5.44
Kiliani et al., J. Appl. Phys. 110 (2011) 054508	9.30	3.60
Dark conductivity analysis (here)	8.25	3.65

zone silicon wafer (1 Ω cm, featuring a silicon nitride surface passivation) was covered by an intransparent sheet with a small circular opening (Ø: 2.5 mm). This pinhole was successively positioned for measurements at 21 equally spaced locations (distance: 3 mm) along two perpendicular axes through the center of the QSSPC sensor region, as also depicted in Fig. B.2. Both thereby obtained radial cross sections started and ended 10 mm beyond the circular flag (Ø: 40 mm) indicative of the boundary of the active measurement area (cf. Fig. B.2). Measurement results at each spot were obtained from the maximum excess photoconductance voltage divided by the maximum voltage of the detector of impinging light intensity (i.e. of excess carrier generation rate). The resulting four radial branches were fitted to a symmetric superposition of two Gaussians

$$\mathcal{S}_R(r) = s_R \sum_{\pm} \exp\left(- \left(\frac{r \pm r_0}{\sigma_S} \right)^2 \right). \tag{B.4}$$

The normalization parameter s_R (which is irrelevant here) was to be determined via Eq. B.2. The best fit of this previous analysis is plotted in Fig. B.1, the fit parameters of interest—r_0 and σ_S—are given in Table B.1.

In an alternative approach [7], Kiliani et al. determined a radial sensitivity function by means of controlled circular removal of a $\mathrm{SiN_x}:\mathrm{H}$ surface passivation from a test wafer via laser ablation. QSSPC lifetime was obtained for a set of QSSPC measurements in different stages of the removal process (each featuring a concentric symmetry) and the shape of the radial sensitivity function—denoted here as $\mathcal{S}_R(r)$—was derived from the measured QSSPC carrier lifetimes via a Gaussian fit, which is also depicted in Fig. B.1 and tabulated in Table B.1. The fact that Kiliani et al. used a mono-Gaussian fit is secondary, as they encountered a vanishing sensitivity at $r = 0$.

In spite of the reasonable agreement between previous results (cf. Fig. B.1), there is a systematic deviation calling for a reassessment—particularly concerning the relative sensitivity in the center of the sensor region. As for potential systematic errors of the applied methods, neither of them can be regarded as error free: our previous setup ([6], SOLMAT 95 (2011) 1011) could particularly suffer from lateral diffusion of excess carriers away from the small illuminated aperture area. However, with an aperture radius of 1.25 mm and an effective minority carrier diffusion length expected to be $L < 1$ mm, the observed deviation cannot solely be attributed to this artifact.

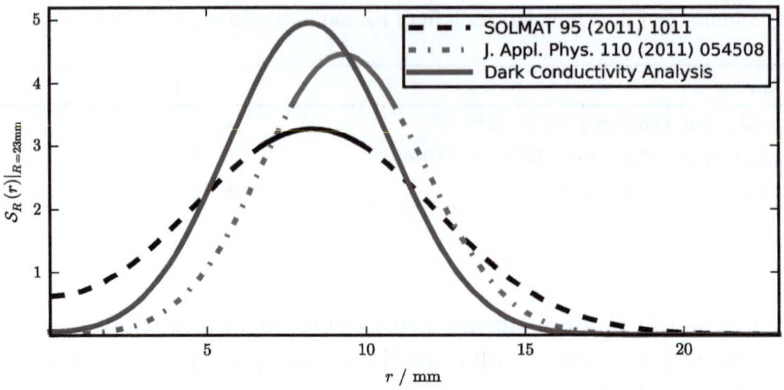

Fig. B.1 Comparison of radial sensitivity functions $\mathcal{S}_R(r)$ as previously reported to the result presented here. The data were normalized according to Eq. B.2 with $R = 23$ mm

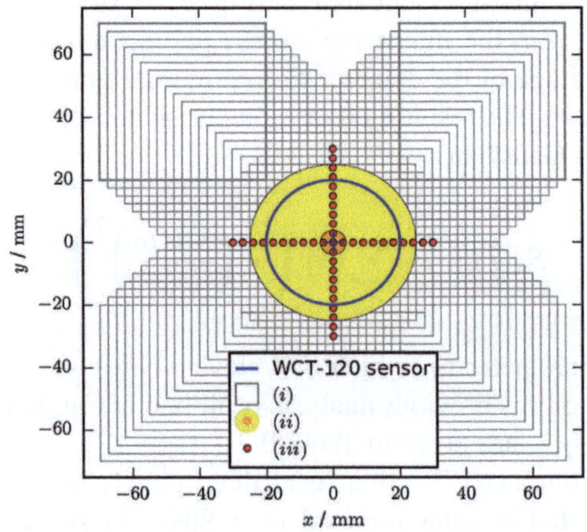

Fig. B.2 Depiction of the discussed measurement designs. Here, x and y indicate distances with respect to the center of the QSSPC sensor region. The blue circle represents the frame of the sensor region as indicated on the instrument. (*i*) Indicates square wafer positions for diagonal cross sections of dark conductivity measurements. (*ii*) Indicates circular wafer positions for the sensitivity check at $r \to 0$ (via dark conductivity). (*iii*) Indicates pinhole positions of a previous investigation of $\mathcal{S}_R(r)$ via excess conductivity [6]

B.4 Sensitivity Analysis via Dark Conductivity

Potential systematic errors of the above explained determination methods of $\mathcal{S}_R(r)$ may originate from any deviation between an ideally assumed and the present lateral distribution of **excess** carriers. Such artifacts could be avoided by measuring net dark conductivity, which is a measure of free majority carrier density rather than of light-induced excess carrier density.

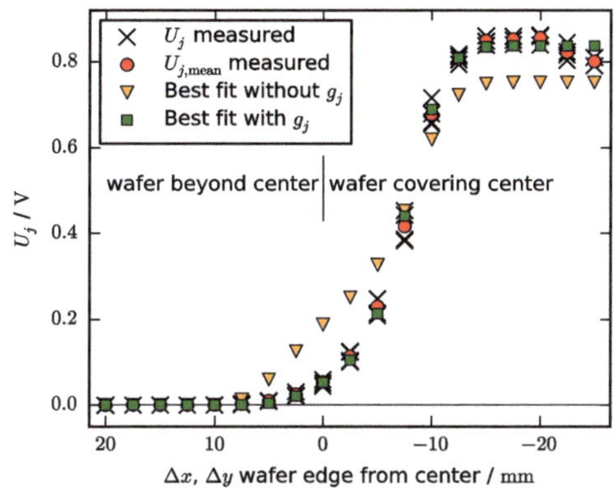

Fig. B.3 Measured dark QSSPC sensor voltage U_j as function of the wafer position represented by the wafer edge distance Δx, Δy from the center of the sensor region (cf. Fig. B.2 (*i*)). Each measurement position features four measurements corresponding to the four symmetrical radial branches of the dark conductivity measurement (*i*). A satisfactory fit to Eq. B.4 was achieved with a geometry factor g_j accounting for the interruption of eddy currents at wafer edges

For the investigation presented in the following, a 330 μm thick 1 Ω cm quadratic $50 \times 50\,\mathrm{mm}^2$ float zone wafer featuring a very homogeneous lateral distribution of boron concentration (dark majority carrier concentration) was used. This wafer was precisely positioned on 37 equally spaced spots (with both horizontal and vertical spot distances of 2.5 mm, respectively) along the two diagonal cross sections of the sensor region, as depicted in Fig. B.2. At each spot, the net dark voltage of the sensor (difference between voltages with and without the wafer in place) was measured. Thereby, four symmetrical radial branches were obtained from these cross sections (cf. Fig. B.3). As can be judged from Fig. B.3, the very small scattering between these branches substantiates the assumption of both, the radial symmetry of the lateral sensitivity distribution as well as the homogeneous lateral distribution of dopant concentration of the test wafer. The measured voltages U_j were to be fitted to Eq. B.4 via a least squares approach involving the parameters r_0 and σ_S. This fit was performed according to

$$U_j = c_U \int dx \int dy\, \Theta_j\,(x, y)\; g_j\,(x, y)\; \mathcal{S}_R\,(x, y)\,. \tag{B.5}$$

Here, the constant c_U ensures normalization of the expression on the right with respect to measured voltages. The function

$$\Theta_j\,(x, y) = \begin{cases} 1 & \forall \{x, y\} \in \text{wafer} \\ 0 & \forall \{x, y\} \notin \text{wafer} \end{cases} \tag{B.6}$$

denotes a two dimensional mask representative of the position of the wafer.

Table B.2 Check of the radial sensitivity function $\mathcal{S}_R(r)$ at $r \rightarrow 0$. The ratio of net dark sensor voltages as measured via concentrically positioned circular wafers (with diameters of 50.0 and 7.8 mm, respectively) was compared to expectations from the three curve parametrizations of $\mathcal{S}_R(r)$ according to Eq. B.4 given in Table B.1 (cf. Fig. B.1)

Measurement and expectation	Dark voltage ratio
Measurement	55 ± 10
Expectation/calculation from	
Giesecke et al., SOLMAT 95 (2011) 1011	29
Kiliani et al., J. Appl. Phys. 110 (2011) 054508	182
Dark conductivity analysis (as described here)	62

The geometry factor g_j weights every location $\{x, y\}$ of the active wafer area by the fraction of a concentric revolution of eddy currents at a distance $r\,(x, y)$ from the istrument's center of symmetry, which lies within the active wafer area. Eddy currents which give rise to the inductance change due to free carriers, are interrupted at wafer edges. Therefore, the reduction of their contribution to conductance might be approximated in a simplified model by a factor of g_j. From Fig. B.3, it can be seen that it was not possible to fit the measured data without the geometry factor g_j.

The resulting parametrization of $\mathcal{S}_R(r)$ is given in Table B.1 and depicted in Fig. B.1. As can be seen, it lies between the curves obtained in previous analyses. Yet, the experimental procedure described herein is judged to feature insufficient sensitivity with respect to the curve shape of $\mathcal{S}_R(r)$ for small radii $r \rightarrow 0$. Therefore, a test measurement was conducted with a series of homogeneously doped $d = 220\,\mu\text{m}$ thick $2.5\,\Omega\,\text{cm}$ circular Czochralski-grown wafers featuring diameters of 7.8 and 50.0 mm, respectively (cf. depiction in Fig. B.2 (*ii*)). These wafers were concentrically positioned on the QSSPC sensor region, and net dark voltage of the sensor was measured. The ratio of the resulting net voltages between the wafers with diameters of 7.8 and 50.0 mm was compared to what one would expect according to the radial sensitivity functions shown in Fig. B.1. The resulting voltage ratio of 55 ± 10 confirms the result of the dark conductivity analysis of 62 (cf. Table B.2).[4]

There is a noteworthy artifact to be observed in the depiction of experimental results in Fig. B.3: It is a decline of measured dark voltage toward the centermost position of the quadratic test wafer (at a wafer edge position of −25 mm with respect to the center of symmetry), which apparently cannot be modeled according to Eq. B.4. This could be an indication of a too simplified geometry factor g_j. The observed effect suggests that the geometry factor would also have to account for a concentration of eddy currents at non-perpendicular incidence of eddy currents on wafer edges. The deflection of eddy current trajectories at wafer edges into sensor regions of higher

[4] Specified errors are related to the reproducibility of the experiment, which was determined by two factors: First, the experiment was repeated for four sets of circular wafer pieces cut out of one wafer in order to ensure a negligible impact of any minor lateral variation of dopant concentration. Second, the measurement precision was limited by the positioning of the small circular wafer piece.

sensitivity is a likely cause of the observed decline. However, considering such effects would much complicate the calculation of the geometry factor g_j.

Given that the radial symmetry of the local sensitivity distribution of QSSPC could be demonstrated here, the radial symmetry of the experimental design by Kiliani et al. [7] is deemed most appropriate for an accurate determination of $\mathcal{S}_R(r)$. In such a straightforward measurement design (adapted to the instrument's symmetry), no geometry factor has to be taken into account, and the deflection of eddy current trajectories at wafer edges is of negligible impact. Therefore, a most accurate test of $\mathcal{S}_R(r)$ would be an analysis of net dark conductivity of a set of (more than two) homogeneously doped and concentrically positioned circular test wafers with different diameters.

References

1. R. A. Sinton, A. Cuevas, M. Stuckings, Quasi-steady-state photoconductance, a new method for solar cell material and device characterization, in *Proceedings of the 25th IEEE PVSC* (Washington, D.C., 1996), pp. 457–460
2. R. A. Sinton, A. Cuevas, Contactless determination of current–voltage characteristics and minority-carrier lifetimes in semiconductors from quasi-steady- state photoconductance data. Appl. Phys. Lett. **69**, 2510–2512 (1996)
3. F. Dannhäuser, Die Abhängigkeit der Trägerbeweglichkeit in Silicium von der Konzentration der freien Ladungsträger—I. Solid-State Electron. **15**, 1371–1375 (1972)
4. J. Krausse, Die Abhängigkeit der Trägerbeweglichkeit in Silicium von der Konzentration der freien Ladungsträger—II. Solid-State Electron. **15**, 1377–1381 (1972)
5. S. Herlufsen, J. Schmidt, D. Hinken, K. Bothe, R. Brendel, Photoconductance-calibrated photoluminescence lifetime imaging of crystalline silicon. Phys. Stat. Sol. (RRL) **2**, 245–247 (2008)
6. J. A. Giesecke, M. C. Schubert, B. Michl, F. Schindler, W. Warta, Minority carrier lifetime imaging of silicon wafers calibrated by quasi-steady-state photoluminescence. Solar Energy Mater. Solar Cells **95**, 1011–1018 (2011)
7. D. Kiliani, G. Micard, B. Steuer, B. Raabe, A. Herguth, G. Hahn, Minority charge carrier lifetime mapping of crystalline silicon wafers by time-resolved photoluminescence imaging. J. Appl. Phys. **110**, 054508 (2011)

Appendix C
Comparative Analysis of QSSPC and Modulated Luminescence

In Sect. 8.3, systematic deviations between effective carrier lifetime calibrations of photoluminescence images via QSSPC and quasi-steady-state harmonically modulated luminescence were encountered. Several potential causes of a mismatch between the calibration methods were addressed therein, including the impact of partial compensation of dopant species on carrier mobility, the resistive impact of crystal defects, and lateral optical blurring of luminescence images. These and other possible causes are investigated in a statistical analysis of effective lifetime calibrations conducted on a specifically prepared $\sim 1\,\Omega\,\text{cm}$ batch of p-type mono- and multicrystalline silicon wafers.

C.1 Aim and Definition of Modus Operandi

The aim of the present comparative analysis is to identify correlations between a mismatch between the compared lifetime calibration techniques and specific characteristics of the investigated wafers, and to draw conclusions thereof. Such characteristics include

- the lifetime level,
- interface morphology,
- the compensation ratio,
- the occurence of locally enhanced recombination,
- the multicrystalline grain structure (i.e. the configuration of grain boundaries)
- and the optical blurring of lifetime information in luminescence images.

The investigated wafer batch contained four different classes of silicon wafer materials, comprising 4 Czochralski-grown monocrystalline 4″ wafers, another 4 monocrystalline float zone wafers (6″ pseudo square), 22 gas-phase-cleaned multicrystalline wafers ($125 \times 125\,\text{mm}^2$) from three different bricks of two directionally solidified ingots, and 25 upgraded metallurgical grade (umg) multicrystalline wafers ($125 \times 125\,\text{mm}^2$) from five different bricks of two directionally solidified ingots.

J. Giesecke, *Quantitative Recombination and Transport Properties in Silicon from Dynamic Luminescence*, Springer Theses, DOI: 10.1007/978-3-319-06157-3,

© Springer International Publishing Switzerland 2014

Multicrystalline wafers were taken from a variety of ingot heights from bottom to top. Wafer interface morphology was varied via different texturing techniques, comprising a microscopically planar physico-chemical polish, an in-house chemical polish (featuring substantial interface roughness on a micron scale for multicrystalline wafers, which is highly dependent on crystal orientation with respect to the surface normal), and an efficient acidic isotropic texture etched upon as-cut wafer interfaces [1], yielding the roughest interfaces investigated here. All multicrystalline wafers were exposed to a phosphorous gettering treatment followed by a subsequent PSG and emitter etch-off. Wafer interfaces were passivated via several silicon nitride and aluminum oxide passivation schemes available at Fraunhofer ISE.

C.1.1 An Adequate Comparative Measure

Harmonically modulated carrier lifetime measurements are not generally comparable to QSSPC lifetime measurements, even if conducted on identical wafer spots and at identical excess carrier generation rates. This is due to differing local sensitivities and differing averaging procedures in the presence of a laterally inhomogeneous carrier lifetime distribution (cf. Sect. 8.3), but deviations could also be caused by differing depth profiles of excess carrier density and by a different weighting of excess carrier decay modes in the case of high surface recombination velocities (cf. Sect. 6.1.2). Therefore, in order to be able to compare harmonically modulated effective lifetime with QSSPC effective lifetime, sufficiently low surface recombination velocities are to be ensured, and adequate averaging procedures to access spatially resolved lifetime are to be applied.

Both, the spatially resolved lifetime calibration procedure introduced in Sect. 8.3 and the quasi-arithmetic averaging[5] of QSSPC given in Appendix B are capable of determining a scaling factor c_φ, which relates measured local luminescence intensity φ_i to local steady-state effective lifetime $\tau_{eff,i}$ via Eq. 8.15. The latter scaling factor is deemed a most appropriate comparative measure for the two investigated measurement techniques, as both techniques claim to access the same quantity—steady-state effective local lifetime $\tau_{eff,i}$—on surface-passivated wafers. Therefore, the pivotal quantity to be investigated is the ratio $\frac{c_{\varphi,\mathrm{PC}}}{c_{\varphi,\mathrm{PL}}}$, with indices PC and PL denoting (quasi-steady-state) photoconductance and (quasi-steady-state harmonically modulated) photoluminescence calibrations, respectively.

C.1.2 Experimental Procedure

Steady-state photoluminescence images of all 55 wafers were conducted at an irradiance equivalent to one sun (cf. Appendix A.6). Five calibration spots were defined

[5] Arithmetic averaging except for the consideration of local detector sensitivity $\mathcal{S}_R(r)$ as of Eq. B.3.

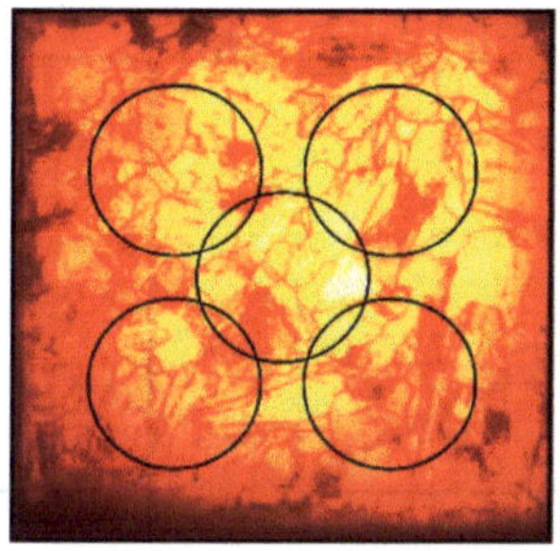

Fig. C.1 Depiction of the designated calibration positions of photoluminescence images in the comparative analysis described herein. The depicted wafer features a sidelength of $12.5 \times 12.5\,\mathrm{cm}^2$

on each wafer, located in the wafer center and (if possible) at a displacement of Δx, $\Delta y = \pm 25$ mm in both lateral directions, as indicated in Fig. C.1.

For best comparability, the circular aperture with a diameter of 40 mm was used for the harmonically modulated calibration, thereby covering the same active measurement area as the designated QSSPC *sensor region*. The total number of calibration spots amounted to 250.

For the photoluminescence analysis, net dopant concentrations as obtained via dark conductivity measurements of QSSPC were adopted. As for wafer reflectivity, reflection spectra were measured with a spectrophotometer in the wavelength range of 300 – 1,200 nm. These spectra were used for a most accurate approximation of the optical constant required for QSSPC measurements (cf. Appendix C.2). For the photoluminescence analysis, the reflectance at the laser wavelength of 880 nm was used in order to accurately determine excess carrier generation.[6]

In order to avoid systematic errors due to light induced lifetime changes, all wafers (except the float zone wafers) were exposed to a Xe light spectrum with an excess carrier generation capacity equivalent to 1/3 sun for 36 hours.

The fact that lifetime calibrations on wafers taken from different block heights of directionally solidified silicon ingots were conducted at predefined spots (cf. Fig. C.1) implies a certain redundancy, or a lack of statistical independence. Yet, this could be exploited to identify a correlation between crystal structure and the departure of $\frac{c_{\varphi,\mathrm{PC}}}{c_{\varphi,\mathrm{PL}}}$ from unity. Beyond, the conduction of up to five calibrations per wafer allows to analyze the characteristic scattering of the calibration procedure, as all calibrations on one wafer are a priori expected to yield the same scaling factor c_{φ}.

[6] As detailed in Sect. 5.2.3, both assumptions about dopant concentration and reflectance are of very minor impact for harmonically modulated lifetime analyses under low-level injection conditions.

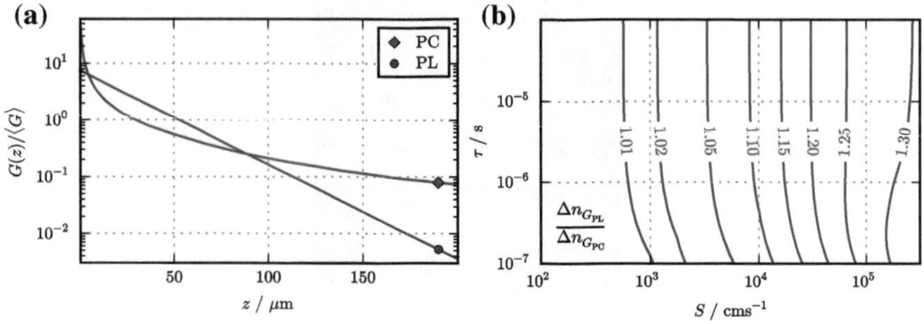

Fig. C.2 On the left, differing depth profiles of excess carrier generation rate corresponding to the Xe flash spectrum of QSSPC (PC) and to monochromatic 880 nm ($T = 300$ K) irradiation (PL) are compared. On the right, the ratio of corresponding integral excess carrier densities $\Delta n_{G_{\text{PC}}}$ and $\Delta n_{G_{\text{PL}}}$ is shown as a function of bulk lifetime τ and (symmetrical) surface recombination velocity S—indicating a negligible departure from unity for $S < 10^2\,\text{cms}^{-1}$

C.2 Preliminary Investigations

In order to ensure comparability of lifetime techniques, several preliminary investigations had to be done. For this purpose, the QSSPC instrument was recalibrated (i.e. the relation between conductance and detector voltage was determined). Further, it was checked and could be confirmed that local QSSPC irradiance was spatially uniform throughout light detector and sensor region. Detailed investigations of local sensitivity of both, QSSPC and harmonically modulated luminescence are to be found in Appendix B and in Sect. 7.1.3, respectively.

C.2.1 Potential Impact of Differing Irradiation Spectra

QSSPC irradiation is carried out via a Xe flash lamp, while the luminescence setup exploits monochromatic laser irradiation. Therefore, the depth profiles of excess carrier generation rate may substantially differ, as also depicted in Fig. C.2a.

Differing depth profiles of excess carrier generation may be reflected in different steady-state effective lifetimes at a given bulk lifetime τ and surface recombination velocity S— particularly for high S. Figure C.2b depicts the ratio of integral excess carrier densities corresponding to the two depth profiles of excess carrier generation rate given in Fig. C.2a—indicating a negligible impact for $S < 10^2\,\text{cms}^{-1}$. Conservatively estimating $S < 50\,\text{cms}^{-1}$ from effective lifetimes achieved with any of the above-mentioned surface passivation schemes allows exclusion of significant artifacts due to differing irradiation spectra.

Note that with $S < 50\,\text{cms}^{-1}$, artifacts related to a different weighting of excess carrier decay modes can be estimated from Fig. 6.3 to amount to the same order of magnitude as artifacts caused by differing irradiation spectra.

C.2.2 Modeling of the QSSPC Optical Constant

Unlike the case of monochomatic irradiation at a room temperature penetration depth $\alpha_\gamma^{-1} \ll d$, QSSPC features a broad spectrum of irradiation wavelengths, which renders modeling of an optical constant f_{PC} corresponding to the characteristics of each specific wafer indispensable. The optical constant f_{PC} is defined as the fraction of an impinging photon current density, which actually contributes to excess carrier generation. f_{PC} depends on the impinging spectrum of irradiation $\frac{dj_{\gamma,\text{PC}}}{d\varepsilon_\gamma}$, on the spectral reflectance of an investigated wafer $R_\gamma\left(\varepsilon_\gamma\right)$, on wafer thickness d, and on the spectral band-to-band absorption coefficient $\alpha_\gamma\left(\varepsilon_\gamma\right)$

$$f_{\text{PC}} = \frac{\int d\varepsilon_\gamma \left(1 - R_\gamma\right)\left(1 - e^{-d\alpha_\gamma}\right)\frac{dj_{\gamma,\text{PC}}}{d\varepsilon_\gamma}}{\int d\varepsilon_\gamma \frac{dj_{\gamma,\text{PC}}}{d\varepsilon_\gamma}}. \tag{C.1}$$

In contrast to the numerical model used for the modeling of f_{PC} here, Eq. C.1 omits internal reflections of irradiation for reasons of simplicity.

Inconsistencies in the common practice of application of the QSSPC optical constant f_{PC} gave rise to an independent modeling within the scope of the present comparative analysis.[7] Optical constants were determined for all investigated wafers according to Eq. C.1. Beyond, a test calibration of the irradiation detector as well as the mismatch between the physical definition of f_{PC} and its implementation in the QSSPC software were taken into account in the remodeled optical constants f_{PC}, thereby enabling the comparative analysis of QSSPC and harmonically modulated photoluminescence lifetime calibrations at physically identical excess carrier generation rates.

C.3 Results and Discussion

Figure C.3 shows average values of the measured mismatch $\frac{c_{\varphi,\text{PC}}}{c_{\varphi,\text{PL}}}$ over all 55 investigated wafers. A ratio $\frac{c_{\varphi,\text{PC}}}{c_{\varphi,\text{PL}}} > 1$ corresponds to smaller lifetimes obtained via QSSPC than via harmonically modulated luminescence. While deviations from unity are

[7] Historically, $f_{\text{PC}} = 1$ was associated with a non-ideal semiconductor absorber, leading to values $f_{\text{PC}} \leq 1.1$ in present look-up tables. The QSSPC measurement software properly takes this inconsistency with regard to the definition of f_{PC} into account. However, the instructions manual of the lifetime tester in use (*Sinton Instruments WCT-120*, revision as of September 2006) explicitly allows an alternative determination of f_{PC} via spectral reflectance and transmittance measurements.

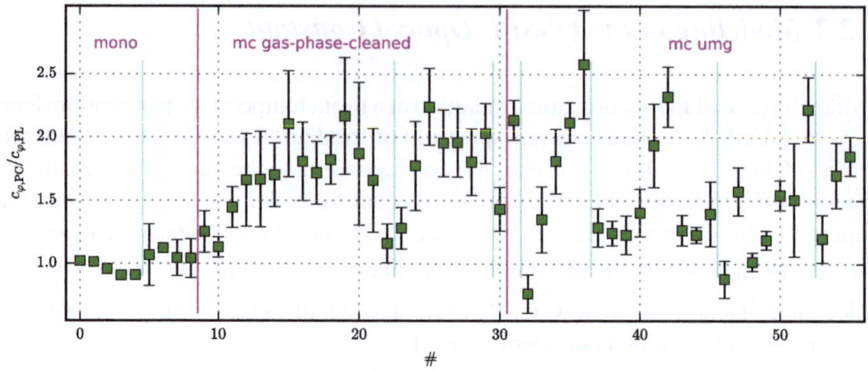

Fig. C.3 Results of the comparative analysis of steady-state effective lifetime calibrations via QSSPC and harmonically modulated photoluminescence. The depicted quantity is the ratio $\frac{c_{\varphi,PC}}{c_{\varphi,PL}}$ of scaling factors of local lifetime in photoluminescence images according to Eq. 8.15. Purple lines separate different material classes. For monocrystalline wafers, the cyan line separates Czochralski-grown wafers from float zone wafers, while for multicrystalline silicon wafers cyan lines separate wafers originating from different bricks. Within each brick, wafers are sorted in ascending order from *bottom* to *top*

minor and tolerable for all investigated monocrystalline wafers,[8] pronounced deviations from unity are to be stated for most multicrystalline wafers.

Considering the fact that the low lifetime multicrystalline wafers from the top and bottom of ingots appear to feature a much higher impact of carrier trapping for QSSPC than for harmonically modulated luminescence lifetime calibrations,[9] the resulting ratio significantly exceeds unity in any case—i.e. QSSPC lifetime is lower than harmonically modulated lifetime throughout all multicrystalline wafers, except for such wafers featuring a dominant trapping artifact in QSSPC calibrations.

Interestingly, a correlation between block height and $\frac{c_{\varphi,PC}}{c_{\varphi,PL}}$ can be stated for both gas-phase-cleaned and umg wafers. Beyond the obvious impact of the compensation ratio in umg wafers, where this correlation is very pronounced, this could also be an indication that the multicrystalline grain structure might give rise to a departure of $\frac{c_{\varphi,PC}}{c_{\varphi,PL}}$ from unity.

As for the variance of scaling factors c_φ deduced from multiple lifetime calibrations on wafers, both calibration techniques can be reported to exhibit a nearly identical overall amount of scattering, as depicted in Fig. C.4. Both calibration techniques could suffer from a lateral inhomogeneity of net dopant concentration, excess carrier generation rate, interface morphology (i.e. roughness), and from optical artifacts such as image vignetting. Beyond, QSSPC is empirically found to suffer more

[8] The specified error in Fig. C.3 relates to the scattering of c_φ upon multiple lifetime calibrations on one wafer. Therefore, if only one calibration per wafer was performed, the indicated error is null. In fact however, it is in any case expected to be finite, and one may expect $\frac{c_{\varphi,PC}}{c_{\varphi,PL}}$ to not significantly differ from unity for all monocrystalline wafers, given the measurement precision.

[9] This relates to wafers 9, 10, 22, 23, 32, 46, 48, and 53.

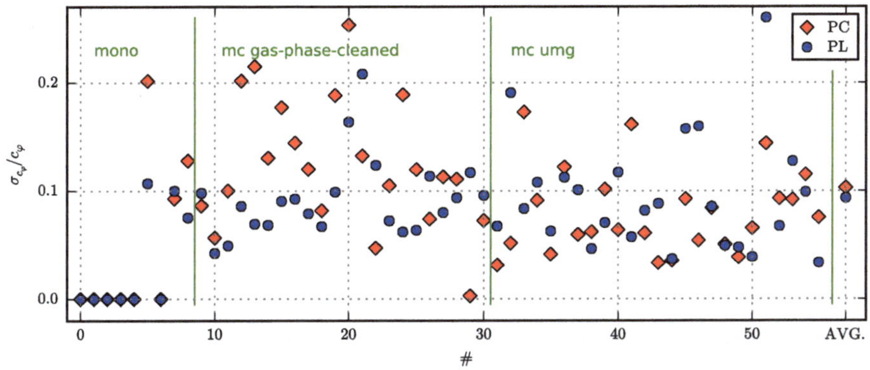

Fig. C.4 Depiction of the relative standard deviation of scaling factors $\sigma_{c_\varphi}/c_\varphi$ obtained via QSSPC (PC) or via quasi-steady-state harmonically modulated photoluminescence (PL). $\sigma_{c_\varphi}/c_\varphi = 0$ implies that only one lifetime calibration per wafer was performed

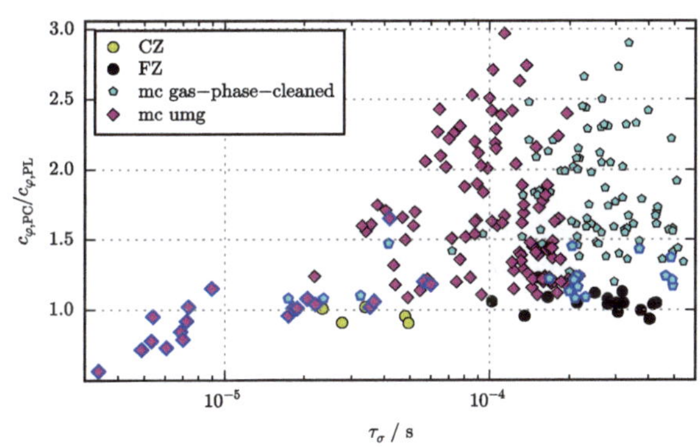

Fig. C.5 Investigation of the correlation between self-consistent quasi-steady-state harmonically modulated carrier lifetime τ_σ and the comparative measure $\frac{c_{\varphi,\mathrm{PC}}}{c_{\varphi,\mathrm{PL}}}$. Bold blue symbol edges are to indicate wafers which are possibly prone to a carrier trapping artifact

severely from carrier trapping, and it could be affected by some errors related to eddy currents (to be addressed later), while the spatially resolved lifetime calibration via harmonically modulated luminescence will be demonstrated to be particularly affected by optical blurring of luminescence images.

In the following, the potential correlations as listed in Appendix C.1 are consecutively addressed.

C.3.1 Correlation with Self-Consistent Lifetime

In order to investigate the correlation between the comparative measure $\frac{c_{\varphi,\mathrm{PC}}}{c_{\varphi,\mathrm{PL}}}$ and carrier lifetime, scaling factor ratios of all 250 lifetime calibrations were plotted over the self-consistent lifetime τ_σ resulting from the harmonically modulated photoluminescence lifetime calibration, as shown in Fig. C.5.

The data shown in Fig. C.5 allow three major conclusions:

1. Carrier trapping appears to affect QSSPC more severely than harmonically modulated luminescence, notwithstanding the fact that both methods are potentially prone to trapping, as detailed in Sect. 3.1.6. A likely cause of this empirical finding could be the fact that the condition for trapping to affect lifetime measurements (cf. Eq. 3.53) is more rigorous for harmonically modulated luminescence than for QSSPC. Illustrations of the impact of trapping on lifetime measurements in Figs. 3.4, 3.6, and 3.7 visualize this.
2. A broad band of scaling factor ratios $\frac{c_{\varphi,\mathrm{PC}}}{c_{\varphi,\mathrm{PL}}}$ near unity over a lifetime range of 20–500 μs can be observed. It is essentially sustained by monocrystalline wafers, but it also comprises a few multicrystalline wafers which are likely to feature compensating lifetime errors.
3. A quasi-correlation-free set of data points is encountered in the range between $30-600\,\mu\mathrm{s}$.

Apart from the strong empirical finding of a more pronounced carrier trapping artifact in QSSPC than in harmonically modulated luminescence calibrations, the given data show no significant correlation between the carrier lifetime level and the comparative measure $\frac{c_{\varphi,\mathrm{PC}}}{c_{\varphi,\mathrm{PL}}}$.

C.3.2 Correlation with Interface Morphology

As mentioned in Appendix C.1, surface morphology was deliberately varied in order to investigate whether there was a correlation with the scaling factor ratio $\frac{c_{\varphi,\mathrm{PC}}}{c_{\varphi,\mathrm{PL}}}$. Wafers were provided with a microscopically planar (physico-chemical) polish, a chemical polish exhibiting some roughness in the case of multicrystalline samples, or an acidic isotropic texture, which yielded the roughest (most Lambertian[10]) interfaces throughout. Wafers 3, 4, 10, 11, 17, 18, 22, 27, 28, 33, 35, 38, 41, 43, 44, 46, and 52 were provided with that isotropic texture. No substantial correlation between surface roughness and the comparative measure $\frac{c_{\varphi,\mathrm{PC}}}{c_{\varphi,\mathrm{PL}}}$ could be found at this stage. This is most reliably substantiated by the multicrystalline umg wafers 38, 43, and 44, none of which are prone to a carrier trapping artifact. Neighboring wafers 37, 39, 40, and 45 were all provided with a chemical polish, and they show no systematic difference in $\frac{c_{\varphi,\mathrm{PC}}}{c_{\varphi,\mathrm{PL}}}$, as can be seen in Fig. C.3. However, a more subtle analysis of the impact of lateral blurring of local luminescence intensity (shown below) reveals a minor—albeit not significant—trend.

[10] A simplified one-parameter *Phong* model of surface roughness [2] is sketched in Appendix E, albeit for a slightly different surface texture.

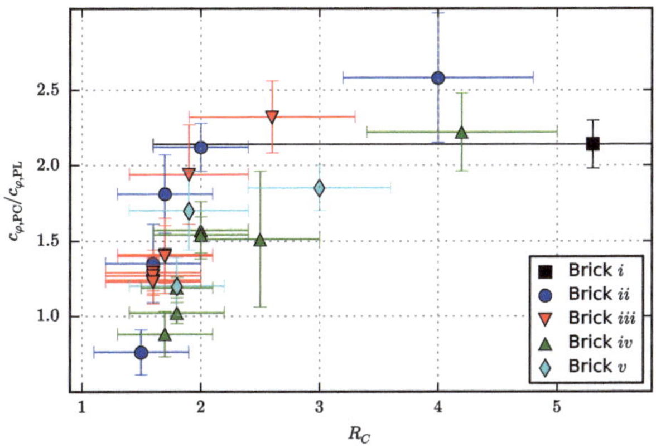

Fig. C.6 Investigation of the correlation between the compensation ratio R_C of the upgraded metallurgical grade silicon wafers of this study and the comparative measure $\frac{c_{\varphi,PC}}{c_{\varphi,PL}}$. The distinction of different bricks of wafer origin reveals a very distinct correlation

C.3.3 Correlation with the Compensation Ratio

Upgraded metallurgical grade silicon commonly features both significant acceptor (N_A) and donor (N_D) concentrations. With ionized or net dopant concentration $n_0 = |N_A - N_D|$, a so-called compensation ratio R_C is defined as

$$R_C = \frac{N_A + N_D}{n_0}. \tag{C.2}$$

QSSPC is sensitive to the product of carrier mobilities and free carrier densities. Therefore, QSSPC lifetime largely depends on assumptions about carrier mobility, to be inferred from a dark conductivity measurement prior to any QSSPC lifetime measurement according to a one-species carrier mobility model [3, 4]. As addressed in Sect. 9.1.1, such interpretation of carrier mobility leads to a systematic overestimation in the presence of a significant amount of complementary dopant species, which again brings about an underestimation of true effective lifetime in QSSPC measurements by the same mismatch factor, and consequently leads to a scaling factor ratio $\frac{c_{\varphi,PC}}{c_{\varphi,PL}} > 1$.

Figure C.6 depicts the scaling factor ratio $\frac{c_{\varphi,PC}}{c_{\varphi,PL}}$ over the compensation ratio R_C of all umg wafers of the present study.[11] A strong correlation between R_C and $\frac{c_{\varphi,PC}}{c_{\varphi,PL}}$ is encountered. The scattering of data is virtually eliminated upon separate investigation of each brick. This is another indication that the characteristic grain structure of the directionally solidified ingots might be a relevant factor for departure of $\frac{c_{\varphi,PC}}{c_{\varphi,PL}}$ from

[11] Acceptor concentrations were determined via measurements of Fe-acceptor repairing according to Macdonald et al. [5].

unity. Any of the wafers shown in Fig. C.6 with $\frac{c_{\varphi,\mathrm{PC}}}{c_{\varphi,\mathrm{PL}}} < 1$ are likely to suffer from a pronounced carrier trapping artifact of QSSPC, which compensates the mobility-induced error.

C.3.4 Eddy Currents and Locally Enhanced Recombination

In order to investigate whether the presence of confined spots of enhanced recombination within the QSSPC active measurement area could impact on Δn and thereby on steady-state effective lifetime as an enhanced excess carrier sink due to lateral eddy currents, the following experiment was conducted: The QSSPC sensor region (*Sinton Instruments WCT-120* measurement stage) was mounted in a photoluminescence imaging setup, and steady-state photoluminescence images of several multicrystalline wafers featuring pronounced local recombination centers were acquired under identical irradiation conditions with the eddy-current driving radio frequency field of the QSSPC inductor switched on and off, respectively. If local photoluminescence intensity (and thereby excess carrier density and effective lifetime) in the vicinity of recombination centers within the QSSPC sensor region was measurably lowered upon switching on the driving radio frequency field, this would be an indication of an enhanced excess carrier sink induced by eddy currents. However, despite the very high sensitivity of the luminescence imaging setup, no such effect could be observed. Therefore, the hypothesis of enhanced recombination due to eddy currents is to be discarded. This result is consistent with the observation of $\frac{c_{\varphi,\mathrm{PC}}}{c_{\varphi,\mathrm{PL}}} \approx 1$ throughout all monocrystalline wafers investigated, regardless whether of laterally homogeneous electronic material quality or not.

C.3.5 Correlation with Multicrystalline Grain Structure

There are several indications suggesting that the grain structure of multicrystalline material (especially certain grain boundaries perpendicular with respect to eddy current paths) may artificially enhance wafer resistivity and thereby lower eddy currents and QSSPC effective lifetime. The dominant causal mechanism for an electrical potential barrier at grain boundaries found in literature is dopant segregation in grain boundaries, leading to a depletion region and a corresponding band bending, which impedes majority carrier transport [6, 7, 8, 9, 10, 11, 12, 13].[12] Recent findings by Spitz and Rein [15] suggest a tremendous resistive impact of grain boundaries upon damage etching (particularly in combination with air drying). This effect is expected

[12] Further complications may arise from the impact of excess carrier density (e.g. induced by illumination or electrical injection) on such potential barriers [14], because such an effect might give rise to substantial inaccuracy of light-induced excess photoconductance.

Table C.1 Results of a QSSPC (*Sinton Instruments WCT-120*) dark conductance analysis involving four 1 Ω cm wafers of identical thickness and with a square area of 50 × 50 mm^2. Unsurmountable electrical potential barriers perpendicular to the eddy current field were simulated through physical wafer edges, which were concentrically aligned such that they covered the entire sensor region

Number of potential barriers	Sensor dark voltage/ mV	Attenuation
0	908 ± 3	1.00
2	368 ± 2	0.41
4	234	0.26

to also affect QSSPC, as lateral eddy currents imposed through a radio frequency inductor are attenuated through electrical potential barriers at grain boundaries.

In the present comparative analysis, no direct proof of an attenuation of eddy currents by grain boundaries could be found. However, several experimental indications suggest such an effect. First, there is an apparent disparity between $\frac{c_{\varphi,\mathrm{PC}}}{c_{\varphi,\mathrm{PL}}} \approx 1$ encountered in monocrystalline material[13] and values of $\frac{c_{\varphi,\mathrm{PC}}}{c_{\varphi,\mathrm{PL}}}$ substantially in excess of unity even in non-compensated (gas-phase-cleaned) multicrystalline material. Another indication can be found in Fig. C.6. Here, the correlation between the compensation ratio R_C of the investigated umg wafers and the measure $\frac{c_{\varphi,\mathrm{PC}}}{c_{\varphi,\mathrm{PL}}}$ is substantially sharpened when accounting for differences in crystal structure and grain boundary configuration by grouping of wafers in blocks of very similar grain boundary configuration.

In order to demonstrate the maximal impact of electrical potential barriers induced by grain boundaries, a systematic analysis of dark photocurrents was conducted on float zone wafers of uniform thickness and dopant concentration. Four 1 Ω cm square wafers with a size of 50 × 50 mm^2 were used to experimentally simulate unsurmountable electrical potential barriers perpendicular to the direction of eddy currents. First, dark conductivity was measured on all four wafers concentrically positioned on the QSSPC sensor region. The corresponding proportional sensor voltage is provided in Table C.1. Subsequently, two wafers were positioned such that their edges were perfectly aligned on the center of the QSSPC sensor region, thereby simulating two unsurmountable electrical potential barriers perpendicular to the eddy current field. Although the complete sensor region was covered by the exact same wafer volume as before, the resulting sensor voltage dropped to a fraction of 0.41 with respect to the case without any potential barrier.[14] Last, all four wafers were positioned such that the position of one of their corners coincided in the exact center of the sensor region, thereby simulating four unsurmountable electrical potential barriers perpendicular to the eddy current field. The resulting sensor voltage dropped to a fraction of 0.26 of the unperturbed system.

[13] A minor (non-significant) trend toward $\frac{c_{\varphi,\mathrm{PC}}}{c_{\varphi,\mathrm{PL}}} > 1$ occurrng exclusively in laterally inhomogeneous monocrystalline material remains to be discussed.

[14] The specified uncertainty in Table C.1 relates to the variation of sensor voltage upon exchange of wafers and wafer edges within the sensor region.

C.3.6 Correlation with Optical Blurring of Luminescence Images

Harmonically modulated (e.g. self-consistent) lifetime was shown to perform an intensity weighted lifetime average rather than an arithmetical lifetime average in the case of a laterally inhomogeneous carrier lifetime distribution within the active measurement area (cf. Sect. 8.3.2). This fact may bring about systematic errors of an effective lifetime calibration of photoluminescence images via harmonically modulated photoluminescence, if photoluminescence images feature lateral optical blurring. The possible impact of optical blurring of lifetime information on the non-arithmetical calibration procedure (represented by Eq. 8.24) is detailed in Sect. 8.3.3.

Optical blurring of spatially resolved lifetime information [16, 17] could explain the weak and non-significant trend toward $\frac{c_{\varphi,\mathrm{PC}}}{c_{\varphi,\mathrm{PL}}} \gtrsim 1$ of the monocrystalline wafers 5–8 of Fig. C.3, which feature some lateral variation of effective lifetime.

In order to evaluate the impact of lateral optical blurring of luminescence images on a harmonically modulated lifetime calibration, 20 spots on 11 wafers with most pronounced deviations of $\frac{c_{\varphi,\mathrm{PC}}}{c_{\varphi,\mathrm{PL}}}$ from unity were subjected to a follow-up investigation. For this purpose, self-consistent quasi-steady-state harmonically modulated lifetime measurements τ_σ were conducted on small but electronically homogeneous spots of $1 \times 1\,\mathrm{cm}^2$ and evaluated at an irradiance equivalent to one sun. The locally measured lifetimes $\tau_{\sigma,i}$ were then compared with local effective lifetime $\tau_{\mathit{eff},i}$ as obtained from spatially resolved lifetime calibrations via both, QSSPC and harmonically modulated photoluminescence. The resulting ratios $\frac{\tau_{\mathit{eff},i}}{\tau_{\sigma,i}}$ are shown in Fig. C.7. An intensity weighted average of all $\tau_{\mathit{eff},i}$ within the designated measurement area of $\tau_{\sigma,i}$ was calculated in order for a most accurate analysis even in the presence of very minor lateral variations of local lifetime (cf. Eq. 8.25).

Here again, a local scattering of $\frac{\tau_{\mathit{eff},i}}{\tau_{\sigma,i}}$ could be due to lateral variations of dopant concentration, irradiance, or surface morphology. Figure C.7 allows the conclusion that lateral optical blurring of luminescence images could cause significant error, but it is unlikely to be the primary cause to deviations between QSSPC and harmonically modulated luminesence lifetime calibrations. Note that the fact that the average value of $\frac{\tau_{\mathit{eff},i}}{\tau_{\sigma,i}}$ coincides with unity for the modulated photoluminescence calibration, whereas it substantially deviates from unity for the QSSPC calibration, implies no direct proof of the *superiority* of the first. Rather, it allows to rule out an optical blurring artifact in the modulated photoluminescence calibration as the primary cause to $\frac{c_{\varphi,\mathrm{PC}}}{c_{\varphi,\mathrm{PL}}} \neq 1$.

It was deemed worthwhile to further investigate the correlation between optical blurring and the ratio $\frac{\tau_{\mathit{eff},i}}{\tau_{\sigma,i}}$. For this purpose, an adequate one-parameter measure of the proneness of each laterally inhomogeneous calibration spot to an optical blurring artifact was to be identified at first. For this purpose, the set of (pixel wise) local luminescence intensities φ_i contributing to the lifetime calibration measurement were sorted in a histogram $\varphi_k \mapsto p_k$, assigning a relative frequency p_k to luminescence intensities in a defined interval around φ_k. Such histograms $\varphi_k \mapsto p_k$ corresponding to depicted calibration areas are shown in Fig. C.8.

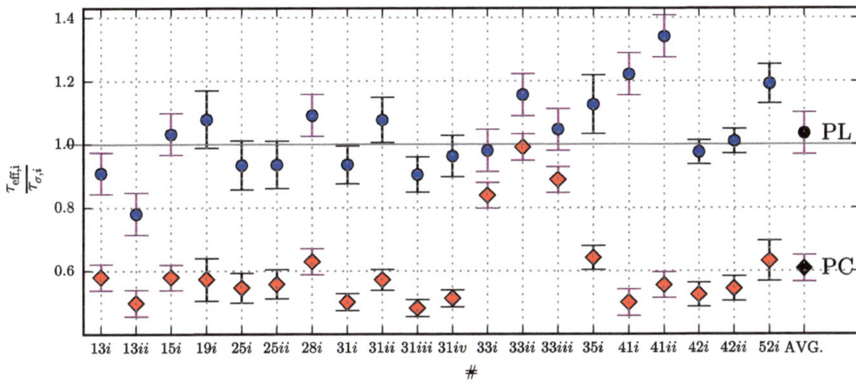

Fig. C.7 Comparison of local effective lifetimes $\tau_{eff,i}$ as obtained with lifetime calibrations via both, QSSPC (PC, rhombs) and quasi-steady-state harmonically modulated photoluminescence (PL, circles) to local self-consistent lifetime measurements on 20 1×1 mm^2 electronically homogeneous spots of 11 wafers featuring most pronounced deviations of $\frac{c_{\varphi,\text{PC}}}{c_{\varphi,\text{PL}}}$ from unity. Wafer numbers accord with Fig. C.3, and measurement spots are denoted with Roman numbers, respectively. Black error bars indicate the variance due to multiple calibrations. Purple errorbars represent an average variance in the case of only one performed lifetime calibration per wafer

Calibration spots are prone to optical blurring, if the characteristic dimension of areas of relatively high lifetime is on the order or less than the characteristic dimension of optical blurring. Therefore, an adequate blurring measure would have to parametrize the characteristic dimension of areas of relatively high lifetime, whose contribution to harmonically modulated lifetime τ_σ is most pronounced. The essential distinction between the distributions of photoluminescence intensities as shown in Fig. C.8 is that the distribution weakly affected by optical blurring (left) is rather uniform, whereas the distribution heavily affected by optical blurring (right) features a very pronounced maximum toward relatively low photoluminescence intensities φ_k and very low relative frequencies p_k at relatively high photoluminescence intensities φ_k. In the latter case, the probability to find two locations (pixels) of relatively high φ_k in the vicinity of each other is accordingly low, and consequently the characteristic dimension of areas of relatively high lifetime can be judged to be rather small from the histogram. Very similar distributions of photoluminescence intensities were encountered on all other calibration spots of interest, depending on their characteristic dimension of areas of relatively high photoluminescence intensity. Therefore, one possible heuristic one-parameter measure for the proneness of a calibration spot to optical blurring is[15]

$$\mathcal{L} = \sum_k p_k^2. \tag{C.1}$$

[15] In miscellaneous interdisciplinary applications of statistics, this measure—otherwise known e.g. as a so-called Herfindahl-Hirschmann index—is used to quantify concentration and deviations from uniformity of statistical distributions.

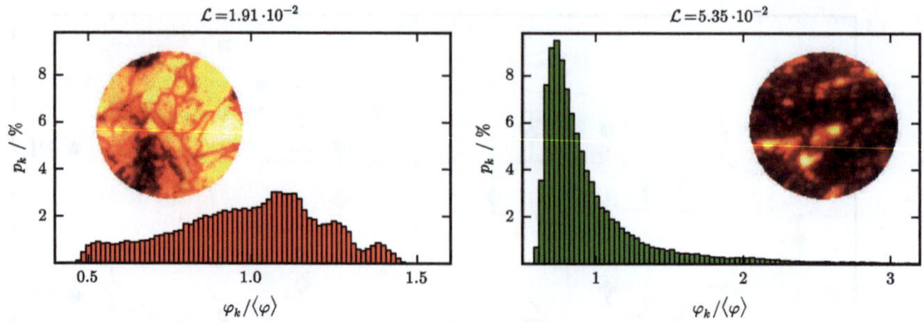

Fig. C.8 Depiction of histograms of the relative frequency p_k of local photoluminescence intensities φ_k corresponding to the lifetime calibration spots shown in the inserted maps. A calibration spot rather weakly prone to optical blurring on the left (featuring a relatively low blurring measure of $\mathcal{L} = 1.91 \times 10^{-2}$) is compared to a calibration spot much prone to optical blurring on the right (featuring a relatively high blurring measure $\mathcal{L} = 5.35 \times 10^{-2}$)

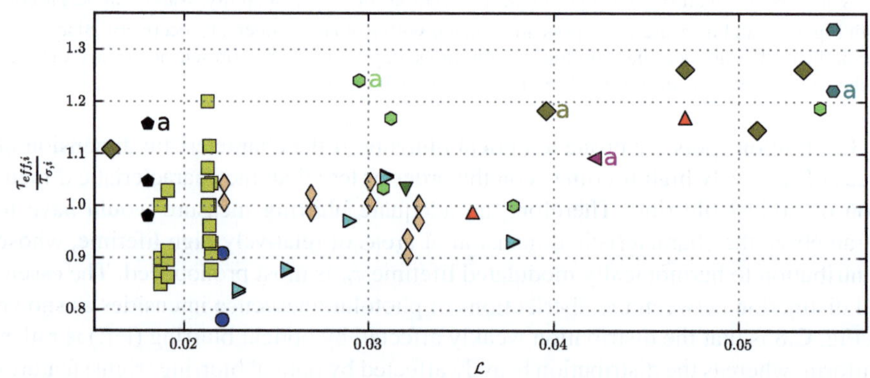

Fig. C.9 Investigation of the correlation between the blurring measure $\mathcal{L}$ as defined in Eq. C.1 and the ratio $\frac{\tau_{eff,i}}{\tau_{\sigma,i}}$ between a harmonically modulated photoluminescence lifetime calibration $\tau_{eff,i}$ and self-consistent lifetime test measurements $\tau_{\sigma,i}$ on designated (laterally homogeneous) spots (cf. Fig. C.7). Each combination of a lifetime calibration spot and a self-consistent test measurement $\tau_{\sigma,i}$ is represented by one data point here. Different symbols and colors denote different wafers. Wafers featuring an acidic isotropic texture are labeled with an acronym **a**, respectively (only one label per wafer). Other wafers are chemically polished

Figure C.9 shows the correlation between the blurring measure $\mathcal{L}$ and the ratio $\frac{\tau_{eff,i}}{\tau_{\sigma,i}}$ for a harmonically modulated photoluminescence lifetime calibration. Indeed, in spite of a significant scattering, which could also be interpreted as an indication of some inaccuracy of the blurring measure $\mathcal{L}$, a significant correlation is yet encountered, particularly when focusing on wafers featuring a relatively high blurring measure $\mathcal{L}$.

As for the question whether surface texture enhances blurring artifacts in modulated luminescence lifetime calibrations, isotropically textured wafers were labeled in Fig. C.9. Although a minor trend toward an enhancement of the mismatch $\frac{\tau_{eff,i}}{\tau_{\sigma,i}}$ due to interface roughness is corroborated by the data shown, interface texture appears not to be necessary for a mismatch to occur. A greater number of test wafers would be required to draw robust conclusions in this respect.

C.4 Interim Conclusions

QSSPC and quasi-steady-state harmonically modulated photoluminescence effective lifetime calibrations of photoluminescence images agree very well in the case of mono-crystalline and non-compensated material featuring a laterally homogeneous lifetime distribution and negligible carrier trapping.

A very pronounced correlation was found between the compensation ratio R_C of partially compensated wafers and the comparative measure, being the scaling factor ratio $\frac{c_{\varphi,\mathrm{PC}}}{c_{\varphi,\mathrm{PL}}}$. This correlation is due to a substantial underestimation of true effective lifetime by QSSPC in the presence of compensating dopant species.

Several independent experimental indications suggest an impact of grain structure (and particularly of the configuration of grain boundaries) on QSSPC lifetime such that some grain boundaries substantially attenuate eddy currents. This finding calls for further systematic validation on the basis of different batches,[16] and for an appropriate experimental design to unambiguously prove or disprove this hypothesis and to quantify the impact of electrical potential barriers induced by some grain boundaries. Note that a direct and unambiguous proof of the hypothesis of grain boundaries acting as significant electrical potential barriers has not been accomplished herein.

Carrier trapping was empirically found to affect QSSPC more severely than harmonically modulated luminescence.

Lateral optical blurring in luminescence images was identified as a significant—albeit not inherently limiting—error source of a lifetime calibration via harmonically modulated luminescence, leading to a slight overestimation of true steady-state effective lifetime $\tau_{eff,i}$. Blurring could be avoided via measures elaborated elsewhere [16, 17].

References

1. R. Einhaus, E. Vazsonyi, J. Szlufcik, J. Nijs, R. Mertens, Isotropic texturing of multicrystalline silicon wafers with acidic texturing solutions, in *Proceedings of the 26th IEEE PVSC* (Anaheim, 1997), pp. 167–170
2. J.D. Foley, A. Van Dam, S. Feiner, J. Hughes, *Computer Graphics: Principles and Practice* (Addison-Wesley, Amsterdam, 1990)

[16] Most prominently, a broad variation of gas-phase-cleaned and non-compensated multicrystalline wafer material is required in order to rule out artifacts specific to the present batch, and in order to obtain a most representative statistical ensemble. Further, it must be noted that both gas-phase-cleaned and umg multicrystalline wafers of the investigated batch were exposed to phosphorous gettering followed by a PSG and emitter etch-off. Prospective investigations should elicit whether any such processes could impact grain boundary induced potential barriers [15].

3. F. Dannhäuser, Die Abhängigkeit der Trägerbeweglichkeit in Silicium von der Konzentration der freien Ladungsträger—I. Solid-State Electron. **15**, 1371–1375 (1972)
4. J. Krausse, Die Abhängigkeit der Trägerbeweglichkeit in Silicium von der Konzentration der freien Ladungsträger—II. Solid-State Electron. **15**, 1377–1381 (1972)
5. D. Macdonald, A. Cuevas, L.J. Geerligs, Measuring dopant concentrations in compensated p-type crystalline silicon via iron-acceptor pairing. Appl. Phys. Lett. **92**, 202119 (2008)
6. T.I. Kamins, Hall mobility in chemically deposited polycrystalline silicon. J. Appl. Phys. **42**, 4357–4365 (1971)
7. J.Y.W. Seto, The electrical properties of polycrystalline silicon films. J. Appl. Phys. **46**, 5247–5254 (1975)
8. G. Baccarani, B. Ricco, G. Spadini, Transport properties of polycrystalline silicon films. J. Appl. Phys. **49**, 5565–5570 (1978)
9. G.J. Korsh, R.S. Muller, Conduction properties of lightly doped, polycrystalline silicon. Solid-State Electron. **21**, 1045–1051 (1978)
10. C.H. Seager, T.G. Castner, Zero-bias resistance of grain boundaries in neutron-transmutation-doped polycrystalline silicon. J. Appl. Phys. **49**, 3879–3889 (1978)
11. A.K. Ghosh, A. Rose, H.P. Maruska, D.J. Eustace, T. Feng, Hall measurements and grain-size effects in polycrystalline silicon. Appl. Phys. Lett. **37**, 544–546 (1980)
12. N.C.C. Lu, L. Gerzberg, C.Y. Lu, J.D. Meindl, Modeling and optimization of monolithic polycrystalline silicon resistors. IEEE Trans. Electron Dev. **28**, 818–830 (1981)
13. N.C.C. Lu, L. Gerzberg, C.Y. Lu, J.D. Meindl, A conduction model for semiconductor-grain-boundary-semiconductor barriers in polycrystalline-silicon films. IEEE Trans. Electron Dev. **30**, 137–149 (1983)
14. H.P. Maruska, A.K. Ghosh, A. Rose, T. Feng, Hall mobility of polycrystalline silicon. Appl. Phys. Lett. **36**, 381–383 (1980)
15. M. Spitz, S. Rein, Impact of potential barriers at grain boundaries of multicrystalline silicon wafers on inductively coupled resistivity measurements, in *Proceedings of the 26th EU PVSEC*, Hamburg, 2011, pp. 338–343
16. B. Mitchell, J.W. Weber, D. Walter, D. Macdonald, T. Trupke, On the method of photoluminescence spectral intensity ratio imaging of silicon bricks: advances and limitations. J. Appl. Phys. **112**, 063116 (2012)
17. D. Walter, A. Liu, E. Franklin, D. Macdonald, B. Mitchell, T. Trupke, Contrast enhancement of luminescence images via point-spread deconvolution, in *Proceedings of the 38th IEEE PVSC* (Austin, 2012), pp. 307–312

Appendix D
Absorption Coefficient Measurement

Existing data compilations of the room temperature band-to-band absorption coefficient of crystalline silicon [1, 2] are inconsistent—calling for an independent validation.[17] For this purpose, the quasi-steady-state harmonically modulated photoluminescence setup was modified in order to determine the light penetration depth of the single mode diode lasers used in this work. This measurement was conducted at a temperature of $T = 301.0 \pm 0.7$ K. The relative transmission of a set of polished p-type 1 Ω cm float zone wafers was measured as a function of accurately determined wafer thicknesses d ranging within $110\,\mu\text{m} < d < 180\,\mu\text{m}$. Laser spectra were individually measured and provided by the manufacturers, respectively. In order to ensure negligible detection of luminescent radiation, the transmission detector was positioned at a distance greater than 10 cm from the unpassivated and bilaterally planar (highly polished) monocrystalline wafers, and an additional optical short pass filter with an edge wavelength of 900 nm and a specified optical density greater than 5 was positioned between wafer and detector (cf. Fig. E.2).

Artificial (e.g. luminescent) radiation was verified to amount to $\ll 1\,\%$ of the detected transmission signal in the most unfavorable measurement configuration[18] (irradiation of the thickest wafer with a wavelength of 791 nm). The resulting band-to-band absorption coefficients are shown in Table D.1, as follows from the linear fit of logarithmized and normalized transmitted light intensity $\ln \frac{I(d)}{I(d_0)}$ shown in Fig. D.1. Here, transmitted irradiance was normalized to a wafer thickness of $d_0 = 100\,\mu\text{m}$.

[17] Parts of the account given here are adapted from the corresponding original publication [3].

[18] This was verified by tilting the detector out of the optical axis of irradiation such that nearly the same amount of wafer luminescence, but no directly transmitted (only scattered) irratiation light would enter the detector.

J. Giesecke, *Quantitative Recombination and Transport Properties in Silicon from Dynamic Luminescence*, Springer Theses, DOI: 10.1007/978-3-319-06157-3,

© Springer International Publishing Switzerland 2014

Table D.1 Band-to-band absorption coefficient α_γ of crystalline silicon at the two irradiation wavelengths used in this work. The results of the transmission measurement presented here are compared to data tabulated by Green and Keevers [1] and by Aspnes [2]

λ_γ (nm)	$\alpha_\gamma^{300\,\mathrm{K}}$ Green/Keevers (cm^{-1})	$\alpha_\gamma^{300\,\mathrm{K}}$ Aspnes (cm^{-1})	$\alpha_\gamma^{301\,\mathrm{K}}$ this work (cm^{-1})
791	920 ± 37	702	946.6 ± 16.0
880	383 ± 15	271	416.6 ± 6.7

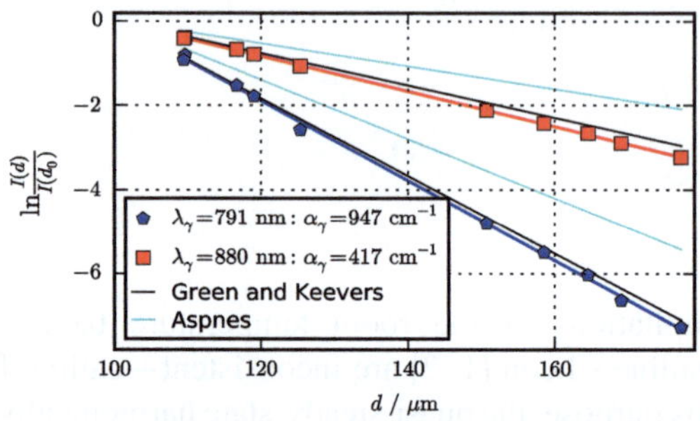

Fig. D.1 Determination of the 301 K band-to-band absorption coefficient of crystalline silicon (p-type 1 Ω cm) at the specified wavelengths via measurement of the transmitted laser intensity I as a function of accurately determined wafer thickness d. The measurement result is compared to 300 K absorption data on intrinsic silicon tabulated by Green and Keevers [1] and by Aspnes [2]. Adapted after [3]

D.1 Discussion

In order to be able to evaluate the deviation of the present absorption data with respect to the most frequently used data by Green and Keevers [1], an elaborate error analysis based on thorough estimates of uncertainties of both thickness d and transmitted intensity I was conducted.

As for thickness measurements, a gauge was used to determine wafer thickness d both at the corners and in the center of each $1 \times 1\,\mathrm{cm}^2$ square measurement spot. The lateral accuracy of the positioning of measurement points (with respect to the exact measurement position) was estimated to amount to 1.5 mm. The best value of d for each measurement spot (average thickness) was obtained via triangulation. Departure from the exact wafer interface topography between the locations of thickness measurements may give rise to an uncertainty of thickness δd, which was approximated via the standard error of the measured 5 thickness values of each spot, respectively.

As for the measurement of transmitted intensity, the encountered precision was found to be very high—yielding a negligible uncertainty compared with a possible error due to the uncertainty of temperature. Temperature was measured via platinum resistance temperature detectors (Pt100). Several sources of temperature uncertainty δT could be identified:

1. **Ex-Situ Measurement:** The temperature detector was in contact with the metal aperture physically connected to the illuminated silicon wafer. Even though the detector was positioned as closely as possible to the wafer, a comparison of in-situ illuminated wafer temperature and ex-situ measurement in the described measurement position revealed a difference of 0.8 K upon irradiation equivalent to 1 sun, which had to be taken into account. This difference proved to be quite robust against light-induced variations of wafer temperature. A remaining uncertainty accounting for the slightly varying detector position (within $\sim$ 1cm^2) and dynamic effects of finite heat dissipation was very conservatively approximated with an uncertainty of $\delta T \leq 0.5K$.
2. **Inherent Uncertainty of Detectors:** Different (Pt100) detectors were found to yield a systematic temperature variation of $\delta T \leq 0.2$ K, which is likely to be attributed to the individual four point contact geometry, but could originate from an inherent uncertainty of detector calibrations as well. It is likely that this source of error could be reduced in the future, but here it must be taken into account.

The estimated uncertainty of temperature was used to calculate a resulting uncertainty of transmitted irradiance $\ln I$. Based on a simplified model of the temperature dependence of the absorption coefficient

$$\alpha_\gamma(T) = \alpha_\gamma(T_0)\, e^{\frac{T-T_0}{\chi_T}}, \tag{D.1}$$

the parameter χ_T was approximated for both laser wavelengths of interest from absorption data $\alpha_\gamma|_T$ as published by Green [4]. For temperatures $T_0 = 300$ K and $T_1 = 363$ K, the encountered parameters were $\chi_T\,(791\,\text{nm}) = 218$ K and $\chi_T\,(880\,\text{nm}) = 248$ K. The temperature uncertainty of $\delta T = 0.7$ K which relates to an uncertainty of $\delta\alpha_\gamma$ according to Eq. D.1 gives rise to an uncertainty of the slope of logarithmized transmitted irradiance $\ln I(d)$. This uncertainty of slope was accounted for by attributing an uncertainty $\delta \ln I(d) = |d - \langle d\rangle|\delta\alpha_\gamma \ln I(d)$ to each measured transmitted irradiance $\ln I(d)$. Measurement results d and $\ln I(d)$ and their estimated uncertainties δd and $\delta \ln I(d)$ were subsequently used for a random generation of normally distributed sets of $\ln I_j^k\left(d_j^k\right)$ [19]—thereby robustly and reproducibly yielding the specified uncertainties of the measured absorption coefficient (cf. Table D.1) upon multiple repetition.

Transferring these results from the best estimate of measured temperature of 301.0 K to the measurement temperature of the data tabulated by Green and Keevers of 300.0 K via Eq. D.1, one obtains $\alpha_\gamma^{300\,\text{K}}\,(791\,\text{nm}) = 942.3 \pm 16.0\,\text{cm}^{-1}$ and $\alpha_\gamma^{300\,\text{K}}\,(880\,\text{nm}) = 414.9 \pm 6.7\,\text{cm}^{-1}$.

The encountered absorption coefficients of this study are greater than the ones tabulated by Green and Keevers. However, with regard to the 4 % uncertainty of α_γ as stated by Green and Keevers (in the wavelength range of interest), the deviation is

[19] Here, the index j denotes each of the 10 wafers of the present study, whereas the index k denotes the k-th random generation of a normally distributed set of thicknesses and corresponding intensities.

only significant at a wavelength of 880 nm. The value reported for 791 nm by Green and Keevers coincides with the more precise result of the present study.

Free carrier absorption can be excluded as an error source of the present study, even for the 1 Ω cm wafers used in the present study. This is because, according to [5], band-to-band absorption can be approximated to exceed free carrier absorption by more than four orders of magnitude even at a wavelength of 880 nm.

Irradiance spectra of the two single mode diode lasers of interest have not been independently validated. However, at the indicated accuracy of the manufacturers' measurements, this appears to be an unlikely error source. Also, the laser diodes were produced (and spectra were measured) by independent competitors, thereby rendering a systematic error of the measured laser spectra of both similar direction and magnitude unlikely.

A remaining possible cause of deviating absorption data with respect to the data tabulated by Green and Keevers [1] could be band gap narrowing due to the dopant concentration of the investigated p-type 1 Ω cm wafers. With $\Delta\varepsilon_G^{1\,\Omega\mathrm{cm}}$ (300 K) $\approx$ 5 meV [6], and with a simplifying assumption of

$$\alpha_\gamma^{1\,\Omega\mathrm{cm}}\left(\varepsilon_\gamma\right) \approx \alpha_\gamma^{\mathrm{intrinsic}}\left(\varepsilon_\gamma + \Delta\varepsilon_G^{1\,\Omega\mathrm{cm}}\right), \tag{D.2}$$

the measurement results obtained here would have to be compared to the intrinsic data of Green and Keevers at slightly shifted wavelengths λ_γ $\alpha_\gamma^{300\,\mathrm{K}}\left(\lambda_\gamma = 788.5\,\mathrm{nm}\right) = 398 \pm 16\,\mathrm{cm}^{-1}$ and $\alpha_\gamma^{300\,\mathrm{K}}\left(\lambda_\gamma = 876.9\,\mathrm{nm}\right) = 940 \pm 38\,\mathrm{cm}^{-1}$, respectively. Considering this effect, the present measurement results agree with the data tabulated by Green and Keevers within the specified measurement uncertainty.

References

1. M.A. Green, M.J. Keevers, Optical properties of intrinsic silicon at 300K. Prog. Photovoltaics Res. Appl. **3**, 189–192 (1995)
2. D.E. Aspnes, Optical functions of intrinsic c-Si for photon energies up to 7.5 eV, in *Properties of Crystalline Silicon*, ed. by R. Hull (INSPEC IEEE, London, 1999), pp. 683–690
3. J.A. Giesecke, F. Schindler, M. Bühler, M.C. Schubert, W. Warta, Accurate determination of minority carrier mobility in silicon from quasi-steady-state photoluminescence. J. Appl. Phys. **113**, 213705 (2013)
4. M.A. Green, *Silicon Solar Cells: Advanced Principles & Practice* (University of New South Wales, Sydney, 1995)
5. D.K. Schroder, R.N. Thomas, J.C. Swartz, Free carrier absorption in silicon. IEEE J. Solid-State Circ. **13**, 180–187 (1978)
6. A. Schenk, Finite-temperature full random-phase approximation model of band gap narrowing for silicon device simulation. J. Appl. Phys. **84**, 3684–3695 (1998)

Appendix E
Impact of Non-Planar Interfaces on Recombination Measurements

A non-planar wafer interface leads to a non-perpendicular transmission of perpendicularly impinging light.[20] This non-perpendicular light transmission reduces the effective perpendicular penetration depth of light. At a given angular probability density distribution of light transmission $p(\vartheta)$ (with the probability $P(\vartheta) = d\vartheta\, p(\vartheta) \sin\vartheta$ of transmission in the polar angle increment $[\vartheta, \vartheta + d\vartheta]$), this effect can be accounted for in the solution of the continuity equation by expressing $\Delta n(z)$ as a superposition of steady-state solutions of the continuity equation corresponding to monochromatic irradiation

$$\Delta n(z) = \sum_{l} \Delta n_l(z)\,\big|_{\vartheta_l}, \tag{E.1}$$

with solutions $\Delta n_l(z)$ corresponding to a perpendicular projection of penetration depth $\alpha_{\gamma,l} = \alpha_\gamma \cos^{-1}\vartheta_l$ and to an absorbed photon current density $j_{\gamma,l} = j_\gamma P(\vartheta_l)$. Therefore, enhanced surface recombination is expected to affect excess carrier density in non-planar wafers more than in planar wafers. For an assessment of the relevance of this effect, a simplified one-parameter model to describe the scattering of light through rough interfaces—the so-called *Phong* model [2]—was exploited in order to roughly estimate the angular probability density distribution of light transmission at wafer interfaces typically encountered in silicon photovoltaic characterization. According to the *Phong* model, the angular probability density distribution of light transmission becomes

$$p_w(\vartheta) = p_0 \cos^w \vartheta. \tag{E.2}$$

The *Phong* model may be referred to as a one-parameter model, as the second parameter p_0 only ensures conservation of probability $\int P = 1$. Special cases of *Phong* coefficients are the planar or specular limit $w \to \infty$, the Lambertian case of $w = 1$, and the isotropic case of $w = 0$. The analysis of the impact of typical non-planar interfaces on recombination measurements comprises two steps:

---

[20] Parts of the account given here are adapted from the corresponding original publication [1].

J. Giesecke, *Quantitative Recombination and Transport Properties in Silicon from Dynamic Luminescence*, Springer Theses, DOI: 10.1007/978-3-319-06157-3,

© Springer International Publishing Switzerland 2014

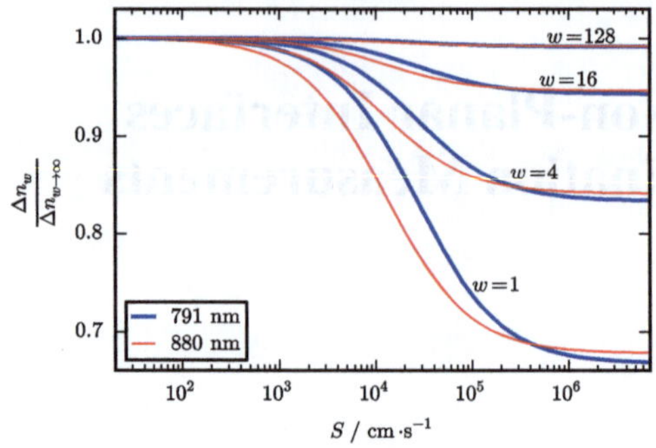

Fig. E.1 Depiction of the impact of non-perpendicular transmission of perpendicularly impinging irradiation on excess carrier density Δn due to a rough wafer interface. Interface roughness was parametrized by a simplified one-parameter *Phong* model. The ratio between Δn_w for an interface characterized by a *Phong* parameter w and $\Delta n_{w\to\infty}$ to be expected for a specular (planar) interface was calculated as a function of surface recombination velocity S for two different irradiation wavelengths available in the harmonically modulated photoluminescence experimental setup at Fraunhofer ISE. For the underlying steady-state solutions of the continuity equation, a minority carrier diffusion length much greater than substrate thickness was assumed ($d = 250\,\mu$m, $\tau = 10^{-1}$ s, $D_a = 27\,\text{cm}^2\text{s}^{-1}$). Adapted after [1]

First, the impact of non-planar surfaces on depth-integrated excess carrier density was modeled on the basis of the steady-state solution of the continuity equation, and for irradiation wavelengths of 791 and 880 nm, respectively. The result is depicted in Fig. E.1, and it shows that a quasi-Lambertian surface morphology has a significant effect on excess carrier density in the case of enhanced surface recombination.

Second, transmission measurements of laser light were performed on wafers showing considerable roughness on the illuminated interface while being planar (highly polished) on the opposite interface with respect to illumination. For this purpose, the harmonically modulated photoluminescence setup was modified such that the luminescence detector would solely capture straightforwardly transmitted laser light at a distance of 10 cm behind a wafer featuring a rough illuminated interface. The measurement setup is schematically depicted in Fig. E.2.

Transmitted intensity in an accurately defined solid angle of detection was measured on both isotropically textured and chemically polished multicrystalline wafers (20 samples each). Transmitted intensity was normalized to a uniform wafer thickness d_0 through multiplication with a factor $e^{\alpha_\gamma(d-d_0)}$. Refraction of non-perpendicular light at the planar interface was accounted for. Remanent luminescence of the (unpassivated) test wafers was suppressed via optical short pass filters in order to solely capture transmitted laser light.

Finally, the *Phong* parameter w was determined from the ratio R_w between transmitted intensities of a unilaterally rough wafer and a bilaterally planar wafer to be expected at a given (setup-specific) solid angle of detection, which is two-dimensionally illustrated in Fig. E.2 as a critical polar angle of detection ϑ_d. In Fig. E.3, R_w is modeled as a function of the *Phong* parameter w, and correspond-

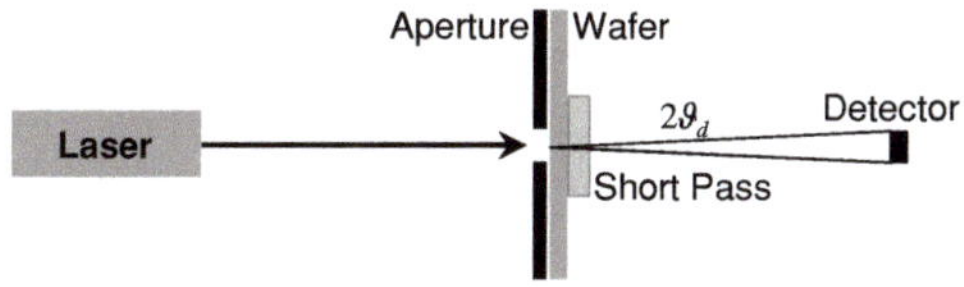

Fig. E.2 Schematic of the experimental setup used for the quantification of interface roughness. A short pass filter enforces effective attenuation of silicon luminescence, allowing exclusive detection of transmitted laser irradiation. An estimate of a remanent luminescent contribution is given in Appendix D. The critical polar angle of detection ϑ_d is used for the calculation of a ratio R_w between detected light intensities of a wafer featuring a rough illuminated interface and a bilaterally planar wafer at a given *Phong* parameter w

Fig. E.3 Depiction of the modeled ratio R_w between expected transmitted light intensities of a unilaterally rough wafer with a *Phong* parameter w and a bilaterally planar wafer, and of the experimentally encountered *Phong* parameters of typical interface morphologies on multicrystalline wafers

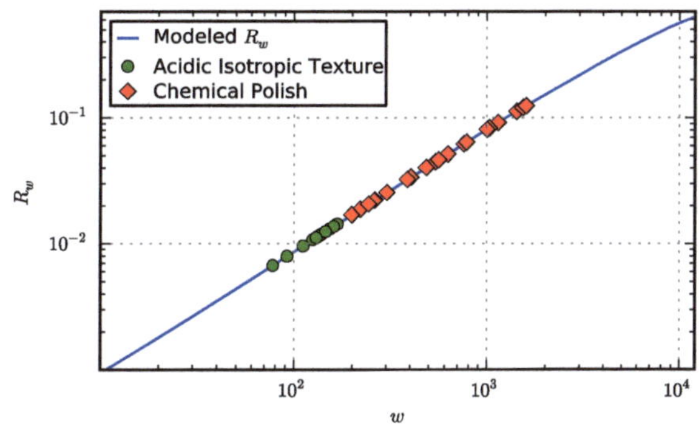

Table E.1 Phong parameters w resulting from the analysis of the angular probability density distribution of light transmission of 2 × 20 multicrystalline wafers featuring one rough interface and another highly polished interface

Interface treatment	$w_{\text{best}} = 10^{\langle \log_{10} w \rangle}$	$\langle \log_{10} w \rangle$
Acidic isotropic texture	136	2.13 ± 0.08
Chemical polish	557	2.75 ± 0.29

ing measurement results of R_w are thereby assigned to w. As the experimentally encountered ratios R_w happen to be symmetrically distributed in $\log_{10} w$, it appears suitable to provide mean values and standard deviations in terms of $\log_{10} w$, as given in Table E.1.

Relating the *Phong* parameters encountered here to the simulation results depicted in Fig. E.1, one might conclude that the investigated non-planar surface morphologies (with $w > 128$) do not significantly affect excess carrier density in the case of enhanced surface recombination. Yet, there is an indication of an improper acidic surface texture of the isotropically textured wafers. Although the acidic texture apparently yields a significantly higher interface roughness than the chemical polish, *Phong* parameters significantly in excess of 10^2 contradict the efficient light trapping effect to be expected from an adequate acidic surface texture. The most likely explanation for a failure of the acidic etching process is a lack of a priori required (as-cut) surface roughness. Therefore, this investigation remains to be repeated with standard acidic surface textures.

References

1. J.A. Giesecke, F. Schindler, M. Bühler, M.C. Schubert, W. Warta, Accurate determination of minority carrier mobility in silicon from quasi-steady-state photoluminescence. J. Appl. Phys. **113**, 213705 (2013)
2. J.D. Foley, A. Van Dam, S. Feiner, J. Hughes, *Computer Graphics: Principles and Practice* (Addison-Wesley, Amsterdam, 1990)

Publications Related to this Dissertation

In the following, original publications on the core topics of this dissertation are listed. Previous publications by the author as well as patents and original publications on derivative applications and related or different topics are excluded. Publications are listed in chronological order.

Publications in Reviewed Journals

- J. A. Giesecke, M. C. Schubert, D. Walter, W. Warta, Minority Carrier Lifetime in Silicon Wafers from Quasi-Steady-State Photoluminescence, Appl. Phys. Lett. 97 (2010) 092109.
- J. A. Giesecke, M. C. Schubert, B. Michl, F. Schindler, W. Warta, Minority Carrier Lifetime Imaging of Silicon Wafers Calibrated by Quasi-Steady-State Photoluminescence, Solar Energy Materials & Solar Cells 95 (2011) 1011–1018.
- J. A. Giesecke, B. Michl, F. Schindler, M. C. Schubert, W. Warta, Minority Carrier Lifetime of Silicon Solar Cells from Quasi-Steady-State Photoluminescence, Solar Energy Materials & Solar Cells 95 (2011) 1979–1982.
- J. A. Giesecke, W. Warta, Microsecond Carrier Lifetime Measurements in Silicon via Quasi-Steady-State Photoluminescence, Prog. Photovoltaics: Res. Appl. 20 (2012) 238–245.
- J. A. Giesecke, T. Niewelt, M. Rüdiger, M. Rauer, M. C. Schubert, W. Warta, Broad Range Injection-Dependent Minority Carrier Lifetime from Photoluminescence, Solar Energy Materials & Solar Cells 102 (2012) 220–224.
- J. A. Giesecke, M. C. Schubert, W. Warta, Self-Sufficient Minority Carrier Lifetime in Silicon from Quasi-Steady-State Photoluminescence, Phys. Stat. Sol. A 209 (2012) 2286–2290.
- J. A. Giesecke, M. C. Schubert, W. Warta, Measurement of Net Dopant Concentration via Dynamic Photoluminescence, J. Appl. Phys. 112 (2012) 063704.

J. Giesecke, *Quantitative Recombination and Transport Properties in Silicon from Dynamic Luminescence*, Springer Theses, DOI: 10.1007/978-3-319-06157-3,
© Springer International Publishing Switzerland 2014

- J. A. Giesecke, S. W. Glunz, W. Warta, Understanding and Resolving the Discrepancy between Differential and Actual Minority Carrier Lifetime, J. Appl. Phys. 113 (2013) 073706.
- J. A. Giesecke, M. C. Schubert, W. Warta, Carrier Lifetime from Dynamic Electroluminescence, IEEE J. Photovoltaics 3 (2013) 1012–1015.
- J. A. Giesecke, F. Schindler, M. Bühler, M. C. Schubert, W. Warta, Accurate Determination of Minority Carrier Mobility in Silicon from Quasi-Steady-State Photoluminescence, J. Appl. Phys. 113 (2013) 213705.
- J. A. Giesecke, R. A. Sinton, M. C. Schubert, S. Riepe, W. Warta, Determination of Bulk Lifetime and Surface Recombination Velocity of Silicon Ingots from Dynamic Photoluminescence, IEEE J. Photovoltaics 3 (2013) 1311–1318.
- J. A. Giesecke, W. Warta, Understanding Carrier Lifetime Measurements at Nonuniform Recombination, Appl. Phys. Lett. 104 (2014) 082103.

Conference Publications

- J. A. Giesecke, D. Walter, F. Kopp, P. Rosenits, M. C. Schubert, W. Warta, Simultaneous Determination of Carrier Lifetime and Net Dopant Concentration of Silicon Wafers from Photoluminescence, in: Proceedings of the 35th IEEE PVSC, Honolulu, 2010, pp. 847–851.
- J. A. Giesecke, B. Michl, F. Schindler, M. C. Schubert, W. Warta, Spatially Resolved Carrier Lifetime Calibrated via Quasi-Steady-State Photoluminescence, Energy Procedia 8 (2011) 64–70.
- J. A. Giesecke, S. W. Glunz, W. Warta, Determination of Actual Carrier Lifetime from Differential Measurements, Energy Procedia 38 (2013) 59–65.

MIX
Papier aus verantwortungsvollen Quellen
Paper from responsible sources
FSC® C105338

If you have any concerns about our products, you can contact us on
ProductSafety@springernature.com

In case Publisher is established outside the EU, the EU authorized representative is:
Springer Nature Customer Service Center GmbH
Europaplatz 3, 69115 Heidelberg, Germany

Printed by Libri Plureos GmbH
in Hamburg, Germany